The Early Mathematics of Leonhard Euler

A young Leonhard Euler

The Early Mathematics of Leonhard Euler

C. Edward Sandifer
Western Connecticut State University

Published and Distributed by
THE MATHEMATICAL ASSOCIATION OF AMERICA

Library of Congress Catalog Card Number 2006933948

ISBN 10: 0-88385-559-3
ISBN 13: 978-0-88385-559-1

Printed in the United States of America

Current Printing (last digit):
10 9 8 7 6 5 4 3 2 1

SPECTRUM SERIES

The Spectrum Series of the Mathematical Association of America was so named to reflect its purpose: to publish a broad range of books including biographies, accessible expositions of old or new mathematical ideas, reprints and revisions of excellent out-of-print books, popular works, and other monographs of high interest that will appeal to a broad range of readers, including students and teachers of mathematics, mathematical amateurs, and researchers.

777 Mathematical Conversation Starters, by John de Pillis

99 Points of Intersection: Examples—Pictures—Proofs, by Hans Walser. Translated from the original German by Peter Hilton and Jean Pedersen

aha! A two volume collection: aha! Gotcha and aha! Insight, by Martin Gardner

All the Math That's Fit to Print, by Keith Devlin

Carl Friedrich Gauss: Titan of Science, by G. Waldo Dunnington, with additional material by Jeremy Gray and Fritz-Egbert Dohse

The Changing Space of Geometry, edited by Chris Pritchard

Circles: A Mathematical View, by Dan Pedoe

Complex Numbers and Geometry, by Liang-shin Hahn

Cryptology, by Albrecht Beutelspacher

The Early Mathematics of Leonhard Euler, by C. Edward Sandifer

The Edge of the Universe: Celebrating 10 Years of Math Horizons, edited by Deanna Haunsperger and Stephen Kennedy

Five Hundred Mathematical Challenges, Edward J. Barbeau, Murray S. Klamkin, and William O. J. Moser

The Genius of Euler: Reflections on his Life and Work, edited by William Dunham

The Golden Section, by Hans Walser. Translated from the original German by Peter Hilton, with the assistance of Jean Pedersen.

I Want to Be a Mathematician, by Paul R. Halmos

Journey into Geometries, by Marta Sved

JULIA: a life in mathematics, by Constance Reid

R. L. Moore: Mathematician and Teacher, by John Parker

The Lighter Side of Mathematics: Proceedings of the Eugène Strens Memorial Conference on Recreational Mathematics & Its History, edited by Richard K. Guy and Robert E. Woodrow

Lure of the Integers, by Joe Roberts

The Magic Numbers of the Professor, by Owen O'Shea and Underwood Dudley

Magic Tricks, Card Shuffling, and Dynamic Computer Memories: The Mathematics of the Perfect Shuffle, by S. Brent Morris

Martin Gardner's Mathematical Games: The entire collection of his Scientific American columns

The Math Chat Book, by Frank Morgan

Mathematical Adventures for Students and Amateurs, edited by David Hayes and Tatiana Shubin. With the assistance of Gerald L. Alexanderson and Peter Ross

Mathematical Apocrypha, by Steven G. Krantz

Mathematical Apocrypha Redux, by Steven G. Krantz

Mathematical Carnival, by Martin Gardner

Mathematical Circles Vol I: In Mathematical Circles Quadrants I, II, III, IV, by Howard W. Eves

Mathematical Circles Vol II: Mathematical Circles Revisited and Mathematical Circles Squared, by Howard W. Eves

Mathematical Circles Vol III: Mathematical Circles Adieu and Return to Mathematical Circles, by Howard W. Eves

Mathematical Circus, by Martin Gardner
Mathematical Cranks, by Underwood Dudley
Mathematical Evolutions, edited by Abe Shenitzer and John Stillwell
Mathematical Fallacies, Flaws, and Flimflam, by Edward J. Barbeau
Mathematical Magic Show, by Martin Gardner
Mathematical Reminiscences, by Howard Eves
Mathematical Treks: From Surreal Numbers to Magic Circles, by Ivars Peterson
Mathematics: Queen and Servant of Science, by E.T. Bell (Out of print)
Memorabilia Mathematica, by Robert Edouard Moritz (Out of print)
Musings of the Masters: An Anthology of Mathematical Reflections, edited by Raymond G. Ayoub
New Mathematical Diversions, by Martin Gardner
Non-Euclidean Geometry, by H. S. M. Coxeter
Numerical Methods That Work, by Forman Acton
Numerology or What Pythagoras Wrought, by Underwood Dudley
Out of the Mouths of Mathematicians, by Rosemary Schmalz
Penrose Tiles to Trapdoor Ciphers . . . and the Return of Dr. Matrix, by Martin Gardner
Polyominoes, by George Martin
Power Play, by Edward J. Barbeau
The Random Walks of George Pólya, by Gerald L. Alexanderson
Reality Conditions: Short Mathematical Fiction, by Alex Kasman
Remarkable Mathematicians, from Euler to von Neumann, Ioan James
The Search for E.T. Bell, also known as John Taine, by Constance Reid
Shaping Space, edited by Marjorie Senechal and George Fleck
Sherlock Holmes in Babylon and Other Tales of Mathematical History, edited by Marlow Anderson, Victor Katz, and Robin Wilson
Student Research Projects in Calculus, by Marcus Cohen, Arthur Knoebel, Edward D. Gaughan, Douglas S. Kurtz, and David Pengelley
Symmetry, by Hans Walser. Translated from the original German by Peter Hilton, with the assistance of Jean Pedersen.
The Trisectors, by Underwood Dudley
Twenty Years Before the Blackboard, by Michael Stueben with Diane Sandford
The Words of Mathematics, by Steven Schwartzman

To order MAA publications, contact:

MAA Service Center
P. O. Box 91112
Washington, DC 20090-1112
800-331-1622 FAX 301-206-9789

Contents

The asterisks in front of the chapter titles below represent the author's rating system: the more asterisks, the more important the article.

Preface

From a scientific standpoint, the 18th century was the Age of Euler. During its early years, Newton and Leibniz were late in their careers, and the Bernoulli brothers, Johann and Jakob, were just passing their peaks. The end of the century featured the likes of Gauss, Lagrange and Laplace. Euler, though, dominated the middle of the century. There were other giants—d'Alembert, Clairaut, Maupertuis, Maclaurin—but Euler alone more than matched their combined productivity. Clifford Truesdell [T] has estimated that Euler made more than a third of the century's discoveries in mathematics and science.

Euler was the preeminent mathematician and scientist during the European Enlightenment, the intellectual revolution of the 18th century that shaped our modern ideas of liberty and human rights. The great thinkers of the Enlightenment—Rousseau, Voltaire, Diderot, Jefferson, Locke and the rest—all wanted to believe that their social and political ideas were rooted in scientific principles, and that, in turn, those scientific principles were based in the firm rigor of mathematics. They worked hard to understand science and mathematics, or at least to gain reputations for understanding.

World leaders, kings and empresses collected scientists in much the way they collected works of art and overseas colonies, partly for power and economic gain, but also partly for show. Euler was a great prize in this game of the powerful. He began his career in the new academy of Peter the Great, though Peter died shortly before the Academy opened. Then Euler moved to Berlin to help reform the academy of Frederick the Great, and finally came back to St. Petersburg during the reign of Catherine the Great. Part of greatness is recognizing and recruiting good help.

Ironically, in many ways Euler was out of step with the Enlightenment. Many of the most charismatic characters of the period were enthusiastic atheists, witty salonists and fawning courtiers. Euler, on the other hand, was religiously faithful. He served on his church council. He was not witty and charming in crowds. Legend has it that he was asked at a party why he spoke so seldom. He supposedly answered something like "Ma'am, where I used to live, people who speak out are hanged." The incident probably never happened, but, as such stories often do, it probably captures the truth that Euler was seldom the life of a large party. He is reported to have been quite charming in small parties, though.

Most readers know some of Euler's discoveries in mathematics and science, especially his pioneering work in number theory. Readers are also likely to have heard that he pioneered the use of $f(x)$ notation for functions and standardized the symbols e and π. We live in an age of specialization, and many of us know Euler's contributions to our own specialty, be it graph theory, fluid flow, Diophantine equations, calculus of variations, mechanics, the physics of music or any of a dozen other fields where the foundational results are due to Euler. But few of us know much of Euler's discoveries outside our own specialty, so we don't appreciate the true depth and breadth of his contributions.

This book is intended as a mathematical biography of Euler. It is an article-by-article description of the fifty or so mathematical articles Euler completed through 1741, the year he left the Imperial Academy in St. Petersburg to join the employ of Frederick the Great in Berlin. I will dwell on the details of his papers far more than on the details of his life or his times, though I will point out connections between the work and the rest of the world when I can.

I also attempt to recapture some of the mathematical and scientific spirit of the 18th century. As the years pass, Euler grows as a scientist and at the same time world changes around him. I will try to show how his early papers were very much in the spirit of the 17th century in ways described in the text, and at the same time they were clearly the work of a young and sometimes naïve scholar of as-yet unproven abilities and potential.

After a good deal of thought, I decided to present the articles in the order in which scholars believe they were written. The authoritative source in this chronology is the Eneström index [E]. That index assigned a number to each Euler work, in the order they were published, but the appendix also listed them in the order they were written. These orders are only approximately the same. For example, Euler's first published article, E-1, was also apparently the first one he wrote. It was written in 1725 and published in 1726. On the other hand, the last article we consider in this book, E-790, "On the usefulness of higher mathematics," was written in 1741, but for reasons we can only guess, it was not published until 1847, over a hundred years later. Other less dramatic permutations in the chronology arise from simple things like publication delay, or for more dramatic reasons, such as late in his career when Euler withheld a result in the calculus of variations because he saw that Lagrange's method was superior.

Euler's mathematical work between 1725 and 1741 has been divided into segments. Most are just a year long, but some cover two or three years. Each segment is introduced with a brief description of Euler's other activities in that time interval, some of the events in his life, and some highlights of then-current events. I hope that these "Interludes" will help illuminate Euler's work in a wider context.

Euler had modest, though by no means impoverished, origins. His father Paul was pastor of a parish near Basel, Switzerland, but had studied with the great Jakob Bernoulli at the University in Basel. When Leonhard went to the university, his teacher was Jakob Bernoulli's younger brother Johann. Euler was an apt student, and Bernoulli's tutelage turned his career ambitions to science. By 1725, when Euler was 18, Bernoulli had him writing papers for *Acta Eruditorum*, one of the most widely read scholarly journals of the era.

In April, 1727, just after his 20th birthday, Euler left Basel to join his friends Nicolas II and Daniel Bernoulli at the St. Petersburg Academy. Euler remained in St. Petersburg until 1741, when political unrest, particularly directed at the German community in Russia, made him welcome an opportunity to join the Academy of Frederick II in Berlin.

Initially, Frederick II and Euler got along very well, but over the next 25 years their relationship gradually deteriorated. In 1766, in the wake of the Seven Years War, Euler persuaded Frederick to let him accept an invitation to return to St. Petersburg, where he was welcomed as a hero.

By the time Euler died in 1783, he had published about 550 books and papers. Another 250 or 300 papers were published posthumously, appearing regularly until about 1820, and then sporadically until 1862, 79 years after his death.

Some remarks are appropriate about the quantity and variety of Euler's publications. They are all cataloged in the Eneström index [E]. Eneström lists 866 publications, and this is usually cited as the "official" number. Closer examination, though, reveals that this is a little too high. For example, about 20 entries late in the list, after E-800, are collections of letters, assembled after Euler's death and probably never intended for publication. Some others are excerpts from the public notebook, the so-called "Day Book" in which Academy members sketched ideas. Still others are clearly "fragments," papers and books started but never finished. Some articles appear twice on the list, in two different languages. Taking all this into account, we have several numbers we could cite as Euler's total number of publications. We could use Eneström's last index number, 866. The number 856 is also often cited, the number of the last article on the list. The entries 857 to 866 are all letters. Here, though, we will prefer the vague but truthful "over 800" published books and papers.

Having settled on the value "over 800," some readers may enjoy asking how this output compares with others. In particular, "Who published more, Euler or Erdős?" It depends on how we make the rules. Erdős authored or co-authored well over 1000 books and papers. Euler wrote about 800, so, in the first analysis, Erdős published more. Moreover, Erdős papers are still being published. Euler hasn't published anything new for over 140 years. If, on the other hand, for some reason we want to penalize Erdős for co-publishing, then Euler comes out ahead. Euler and Erdős combined, though, are more than outdone by Christian Wolff. Today Wolff is mostly forgotten, but in his own time (1679–1754) he was a very influential scientist, philosopher and mathematician. Wolff was the last defender of Leibniz's philosophy of monads. His collected works run to over 300 volumes. Should Wolff be disqualified just because his work is mostly forgotten and was not entirely mathematics? Such questions make pleasant conversation, but seldom lead to valuable conclusions.

Some remarks are appropriate on the philosophy of history behind this book. I focus primarily on the mathematical details of Euler's work up to the time he left St. Petersburg for Berlin in 1741. As such, the work is prima facie what is called internalist.

Moreover, it is chronological, describing Euler's mathematical work in the order he wrote it, which, as pointed out earlier, is only approximately the order in which it was published. The decision to proceed chronologically, rather than thematically, was a difficult one. Eventually, though, it became clear that the many threads of Euler's work cross so often, forming more a web than a bundle of threads, that they would not retain their shape if anyone tried to untangle them. The chronological exposition also lets the reader better appreciate the broad number of projects Euler pursued at any given time.

At least one person who has seen this manuscript has called the work "crypto-internalist." Though I am not exactly sure what the word means, and suspect it was coined just for the occasion, I choose to take it as a compliment. I think and hope it means that the book describes the mathematical details themselves, but that it also makes connections among the various papers, even if they are not obviously part of the same thread. Moreover, it relates the mathematical details to the events in Euler's life, the rest of the scientific community and the world at large.

These considerations on the way this book was written leave the reader a good deal of flexibility in how to read the book. Those with plenty of time and broad interests might want to read it straight through and enjoy its many plots and subplots. Those with particular interests, say number theory or the calculus of variations, can read just those threads. The annotated table of contents should help such readers. Alternatively, there is a collection of some topically related lists of papers at the end of this Preface. Those with broader interests but less time will find that some articles are marked with one, two or three stars. These are the ones that I thought were most important or more interesting. Such judgments are, of course, subjective and vary with individual tastes.

There is a particular pleasure reading Euler in the original language, usually Latin. Euler's writing is as clear and carefully crafted as his mathematics. He seems to make a deliberate effort to keep the language itself relatively simple, knowing, perhaps, that Latin was nobody's first language. I hope that some readers will use this book as a guide to choose what Euler they would like to read in the original.

The reader deserves to be warned of some conventions followed in this book. First, I often refer to Euler's individual works by their numbers in the Eneström index. Certainly "E-20" is a good deal more concise than "De summatione innumerabilium progressionum." Eneström numbers are chronological in order of publication. We sometimes refer to an article by its original title, usually in Latin, and sometimes by an English translation of that title. If this becomes confusing, citations at the end of each chapter should help.

Moreover, it has become standard practice to use two or more dates to cite Euler's works. For example, Euler wrote E-20 in 1731, and it appeared in volume 5 of the journal of the St. Petersburg Academy. That volume contains papers presented to the Academy for the years 1731 and 1732, and was published in 1738. Moreover, E-20 appears in Series I volume 14 of Euler's *Opera Omnia*, and facsimiles of this, and almost all of Euler's works are available online at EulerArchive.com. The bibliographic citation in the footnote at the beginning of the chapter on E-20 contains much of this information. It says

Comm. Acad. Sci. Imp. Petropol. 5 (1730/31) 1738, 91–105; *Opera Omnia* I.14, 25–41

The abbreviation "*Comm. Acad. Sci. Imp. Petropol.* 5" tells us that this article appeared in volume 5 *Commentarii academiae scientiarum Peteropolitanei*, the journal of the St. Petersburg Academy. The years 1730/31 and 1738 are the dates described above, and "*Opera Omnia* I.14, 25–41" tell us where to find the article in the *Opera Omnia*.

In Euler's time, Russia still used the Julian calendar. His correspondents in the rest of Europe mostly used the newer Gregorian calendar, so when it was November 12, 1739, it was already November 23 in Berlin. Eighteenth century mail services were much better than most people would expect, so occasionally it was possible for a letter to seem to be answered before it had been written! We make note of these calendar problems whenever they arise.

As much as possible I try to use the same notation that Euler did. Euler wrote "$l\,x$" where we would write "$\ln x$." Later in his career, he would write $\int n$ where we would write $\sum n$, and he usually wrote xx rather than x^2, but never xxx for the cube. He had no subscript notation. (There are many times it would have been very helpful.) I will explain his notation for definite integrals when the time comes.

There are two main exceptions to this faithfulness to Euler's notation. First, occasionally Euler uses the old German Fraktur alphabet. I find this font very difficult to read, and the printers I was using didn't have the font installed, so I made substitutions. Second, Euler's radical signs don't usually have bars on them. They look like $\surd$ instead of $\sqrt{\ }$. It simply became too tedious to untangle his radical signs over and over again, and then to try to explain what the correct reading of the expression was, so I finally gave up and switched to the modern notation.

The scholarship in this book is based almost entirely on original sources. This lets us avoid repeating the errors of earlier scholars. One is reminded of one reader's criticism of Descartes: "He wiped away the errors of the Ancients and replaced them with his own."

Moreover, there are inevitably a great number of loose ends and interesting threads that could be tied up, perhaps should be, but aren't. I am aware of some of these and blissfully ignorant of many others. It would be interesting, for example, to learn more about Henry Pemberton and the origins of the problems of reciprocal trajectories. Euler studied these in several papers beginning with E-3. One wonders why Pemberton would pose such problems. Also, I suspect there is something more in E-44 and E-45 than I see.

I have been extraordinarily fortunate to be able to read much of Euler from the original 18th century volumes in the library collections of Yale University. Most readers will not be so lucky, but they have at least two recourses. First, in 1907 the Swiss Academy of Sciences began to publish the complete works of Euler under the title *Opera Omnia*. The first volume, *Algebra*, appeared in 1911, and the series is not yet complete. All of Euler's published works in mathematics and physics have been republished in this series, but most of his correspondence and all of his notebooks are still in preparation. The scholarship that has gone into the preparation of the *Opera Omnia* includes much of the 20th century's best historical and mathematical work on Euler. I refer frequently to the *Opera Omnia*.

Second, more and more of Euler's original work is becoming available on line. I regularly use the resources at EulerArchive.org, which has over 95% of Euler's publications scanned and on line. Additional Euler is available at the Berlin Academy of Sciences, the National Library in France and at the Cornell Digital Library, and certainly more is being added all the time.

There are surely errors in this work and places I did not understand what Euler was doing. For this, I apologize and take full responsibility. I hope that they do not obscure the picture of Euler, his early work, and its role in the 18th and 21st centuries that I am trying to paint.

Acknowledgements

Finally, I must thank people. For getting me off to a good start with the Institute for the History of Mathematics and Its Uses in Teaching, and all those hot summers in Washington, DC, I thank Fred Rickey, Victor Katz, Ron Calinger, Florence Fasanelli, and all my friends and classmates at the Institute. Further, thanks go to all of the above for their continued friendship and support, now that I've grown up a bit as an historian of mathematics. Rob Bradley, John Glaus, Mary Ann McLoughlin, Rdiger Thiele and, again, Ron Calinger have been helpful and supportive with their work with The Euler Society. Dominic Klyve, Lee Stemkoski, Rachel Esselstein, Erik Tou and the rest of the people at Dartmouth have performed a technological miracle in creating and maintaining The Euler Archive. I thank my Library Director at Western Connecticut State University, Ralph Holibaugh for generously arranging permissions for me to use the wonderful collections at Yale University. I sincerely appreciate my wife and family for putting up with and even encouraging all of this. The Editors at the MAA have shown that they deserve their reputations as conscientious and helpful people. Finally, I thank the Readers, the ones for whom we write.

Ed Sandifer
Newtown, CT
June 2006

References

[E] Eneström, Gustav, "Verzeichnis der Schriften Leonhard Eulers," *Jahresbericht der Deutschen Mathematiker-Vereinigung*, Ergnzungsband 4, Leipzig, 1919–1913.

[T] Truesdell, C., *An Idiot's Fugitive Essays on Science: Methods, Criticism, Training, Circumstances*, Springer-Verlag, New York, 1984.

Interlude: 1725–1727

World events

In January 1725, Peter I "the Great," Czar of Russia (1682–1725), died. He had organized the Imperial Academy of Sciences of St. Petersburg, where Euler would live and work for much of his life, but Peter did not live to see the Academy open.

Later that year, Euler's friends Nicolas II and Daniel Bernoulli, sons of Euler's teacher Johann Bernoulli, left Basel to join the new Academy in St. Petersburg. Nicolas took the position in the Mathematics and Physics division of the Academy, and Daniel was in the Physiology division. In 1726, Nicolas died and Daniel moved to his position in the Mathematics and Physics division, leaving the position in Physiology open.

In 1725, composers Handel, Telemann and François Couperin were near the peaks of their careers.

Also in 1725, Sir Robert Clive, "Clive of India" (1725–1774), was born. During the Seven Years War (1756–1763) he seized India from the French on behalf of the British East India Company.

In 1726, Benjamin Franklin (1706–1790) moved to Philadelphia, where he spent most of the rest of his life. After growing up in Boston, he went to England to purchase printing equipment. Franklin then returned to Philadelphia, where he opened his famous printing shop.

François-André Danican Philidor, French composer and chess champion, was born in 1726. In the 1750s, Euler studied chess and got a reputation as quite a good player. Apparently Euler never got the chance to play against Philidor.

On March 31, 1727, Sir Isaac Newton (1642–1727) died in London. Also that year, Empress Catherine I of Russia and King George I of England died.

In Euler's life

In 1726, Euler completed his studies at the University of Basel.

On April 5, 1727, ten days before his 20th birthday, Leonhard Euler left Basel to join his friend Daniel Bernoulli at the Academy of Sciences in St. Petersburg. He arrived there on May 24 and stayed until 1741.

Euler had expected to fill the vacant position in the Physiology division of the Academy but his initial appointment in St. Petersburg was as a physician in the Imperial Navy. The recent death of Empress Catherine I caused a "hiring freeze" at the Academy.

Euler's other work

In 1726, Euler wrote E-4, Meditationes super problemate nautico, de implantatione malorum, "Thoughts about a nautical problem on the positioning of masts." The essay earned him an *accessit*, or Honorable Mention in the 1728 Paris Prize competition.

In 1727 he wrote Dissertatio physica de sono, "Essay on the physics of sound." It is sometimes described as Euler's "masters' thesis," though it is better described as a sample essay, part of Euler's unsuccessful application to join the faculty at the University of Basel.

Euler's mathematics

During the years 1725–1727, Euler wrote two articles that we consider in detail in this book. The first, E-1, Constructio linearum isochronarum in medio quocunque resistente, "The construction of isochronal curves in any resistant medium," is a flawed application of the calculus of variations to a mechanical problem.

The second, E-3, Methodus inveniendi trajectorias reciprocas algebraicas, "Method for finding reciprocal algebraic trajectories," is the first of several articles Euler wrote on the now-forgotten problem of reciprocal trajectories.

1

E-1: Constructio linearum isochronarum in medio quocunque resistente*

Construction of isochronal curves in any kind of resistant medium

Euler published his first article when he was only 18 years old. He may have written some of his other works before this one. People wonder particularly about E-2 and E-4, but this was the first to see print, and Gustav Eneström, the authority we are following, believed this was the first one he wrote as well. In some ways, it is like much of Euler's other work, but in many ways, it is different.

For one thing, this paper is very short, only three pages long, and most of Euler's papers run to about 20 pages. It appeared in the *Acta Eruditorum*, and most of his papers were in the journals of the Petersburg Academy. The other articles that Euler published in the *Acta Eruditorum* are also very short, as we shall see when we get to E-10.

This paper is written in Latin. About 80 percent of Euler's 800 or so published works are in Latin, 20 percent in French and about 2 percent in German.

The editors of the *Opera Omnia* classify this paper as "Mechanics" rather than "Mathematics" and put it in Series II rather than Series I. We include it here because it was Euler's first and because the results in E-42 depend in part on it. E-6 is included for the same reason. The editors classify 470 out of about 800 of his published works, that is almost 60 percent, as "Mathematics." Most of what remains depends heavily on mathematics, and today would perhaps be called "mathematical physics" or some such. Conversely, much of Euler's mathematics is motivated by physical applications.

There is another way in which this paper is not typical of Euler. Some of the mechanical assumptions he makes here are incorrect, so the main results of the paper are tainted.

In this paper, Euler studies isochronal curves. A curve is isochronal if it takes a body sliding along the curve under the force of gravity the same amount of time to reach the bottom regardless of where on the curve the body begins.

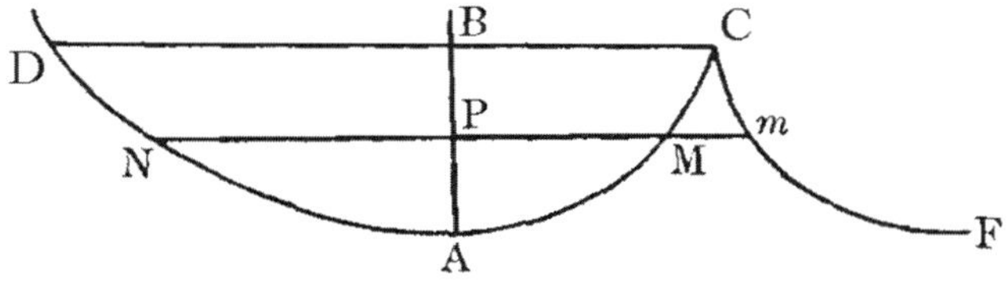

Fig 1

Euler starts the paper with these words: "It is known among Geometers that the ordinary cycloid is an isochrone or tautochrone in a medium that is not resistant and if the force of gravity acts uniformly towards a center that is infinitely far away." He went on to say that Isaac Newton showed

Acta Eruditorum, 1726, 361–363; *Opera Omnia* II.6, 1–3
These notations indicate that this article first appeared in the 1726 volume of *Acta Eruditorum*, pages 361–363. It is reprinted in the *Opera Omnia*, Series II, Volume 6, pages 1–3. Notations like these appear at the beginning of each chapter.

that the isochrone is also a cycloid if there is a resisting force that is proportional to the speed of the object. Euler seeks to generalize this result to other forms of resistance.

Let AB be a vertical line (Figure 1) and *AND* be a curve, which we are to take, without being told, to have its minimum at A. Let $x = AP$. Draw a perpendicular PN and call its length z. Let Z be the function of z that describes the resistance. Then Euler claims that $dx = z\,dz + Z\,dz$. This is where he makes his mistake. Without friction, the appropriate differential equation would be $dx = z\,dz$, where z denotes arc length. He means to be considering wind resistance, so, with friction, he should consider Z to be a function of speed, $\frac{dz}{dt}$, and not a function of z. This invalidates the rest of his analysis.

This done, construct another curve *AMC* on that same axis AB, and extend NP to M so that the arc length AM equals PN. Euler seeks conditions on the curve *CMA* that make it isochronal.

This should be pretty confusing. Euler is graphing two things on the same axis, and both PN and PM, have to be taken as positive, even though they go in opposite directions. The curve on the left is a graph of the arc lengths of the curve on the right. In an order that seems unnatural to us, Euler starts with the graph of the arc length, *AND*, and constructed the curve *AMC* that has *AND* as its arc length function. Euler has also taken the x axis to be vertical, where we usually expect it to be horizontal, and he has chosen z and dz to denote arc length and speed, where we tend to expect s and ds, respectively.

Euler is interested in the location of the point C, the most remote point from which it is possible to reach the point A in the given time. He concludes that C must be a cusp and that at C, the curve *AMC* must have a vertical tangent. He captures this in an equation by asking that $\sqrt{dz^2 - dx^2}$ "debet aequari ziphrae", or ought to be zero. It seems to be a youthful flourish of language for Euler to use this exotic and obsolete word *ziphrae* for "zero."

It would have been easier to define $y = PM$, so that $dz^2 = dx^2 + dy^2$, but Euler does not introduce the extra variable. Because Euler thinks that $dx = z\,dz + Z\,dz$, it follows that

$$\sqrt{dz^2 - dx^2} = dz\sqrt{1 - (z + Z)^2}.$$

At the cusp, then, $z + Z = 1$.

In his typical manner, Euler provides an example, and his first example is chosen to verify previously known results. Suppose that the resistance is proportional to the speed, that is, that $Z = z$. Then $dx = 2z\,dz$ and $x = zz$. Then the curve *AND* is a parabola, and so, by a fact Euler apparently expects his readers to know, the curve *AMC* will be a cycloid. This is the result of Newton that Euler mentions above.

Euler takes as a second example the resistance proportional to the square of the speed, that is $Z = azz$ and so

$$dx = z\,dz + azz\,dz \quad \text{and} \quad x = \frac{1}{2}zz + \frac{1}{3}az^3.$$

Note that this equation describes the companion curve *AND* and not the isochronal curve itself, *AMC*.

Euler closes with a remark that his mentor Johann Bernoulli has posed some questions on brachistochrone curves in a resistant medium and that an anonymous Englishman—now known to be Henry Pemberton (1694–1771)—has posed some interesting questions about inverse tangents. As we shall see, Euler addresses Bernoulli's problem several years later in E-42, but he solves Pemberton's problem almost immediately, in E-3.

There are some technical aspects of this paper that mark it as an early work. In the graph, there are no negatives on the axes. The variable x is measured along the vertical axis. The variable z is measured one direction from that axis, and another variable, not named but we would call it y, is measured in the other direction. Neither a z-axis nor a y-axis appears as part of the drawing. The x-axis measures height above the minimum of the curve and not horizontal displacement, as our modern conventions might lead us to assume.

Moreover, the "answer" is a relation between height and arc length. From this, we do not get an explicit formula for the isochronal curve.

All of these aspects are typical of the early 1700s. As we will see, many of these customs change over the next few decades, and a great many of these changes will be due to the work of Euler himself.

The real surprise, though, of Euler's first published paper is that it was wrong. Even Euler made mistakes, and only sometimes did he find and correct them. In this case, he did find his error just two years later and corrected it in E-13. We won't be examining E-13; it is too mechanical, but the "correction" involves, essentially, solving the right differential equation instead of the wrong one.

References

[E] Eneström, Gustav, *Die Schriften Eulers chronologisch nach den Jahren geordnet, in denen sie verfasst worden sind, Jahresbericht de Deutschen Mathematiker-Vereinigung*, 1913. Available on line at `www.EulerArchive.org`.

2

E-3: Methodus inveniendi traiectorias reciprocas algebraicas*

Method of finding reciprocal algebraic trajectories

Today, almost nobody knows what "reciprocal trajectories" are. People writing papers on Euler dutifully report that he wrote an article on the topic because it was one of just four papers assigned to Euler's "Basel period." They give no clue to what reciprocal trajectories may be.

Euler does only a little better. He omits any description of the problem, though in its sequel, E-5, he does give a brief explanation of it. It seems that in Euler's time, everybody knew the problem of reciprocal trajectories so well that Euler did not feel he had to remind them. Then, over the next fifteen years, Euler, in his four papers on the subject, E-3, E-5, E-23 and E-173, solved it so thoroughly that it was finished and forgotten.

Before we begin looking at Euler's work, we will try to explain the problem of reciprocal trajectories. Apparently Nicolaus Bernoulli (1695–1726) first described the problem, and he attributes it to an "anonymous Englishman." Bernoulli asks "Between two parallel axes *MN* and *FG*, to find and to construct a curve *ABC* and that same curve again, *DBE*, but drawn in an inverse position" (Figure 1) such that the motion of the second one along the parallel axes keeps the curves cutting each other at a constant right angle.

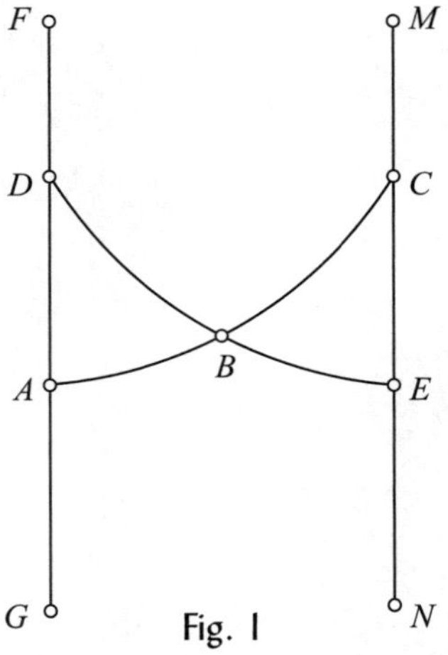

Fig. 1

In Figure 1, *MN*, *FG* are parallel axes, and *ABC* is the curve to be found. The second curve, *DBE* is the mirror image of *ABC*.

In Figure 2, the second curve, *DBE* is moved up (and down) along the parallel axes to form $D'B'E'$ and $D''B''E''$. If the curve were a reciprocal trajectory (as shown, it isn't), then the angles $AB''E''$, *ABE* and $AB'E'$ would all be right angles.

Acta Eruditorum, 1727, 408–412; *Opera Omnia* I.27, 1–5

There are other ways to characterize reciprocal trajectories, for example

$$f'(-x) = \frac{1}{f'(x)}$$

and $f'(0) = \pm 1$, but Euler did not use any such characterizations in this paper.

Reciprocal trajectories get steeper, as measured by their angle, not their slope, in one direction at a rate that balances the rate at which they become less steep in the other direction. Pemberton and Bernoulli do not seem to have had any applications in mind when they posed and re-posed the problem. It seems to have been nothing more than a challenge for the rapidly developing new field of calculus.

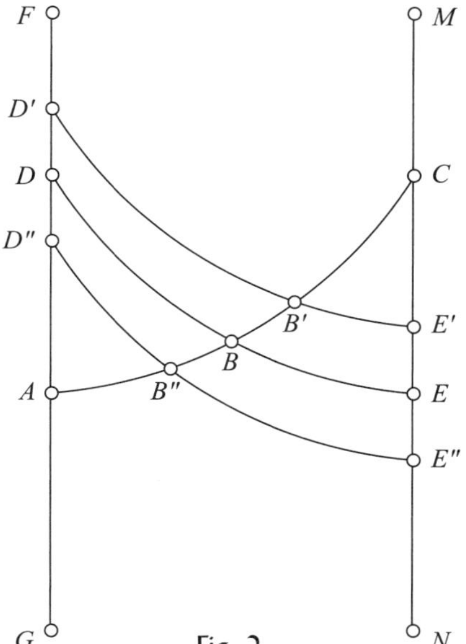

Fig. 2

Reciprocal trajectories are not required to extend beyond the two axes FG and MN. It is interesting to note how the word "axis" is used differently here than in modern times.

When Euler wrote E-3, he was only twenty years old. He was not yet ready to pose and solve problems of his own devising, so he had to find his problems among those suggested by others. As we saw, E-1 extended a problem solved by Newton. His next paper, E-2, "Dissertation on sound" was written as an assignment while he was a student at the University of Basel. He wrote the paper after this one, E-4, for the Paris Prize competition on the best positions for masts on a ship. We will see that E-9 was a homework assignment from his mentor Johann Bernoulli.

This paper also seems to have been homework. Euler explains. "The Celebrated Master Johann Bernoulli, my teacher and patron in these things, has commended this problem to me." Euler left Basel for St. Petersburg on April 5, 1727. This article appeared in the *Acta Eruditorum* later that year, after Euler left, but most historians, du Pasquier, for example, count it among Euler's Basel publications. Surely he had done most, if not all, of the work on the paper before he left Basel, though the article did not appear until after he reached Russia on May 17, 1727.

The problem is not so much finding reciprocal trajectories, but finding reciprocal trajectories that have forms simple enough to be written down. Euler sought reciprocal trajectories that are *algebraic*, that is, that can be expressed in terms of polynomials. Euler finds his reciprocal trajectories by deriving them from algebraic curves for which he can write arc length formulas.

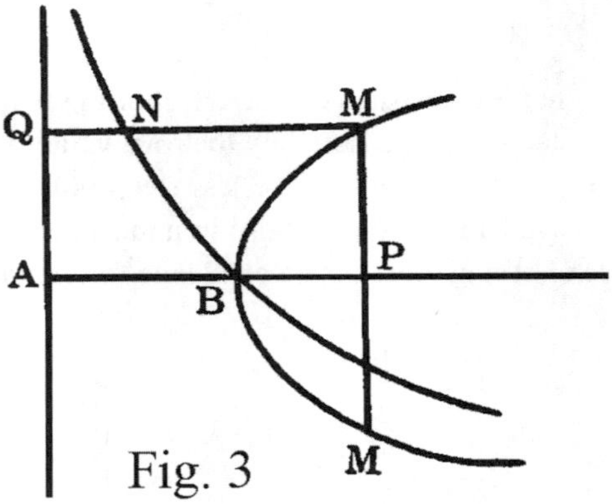

Fig. 3

In Figure 3, *MBM* is the curve $yy + \frac{2}{3}aa = a\sqrt[3]{axx}$. The line *ABP* is called the "diameter" and will be used as an x axis measuring from A. Let $AB = a\sqrt{\frac{8}{27}}$, $AP = x$ and $PM = y$. As is typical of 17th century and early 18th century diagrams, it is uncertain whether y can ever be negative, but because the curve is symmetric, it does not really matter. Clearly, x will always be positive.

Further, let the arc length of *BM* be s. Draw *QM* parallel to *AP* and locate N on *QM* so that *MN* = *BM*. This defines a new curve, *NB*, which Euler says is the "traiectoria quaesita," the curve that is sought.

To justify this, Euler first derives an equation to describe the curve *NB*. Take $AQ = t = y$ and $QN = z = x - s$. Then

$$x = \frac{\left(yy + \frac{2}{3}aa\right)^{3/2}}{aa}$$

so that

$$dx = \frac{3y\,dy\sqrt{yy + \frac{2}{3}aa}}{aa}.$$

Thus

$$ds^2 = dx^2 + dy^2 = \frac{9y^4\,dy^2 + 6aayy\,dy^2 + a^4\,dy^2}{a^4}$$

and

$$ds = \frac{3yy\,dy + aa\,dy}{aa}.$$

Consequently

$$s = \frac{y^3}{aa} + y.$$

Putting t for y, the equation $z = x - s$ becomes

$$z = \frac{\left(tt + \frac{2}{3}aa\right)^{3/2}}{aa} - t - \frac{t^3}{aa}.$$

Rearranging this gives

$$a^4zz + 2a^4tz + 2aat^3z = \frac{1}{3}a^4tt + \frac{8}{27}a^6.$$

Dividing by aa, then substituting $aa = \frac{3}{2}bb$ gives the equation for curve NB, the curve that Euler says is the reciprocal trajectory he seeks:

$$12t^3z + 18bbtz + 9bbzz - 3bbtt - 4b^4 = 0.$$

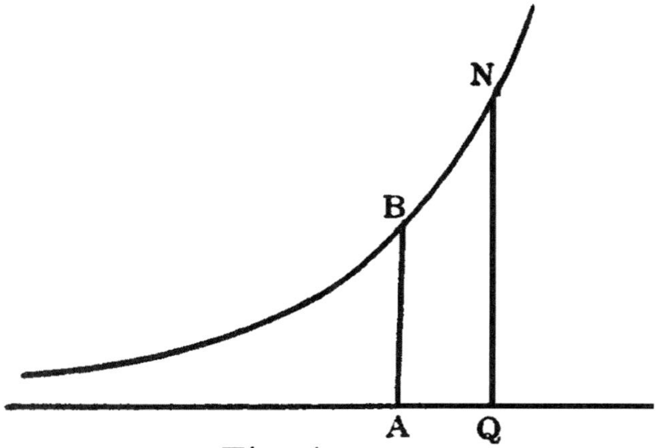

Fig. 4

He turns to trying to show that this curve is a reciprocal trajectory. He doesn't do this directly but instead considers reciprocal trajectories in general. In Figure 4, he sets $x = AQ$, $y = QN$, and, a little mysteriously, supposes

$$dy = dx\left(x + \sqrt{xx+1}\right)^n$$

where n can have any value. The reflected trajectory to this can be found by replacing x with $-x$, so that

$$dy = -dx\left(-x + \sqrt{xx+1}\right)^n.$$

Multiplying these gives $dy^2 = -dx^2$, "which is the essence of reciprocal trajectories." This doesn't quite make sense, and the nonsense is due to the notation. There are two different y's in this equation, one for the curve and one for its reflection. The interested reader can try parameterizing in terms of some variable t and calculating the appropriate dot product. A term $(y'(t)\,dt)(y'(-t)\,dt)$ arises, and Euler represents it as dy^2. Without more modern notation, though, Euler did not have the tools available to him to write this "essence of reciprocal trajectories" any more clearly than he did.

Euler does not explain that the mysterious equation $dy = dx\left(x + \sqrt{xx+1}\right)^n$ was carefully chosen to have two particular properties. First, it has an algebraic integral, except when n is 1 or -1, in which case the integral involves logarithms. Second, it satisfies the "essence of reciprocal trajectories," that is $dy^2 = -dx^2$.

We have already seen why $dy^2 = -dx^2$. To see why the integral is algebraic, take $t = x + \sqrt{xx+1}$, so that

$$x = \frac{tt-1}{2t} \quad \text{and} \quad dx = \frac{dt}{2} + \frac{dt}{2tt}$$

and then

$$2\,dy = t^n\,dt + t^{n-2}\,dt$$

so that

$$2y = \frac{1}{n+1}t^{n+1} + \frac{1}{n-1}t^{n-1} = \frac{1}{n+1}\left(x+\sqrt{xx+1}\right)^{n+1} + \frac{1}{n-1}\left(x+\sqrt{xx+1}\right)^{n-1}.$$

Euler expands and simplifies this for the cases $n = 2$, 3 and 4. For example, in the case $n = 2$, he gets

$$12yx^3 + 18xy - 9yy + 3xx + 4 = 0.$$

He further remarks that if $n = 3$, then he would get an equation of degree 5, and that, in general the degree will be $n + 2$. If $n = \frac{p}{q}$ is a fraction in lowest terms, then the order will be the numerator of the fraction

$$n + 2 = \frac{p+2q}{q}.$$

Euler illustrated this by taking $n = \frac{1}{2}$, so that $n + 2 = \frac{5}{2}$. The resulting equation is

$$72yyx^3 - 81y^4 + 144x^4 - 216yyx - 96xx + 16 = 0,$$

which is of degree 5, as promised.

Euler was just about finished here. He concluded the paper by explaining how a reciprocal trajectory can be constructed even if the original rectifiable algebraic curve has a cusp. The illustration (Figure 5) is interesting, so we include it without much explanation. The curve CAM is the rectifi-

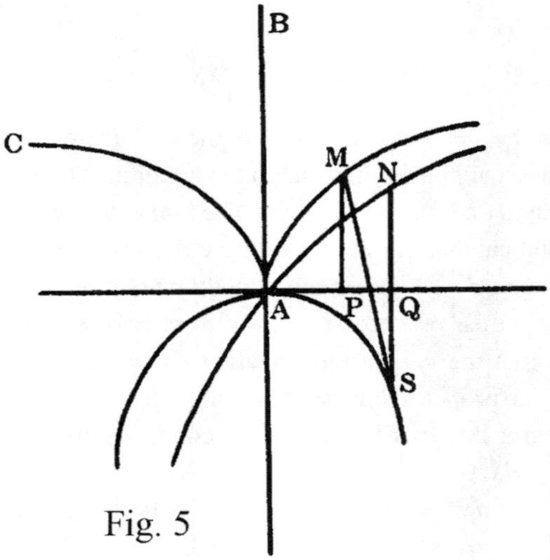

Fig. 5

able algebraic curve. The curve *AS* is its rectifying curve, and, as it turns out, *CAM* is the evolute of *AS*. Then, the curve *AN* is the reciprocal trajectory he was looking for.

He has shown how to use rectifiable algebraic curves to construct reciprocal trajectories, and he has verified that the trajectories thus constructed are indeed reciprocal. He has not given us much of a clue to why he needed rectifiable algebraic curves to begin with or how he came upon his construction. He will address this problem three more times, though, and each time he will improve his material and his presentation.

Interlude: 1728

World events

In February of 1728, Grand Duke Peter Romanov (1728–1762) was born. He would grow up to be Emperor Peter III of Russia and husband of Empress Catherine II "the Great."

The Swedish Academy of Sciences was founded in Uppsala in 1728.

Johann Heinrich Lambert (1728–1777) was born this year. Euler and Lambert exchanged a number of letters, at least 16 of which survive. Lambert is usually credited with proving that the constant e is irrational, but, as we will see in E-71, De fractionibus continuis, Euler probably has a claim that is as strong or stronger.

In Euler's life

In 1728, Euler wrote only once to his mentor in Basel, Johann Bernoulli. He wrote a letter in December, replying to one Bernoulli had sent him in January. Later in his life, Euler would become a much more conscientious correspondent.

Also in 1728 the Academy acquired a printing press and began publishing its first journal, the *Commentarii academiae scientiarum imperialis Petropolitanae*. The journal and its successors were published continuously until the Russian Revolution in 1914 and contained most of Euler's most important work. Euler had articles in every issue of the journal from 1729 to 1830, with the single exception of 1793, when the year's volume was entirely a memorial for Catherine II "the Great" and had no scientific content.

Euler's other work

In 1728, Euler wrote E-7, Tentamen explicationis phaenomenorum aeris, "Essay explaining the phenomena of air" in which he studies some of the properties of the atmosphere that are consequences of the observation that air pressure decreases with altitude. He estimates the thickness of the atmosphere and conjectures some of the effects that its change in density must have on astronomical observations.

He also wrote E-6, Dissertatio de novo quodam curvarum tautochronarum genere. "Dissertation on a new kind of tautochronal curve." This is related to E-1. The Editors of the *Opera Omnia* classify this as a paper in mechanics, so we do not treat it in this book.

Euler's mathematics

E-5, Problematis trajectoriarum reciprocarum solutio, "Solution to a problem in reciprocal trajectories," is an expansion on the material published in E-3.

Euler's other mathematical paper of 1728 is E-10, Nova methodus innumerabiles aequationes differentialis secundi gradus reducendi ad aequationes differentialis primi gradus, "A new method for reducing infinitely many different differential equations of the second degree to differential equations of the first degree." It is a classic in the development of differential equations and one of the first general methods in the subject.

3

E-5: Problematis traiectoriarum reciprocarum solutio*

Solution to problems of reciprocal trajectories

Euler arrived at his new job in Russia on May 17, 1727. His twentieth birthday had been a month earlier. His only claims to fame were his second prize in the Paris Academy competition for his essay "Meditationes super problemate nautico," E-4, on the masting of ships, and the enthusiastic recommendations of several Bernoullis. Euler was officially the Physician for the Imperial Navy, but he never actually did any work either as a physician or as a navy officer. Instead, he worked on mathematics and mechanics. His first project was to continue the work on reciprocal trajectories he had begun in E-3, and this paper was the immediate result. The notation "M. Jul. 1727" in the volume of the *Commentarii* in which E-5 appeared apparently means that it was read before the Academy in July, 1727, about two months after Euler arrived in St. Petersburg. This was probably the first time Euler officially addressed the Academy, though there would be hundreds of such appearances. Though he may have been nervous, only twenty years old and addressing the Academy for the first time, he finds a way to be droll and modest at the same time, beginning his paper "The problem on which I have endeavored to work in this paper that I have thrown together hurriedly is a celebrated one."

Euler briefly restated the problem, attributing it to the now deceased Nicolas Bernoulli. He relaxed the condition that the curve and its reflection intersect at right angles and now required only that they intersect at a constant angle. He noted that there are "an infinity of curves, both algebraic and transcendental, which satisfy" the problem. He said that he would try to find the simplest ones. He then re-cast the problem in a more algebraic and less geometric way, as illustrated in Figure 1.

The problem is to find a curve, *CBD*, around an axis *AB*. If *MP* and *NQ* are parallel to and equidistant from the axis *AB*, then the sum of the angles *PMB* + *QND* should be a constant, and should be equal to double the angle *DBA*.

To begin to derive some equations that describe these conditions, Euler asks us to "invert the curve *CBD* about the axis *AB*, and apply *QN* over *PM*, then move it so that the point *N* coincides with the point *M*, and let the inverse curve in this position be *cbd*." Because this amounts to a rigid motion without rotation, the angles *PMd* and *QND* will be equal, and so the sum of the angles *PMB* + *QND* ought to be constant and equal to *PMB* + *PMd*, which equals *BMd*.

Now introduce another axis, *PQ* normal to *AB*. That makes $AP = AQ$. Augment those distances a bit so that $Aq = Ap$, and erect ordinates *pm*, *qn* up to the curve *CBD*. Also, draw *MR* and *NS* tangent to the curve *CBD* at *M* and *N* respectively. Then the angle *RMm* is the amount by which angle *PMB* decreases, and the angle *SNn* is the angle by which *QND* increases. The condition that the sum of the angles must remain constant requires that $RMm = SNn$.

On the other side of the axis *PQ* Euler draws another curve *FEG*. He extends the segment *MP* to *F* so that *PF* is proportional to the angle *RMm*. This is supposed to measure the rate at which the angle of the curve *CBD* increases as it moves from *P* to *p*. There are some problems with this idea

Comm. Acad. Sci. Imp. Petropol., 2 (1727) 1729, 90–111; *Opera Omnia* I.27, 6–23

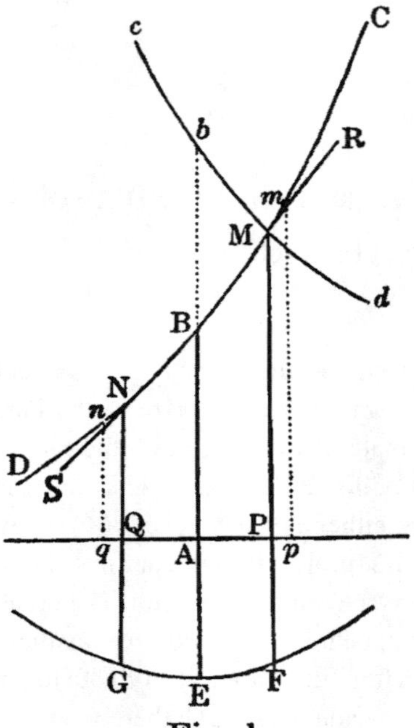

Fig. 1

as stated. First, *MR* is tangent to *CBD* at *M*, and *Mm* is very small, so angle *RMm* should be zero. Second, we are not supposed to be able to take proportions between unlike magnitudes. Proportions should compare areas to areas, lengths to lengths, angles to angles, etc., and not lengths to angles. He really means to measure how the second derivative changes with arc length, and he will make this more explicit in a few lines.

Likewise, he extends *NQ* to *G* so that *QG* is proportional to angle *SNn*. From the work above, $QG = PF$. Now "all the difficulties here are reduced to finding, from the curve *FEG*, that other curve *CBD* in which the angle *BMP* will respond in proportion to the length *PF*, and then the curve *CBD* will be a reciprocal trajectory."

Euler sets to work on those "difficulties."

Let $AP = x$, $PM = y$, $PF = u$ and $Pp = dx$. Euler asks us to "hold dx constant," something he often does. It amounts to letting x be the independent variable and letting it vary at a constant rate so that he can ignore terms containing ddx. Euler tells us

$$\text{angle } RMm \text{ is as } ddy : (dx^2 + dy^2).$$

Euler is using the older notation of proportions. This also shows that Euler is measuring angles as arc lengths, and that he is approximating the arc length as the sine of a differential triangle where ddy is the square of the opposite side and $(dx^2 + dy^2)$ is the square of the hypotenuse. Then

$$u = ddy : (dx^2 + dy^2).$$

Euler is a magician with substitutions. He takes $dy = p\,dx$ so that $ddy = dp\,dx$. Note that actually, $ddy = dp\,dx + p\,ddx$, but because Euler has asked us to "hold dx constant," $ddx = 0$, so he can neglect the second term here. These substitutions make

$$u = \frac{dp}{dx + pp\,dx} \quad \text{and} \quad u\,dx = \frac{dp}{1 + pp}.$$

From this,

"$\dfrac{dp}{1+pp}$ is the double of the element of the sector of a circle with radius 1 and tangent p."

Euler had not yet invented functions, so he couldn't say this in terms of arctangents. This tells Euler that

$$\frac{1}{2}\int u\,dx = \text{the sector of that circle,}$$

and

$$\int u\,dx = \text{area of } APEF.$$

Euler now turns his attention to curves like *FEG*, but he wants results that will apply to curves other than this particular one. He asks us to consider any curve *IEK* (Figure 2). Here, *OA* is one axis and *EA* another, perpendicular to *OA*. Euler follows ancient custom and calls *EA* a diameter, because he is taking the curve *IEK* to be symmetric about *EA*.

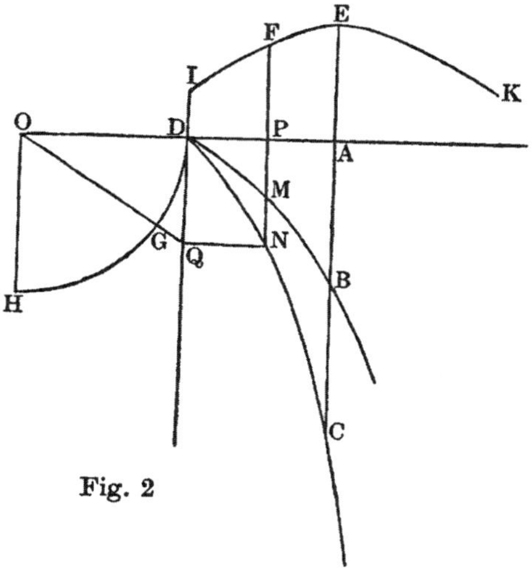

Fig. 2

Draw the line *ID* perpendicular to *OA*, and from the point O, draw circle *DGH* with radius *OD*. Draw any ordinate *FP* and find the point G on the arc of the circle so that the area of the region *PFID* equals the area of the sector *DOG*. Euler does not mention it, but if the distance *OD* is too small, then the point G might not exist. Now, extend the radius *OG* to the point Q on the line *ID*.

Locate the point N so that *QN* is parallel to *DP* and *PN* is parallel to *DQ*. This defines the curve *DN* as a function of *AP*. Assign variable names $x = AP$ and $p = PN$. We see that Euler has made a beautiful geometric construction of the curve *DN*, starting from the curve *IEK*. An algebraic version would be quite formidable.

Once the curve *DN* is constructed, Euler is ready to construct the reciprocal trajectory he calls *DM*. Because $dy = p\,dx$, take *PM* proportional to the area *DPN*. Then *DM* is a reciprocal trajectory with axis *AB* because it satisfies the conditions he developed when he was explaining Figure 1.

The mechanics of this construction are magnificent, a process that would never be seen today. We have orthogonal axes *OA* and *AE*, with a graph above the axis *OA*. We have the arc of a circle drawn from one end of the axis *OA*, and another two graphs drawn below the axis *OA*. One side of the axis is the "positive" direction for one curve, whereas the other side is the positive direction for two more.

Moreover, the construction works as a delicately designed geometry "machine" that converts the curve *IEK* with an axis of symmetry into a curve that has a kind of angle symmetry. It is the key insight in this paper and takes the place of the observation $dy^2 = -dx^2$ in the previous article. Because calculus is well suited to translate between areas and slopes, but is not well adapted to measurements involving angles, Euler invents this circular arc that works like polar coordinates to do the angle work for him. It is a most wonderful geometry machine.

Euler turns to special cases. If the area *DPFI* equals the area of the quadrant of the circle *ODH*, then *DQ* and *PN* become infinite and the curve *DN* has a vertical asymptote. If the area becomes greater than the quadrant, then *PN* becomes negative and the curve *DM* will have asymptotes as well.

It is time for an example. Take

$$u = \frac{b}{xx + aa}.$$

Perhaps by now the reader has noticed that typesetters in the 17th and 18th centuries often, but not always, put xx for x^2 because it was easier to set up. (Now that we do so much mathematics with keystrokes instead of chalk and pencil, this might be a good habit to revive.)

This expression for u can be made into an equation involving x and y by taking $u = ddy : (dx^2 + dy^2)$. Then, setting $dy = p\,dx$ and $ddy = dp\,dx$, we see that, as before, a few steps lead to

$$\frac{b\,dx}{aa + xx} = \frac{dp}{1 + pp}$$

and these integrate to give inverse trigonometric forms, or, as Euler says "depend on the quadrature of the circle." Euler has already shown his discomfort with inverse trigonometric forms and instead chooses to use complex numbers to decompose these using partial fractions. This is apparently Euler's first use of complex numbers. He reduces this equation to

$$\frac{b}{a}\left(\frac{dx}{a + x\sqrt{-1}} + \frac{dx}{a - x\sqrt{-1}}\right) = \frac{dp}{1 + p\sqrt{-1}} + \frac{dp}{1 - p\sqrt{-1}}.$$

He integrates both sides, getting logarithms, then uses laws of logarithms, takes $a = b$, and sets his constant of integration, which he calls h, equal to $\sqrt{-1}$ to get

$$\frac{a + x\sqrt{-1}}{a - x\sqrt{-1}} = \frac{dx\sqrt{-1} - dy}{dx - dy\sqrt{-1}}$$

which reduces to

$$dy = \frac{x - a}{x + a}dx.$$

The reader can check that this integrates to give $y = x - 2a\ln(a + x)$, the derivative of which is

$$y' = 1 - \frac{2a}{a + x}.$$

Since the derivative evaluated at x is the reciprocal of the derivative evaluated at $-x$, the curve $y = x - 2a\ln(a + x)$ is, as hoped, a reciprocal trajectory.

Euler finds another reciprocal trajectory by taking $b = 2a$ and says that more examples can be generated in this way.

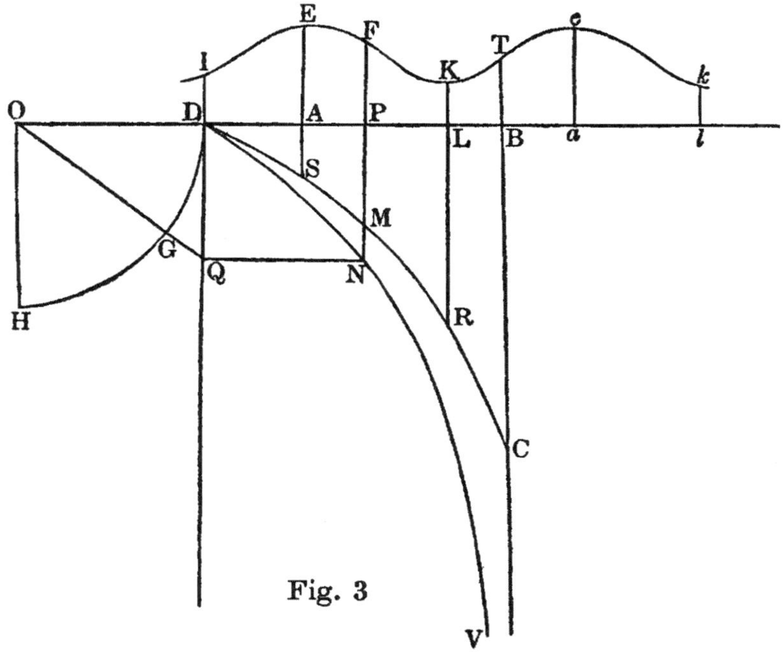

Fig. 3

Note that the definition of a reciprocal trajectory requires an axis, a line about which the curve is to be reflected. In the examples just given, the axis is the line AB, and it corresponds to the diameter, or axis of symmetry, of what Euler now calls the *curva genetrix* or generating curve. He says that if the generating curve has several diameters, then the reciprocal trajectory can have several axes as well.

This idea is developed in Figure 3. Readers who made sense of the work involved in Figure 2 will be able to see how this analysis must go. Because Euler takes the generating curve $IEKek$ always on the same side of the axis OA, there will be periodic asymptotes in the resulting reciprocal trajectory.

Euler discusses the simple case in which the curve $IEKek$ is a straight line parallel to the line DB. He tells us that this gives rise to a curve that Johann Bernoulli had named the *Patagonian curve* when he discovered it in 1726.

The Patagonian curve, it turns out, is given by $y = \ln\sqrt{\sec^2 x} = \frac{1}{2}\ln(1 + \tan^2 x)$, and it has the special property that it is a reciprocal trajectory around any axis, though if the axis is taken to be one of its vertical asymptotes or passing through one of its minima, then there are no intersections, so the condition is satisfied only vacuously.

Wonderful as this is, Euler admonishes us that "this method of duplicate quadrature is not the end of it" if simple or algebraic trajectories are desired and says that he will "move on to a method that is extremely fertile in producing simple and even algebraic trajectories."

Because much of the analysis of reciprocal trajectories involves the substitution of $-x$ for x, it is clear to us that reciprocal trajectories will be related to even functions. Euler has gone a long way toward explaining that relationship, but he has not yet offered a word to distinguish the functions that describe curves like *IEK* in Figure 2. To do this, Euler coins the term "even functions" and "odd functions" as well. We give a translation of Euler's words. The numbers 17, 18, 19 are section numbers, in common use at the time. This paper has 50 sections.

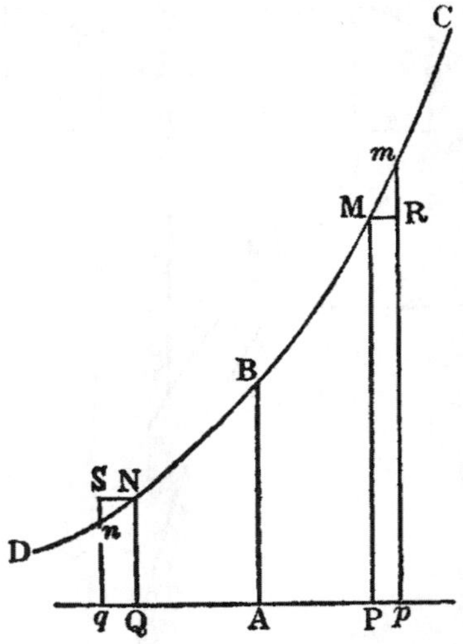

Fig. 4

17. In the first place are noted functions which I call even, of which there is this property, that they remain unchanged if, in the place of x is put $-x$. The exponents of such functions are even numbers, or fractions with numerators that are even numbers and denominators that are odd. Then, functions composed of functions of this kind, by addition or subtraction or multiplication or division or by elevation to any power will also be even, like $x^{4/5}$ or $(ax^2 + bx^{2/3})^n$.

18. Secondly, I observe odd functions, which produce their own negatives if in place of x goes $-x$. Functions of this kind are x itself, x^3, x^5, etc., all powers the exponents of which are odd numbers, or fractions the numerators and denominators of which are odd numbers. Also, functions which are formed by addition or subtraction of such powers, or their elevation to odd powers, like $x^{3/5}$ and $(ax^3 + bx^{5/7})^3$.

19. If an odd function is multiplied by an odd function, it will always make an even function, as x^3 in $x^{1/3}$ gives $x^{10/3}$. ["In" and "drawn in" are old words for multiplication, found from Euclid to Descartes. By Euler's time, their use was dying out, and this may be Euler's last use of the terms.] And an even function in an odd one always produces one that is odd. Sometimes it is possible to have a function that is both odd and even, like $x\sqrt{aa + xx}$ is a function that is an odd times an even, and it is equal to $\sqrt{aaxx + x^4}$, which is even.

Euler now asks us to consider functions that give their reciprocals when $-x$ is put in the place of x. He gives several examples,

$$\left(\frac{a+x}{a-x}\right)^n,$$

exponentials like a^x, and $(aa+xx)^{x^3}$, and all even functions raised to powers that are odd functions.

Because we are more familiar with the properties of odd and even functions than were Euler's readers, we can condense several pages of Euler's work into just a couple of paragraphs. If Q is odd, then the function

$$p = \left(Q + \sqrt{1+QQ}\right)^n$$

has the property that $pq = 1$, where q is, as before "p with $-x$ put into the place of x." This leads to reciprocal trajectories by taking $dy = p\,dx$, as outlined above.

In particular, taking $Q = x$ leads to the class of reciprocal trajectories given by

$$dy = dx\left(x + \sqrt{1+xx}\right)^n$$

that Euler studied in E-3, the previous paper on this subject. Here, taking $n = 1$ leads once again to the curve

$$12yx^3 + 18xy - 9yy + 3xx + 4 = 0$$

that Euler also found in E-3. Euler is sure that this is the simplest example of an algebraic reciprocal trajectory, after the "ordinary third order paraboloid" mentioned in E-3 and so this solves the problem that the "anonymous Englishman proposed to the Celebrated Bernoulli."

Euler finds other solutions, taking $n = 3$, and by using various odd functions like $x^{1/3}$, $x^{1/5}$, etc. for Q. Working through such examples consumes the rest of the paper and produces algebraic solutions as high as order 6.

This is a mostly forgotten paper about a mostly forgotten problem, yet it includes some gems. One is the magnificent geometric construction. Another is Euler's first published use of complex numbers. Probably the most important is Euler's first description of even and odd functions, a topic he will explain in more detail in 1748 in his *Introductio in Analysin Infinitorum*, (E-101 and E-102) book I, pages 11–15.

4

E-10: Nova methodus innumerabiles aequationes differentialis secundi gradus reducendi ad aequationes differentialis primi gradus*

A new method of reducing innumerable differential equations of the second degree to equations of the first degree

In 1728, the only general technique for solving differential equations was the method of separation of variables. In this paper, Euler develops two others. The first gives a substitution that reduces certain differential equations to a form where separation of variables will work. The second technique is much like what we now call the Method of Undetermined Coefficients.

Buried in §16, though, is a real surprise: Euler discovers the method of solving first-order differential equations by an integrating factor. This is such a basic tool that we may forget at times that somebody had to discover it. That "somebody" was Euler, and in this article, we get to watch him do it. He doesn't initially recognize the importance and generality of the method. In the flurry of activity in differential equations at the time, Euler's work was mostly overlooked. About ten years later, in 1739 and 1740, Alexis Clairaut did a more thorough job of explaining integrating factors. Because his presentation was so thorough, Clairaut is often credited with the discovery of the technique. Though there seems to be no evidence that Clairaut was aware of Euler's work on the topic, nonetheless, Euler predated Clairaut by a full decade.

We mostly learn about integrating factors to solve linear differential equations, though they also work for certain non-linear equations. In 1728, there weren't that many differential equations, and people did not yet recognize linear differential equations as an important class. Many of the problems that give rise to linear differential equations come from mechanics, vibrations and circuits. People had not yet developed the links between mechanics and mathematics, and circuits were still in the distant future. In fact, Euler himself would take one of the earliest major steps to link mechanics and mathematics in 1736 with his first important book, *Mechanica*, E-15 and E-16.

Much of this article has been translated by Florian Cajori and is reprinted in a number of places, including Smith's *A Source Book in Mathematics*, pages 638–643 [E]. Ironically, either Cajori did not translate or Smith did not choose to include the key §16.

Euler begins this paper telling us how few general methods there were to solve differential equations. He asks us to consider differential equations of the form $P\,dy^n = Q\,dv^n + dv^{n-2}\,ddv$, where P and Q are functions of y, and, as Euler says, "dy is held constant." Euler chooses this particular equation because one of the variables, in this case v, does not appear in the equation except as a differential. The notation here may be confusing. Here, ddv is a second differential, that is to say, the differential of dv, and it will be zero if "dv is held constant". On the other hand, dv^2 would be the square of dv.

To Euler, "differential equations" really were equations of *differentials*, not equations of derivatives.

**Comm. Acad. Sci, Imp. Petropol.* 3 (1728) 1732, 124–137; *Opera Omnia* I.22, 1–14
Partial translation in Smith, *A Source Book in Mathematics*, pp. 638–643

Note that here, contrary to modern custom, y is the independent variable. Euler indicates this in two different ways. He asks us to consider P and Q as functions of y and asks that dy be held constant. This means that we are to consider the variable y as changing uniformly and not with a rate that depends on some other variable, say time.

Euler substitutes $dv = z\,dy$, "since v is not an ingredient of the equation." Then $ddv = dz\,dy$. This calculation arises from the product rule, $ddv = dz\,dy + z\,ddy$. Because dy is held constant, $ddy = 0$, so the second term on the right vanishes and $ddv = dz\,dy$, as promised. This substitution gives the equation

$$P\,dy^n = Qz^n\,dy^n + z^{n-2}\,dy^{n-1}\,dz.$$

This is what Euler calls a "simple differential" because dividing through by dy^{n-1} leaves a differential equation of the first "degree." Euler uses the term "degree" for what we now call the "order" of a differential equation, that is, the highest derivative that appears. This problem started as a differential equation of degree, or order, 2 and reduced to one of degree 1.

We see that Euler is already comfortable working with fairly general forms. He regards this problem as solved because he has reduced the form of the second degree differential equation to a first degree form.

Euler often structures papers the way he organizes this one. He poses a problem and begins his analysis with some fairly easy special cases before attacking more complicated or interesting forms of the problem. In this case, his first differential equations were easy enough that they did not require the new techniques he will develop later in the paper. For example, in E-9, his first problems in shortest paths were on spheres and planes. Only after he had showed how to understand these paths with the idea of a taut string did he begin to develop his more general and difficult techniques.

Euler now turns his attention to other forms of differential equations. He follows a pattern that will become common in his work and begins with the steps that led him to his new idea. In this paper, Euler makes a number of exponential substitutions. He uses c as the base of the natural logarithm, the value we now denote by e. He knows well that the differential of c^x is $c^x\,dx$, and that, by the product rule, its "differentio-differential" or second differential is $c^x(ddx + dx^2)$.

To prepare us for the substitutions he will make, Euler tells us, following the Cajori translation, that "This being understood, these quantities must be so adapted, when substituted in place of the indeterminates that, after the substitution is made, they do not resist being removed by division; in this manner, one indeterminate or the other is eliminated and only its differentials remain."

Euler admits that the technique he develops does not work in all situations, but promises to show us three cases in which his substitutions reduce second degree (order) equations to first degree equations. His first solution applies to equations that contain only two terms. His second is that in which "each term is of the dimension that the variable in the differential has." That is to say, the equation satisfies a homogeneity condition. His third case is that in which "one or the other of the variables in the separate terms hold the same number of dimensions," another homogeneity condition.

In Euler's first example the equation is

$$ax^m\,dx^p = y^n\,dy^{p-2}\,ddy. \tag{1}$$

Here Euler takes dx to be constant and substitutes $x = c^{\alpha v}$ and $y = c^v t$, so that $dx = \alpha c^{\alpha v}\,dv$ and $dy = c^v(dt + t\,dv)$. Moving on to second differentials, Euler finds that

$$ddx = \alpha c^{\alpha v}(ddv + \alpha\,dv^2) \tag{2}$$

and

$$ddy = c^v(ddt + 2\,dt\,dv + t\,ddv + t\,dv^2). \tag{3}$$

These simplify a bit because dx is constant, so $ddx = 0$ and equation (2) becomes $ddv = -\alpha\,dv^2$. Substituting that value into the equation (3) he gets

$$ddy = c^v\big(ddt + 2\,dt\,dv + (1-\alpha)t\,dv^2\big). \tag{4}$$

Now putting these values for x, y and ddy into equation (1) gives

$$ac^{\alpha v(m+p)}\alpha^p\,dv^p = c^{(n+p-1)v}t^n(dt + t\,dv)^{p-2}\big(ddt + 2\,dt\,dv + (1-\alpha)t\,dv^2\big). \tag{5}$$

This doesn't look like progress, but in the next section, Euler shows us how it is useful.

For the two sides of this differential equation to be equal, it is sufficient, though not necessary, that the parts match. As we shall see, Euler does not always seem to have a clear idea of the distinction between necessary conditions and sufficient ones, and several times he treats necessary conditions as if they were also sufficient. This particular case is unusual for Euler because, for a change, he seems to treat a sufficient condition as if it were also necessary.

Euler now begins to match the parts on the two sides of equation (5), believing that the exponential parts must be equal. That is, he matches the factor $c^{\alpha v(m+p)}$ on the left with the corresponding factor $c^{(n+p-1)v}$ on the right, so, in particular,

$$\alpha v(m+p) = (n+p-1)v.$$

Solving for α, we get

$$\alpha = \frac{n+p-1}{m+p}.$$

Substituting this into equation (5) gives

$$a\left(\frac{n+p-1}{m+p}\right)^p dv^p = t^n(dt + t\,dv)^{p-2}\cdot\left(ddt + 2\,dt\,dv + \frac{m-n+1}{m+p}t\,dv^2\right).$$

Euler continues these substitutions and matchings until he reaches the form

$$\begin{aligned} a&\left(\frac{n+p-1}{m+p}\right)^p z^p\,dt^p \\ &= t^n\,(dt + tz\,dt)^{p-2}\left(\frac{1-n-p}{m+p}z\,dt^2 - \frac{dz\,dt}{z} + 2z\,dt^2 + \frac{m-n+1}{m+p}tz^2\,dt\right), \end{aligned}$$

where the new variable $z = \frac{dv}{dt}$. Interested readers can follow the details in the Cajori translation we mentioned above [E].

Euler multiplies both sides of this equation by z to get

$$a\left(\frac{n+p-1}{m+p}\right)^p z^{p+1}\,dt = t^n(1+tz)^{p-2}\left(\frac{1+2m-n+p}{m+p}z^2\,dt + \frac{m-n+1}{m+p}tz^3\,dt - dz\right) \tag{6}$$

which is again a differential equation of the first degree.

Finally, Euler tells us that there is a shortcut. He could have obtained this equation from equation 1 "in just one step" by substituting $v = \int z\,dt$.

Characteristically, Euler gives us an example. He considers the equation

$$x\,dx\,dy = y\,ddy$$

which he rewrites as

$$x\,dx = y\,dy^{-1}\,ddy$$

so that it is in the form of the original equation, with $a = m = p = n = 1$. Euler doesn't choose this example at random. Rather, as Euler will mention in his closing remarks, it is an equation that had arisen in some physical problem (he doesn't tell us which one!) and which "could not be reduced in any way, either by me or by those with whom I was communicating." After a few steps and armed with his new method, though, Euler presents us with the solution

$$x = c^{\int z\,dt:2} \quad \text{and} \quad y = c^{\int z\,dt}t.$$

Note the use of the ratio rather than a fraction in the exponent $\int z\,dt : 2$. Euler does this occasionally, even though in his time people usually made a clear distinction between ratios and fractions. In this case, Euler surely means to use a fraction rather than a ratio. Perhaps it is only for easier typesetting.

Euler now turns to another kind of second degree equation which this method reduces to the first degree. He gives the form as

$$ax^m y^{-m-1}\,dx^p\,dy^{2-p} + bx^n y^{-n-1}\,dx^q\,dy^{2-q} = ddy.$$

The left-hand side of this equation is homogeneous of degree one in the expressions x, y, dx and dy. He notes that this particular form has only three terms, but that there could well be more terms as long as they are all of the form $ex^r y^{-r-1}\,dx^q\,dy^{2-q}$.

Euler applies a slightly different substitution than before,

$$x = c^v \quad \text{and} \quad y = c^v t$$

and, after a few steps, finds that the equation is transformed into

$$at^{-m-1}\,dv^p(dt + t\,dv)^{2-p} + bt^{-n-1}\,dv^q(dt + t\,dv)^{2-q} = ddt + 2\,dt\,dv.$$

Since the variable v is absent, he substitutes $dv = z\,dt$, and a few more steps yield

$$at^{-m-1}z^p\,dt(1+zt)^{2-p} + bt^{-n-1}z^q\,dt(1+zt)^{2-q} = z\,dt - \frac{dz}{z},$$

and this, again, is an equation of the first degree. As before, Euler tells us that there is a shortcut, namely to substitute

$$x = c^{\int z\,dt} \quad \text{and} \quad y = c^{\int z\,dt}t$$

into the original equation.

Euler gives us the following example of this second case:

$$x\,dx\,dy - y\,dx^2 = y^2\,ddy.$$

This is not in the form of the original equation, so Euler divides by y^2 to get

$$xy^{-2}\,dx\,dy - y^{-1}\,dx^2 = ddy.$$

In the terms of the original equation, $a = m = p = 1$, $b = -1$, $n = 0$ and $q = 2$. Making the substitutions Euler outlined gives

$$t^{-2}z\,dt(1+zt) - t^{-1}z^2\,dt = z\,dt - dz : z,$$

where he has introduced that ratio notation again. Euler multiplies by t^2z, then simplifies, and finds that in the resulting equation the variables can be separated:

$$dz : z^2 = dt(t^2 - 1) : tt.$$

Because the variables are separated, he can use the classic technique and integrate both sides to get

$$atz - t = t^2z + z,$$

where a is the constant of integration, not to be confused with the value a from the original differential equation.

Knowing that $z = dv : dt$, $c^v = x$ and $t = y : x$, Euler substitutes and solves to get the first degree equation

$$y\,dy + x\,dx = ay\,dx.$$

In case $a = 0$ this is the differential equation of a circle.

Perhaps recognizing that he had given only a short justification of his claim that the form of this equation could have more than three terms, Euler now does a longer example. He introduces the equation

$$yy\,dx^3 + xx\,dy^3 - yx\,dx\,dy^2 - yx\,dx^2\,dy + yx^2\,dx\,ddy - y^2x\,dx\,ddy = 0.$$

Again making the substitutions $x = c^v$ and $y = c^vt$, Euler observes that $ddv = -dv^2$ and finds that

$$dt^3 + 2t\,dt^2\,dv - tt\,dt\,dv^2 + t\,dt\,dv^2 + t\,dv\,ddt - tt\,dv\,ddt = 0.$$

Again, because there is no variable v here, Euler substitutes $dv = z\,dt$ so that $ddt = -z\,dt^2 - dz\,dt : z$ and he gets

$$dt + 2tz\,dt - t\,dz + tt\,dz = 0,$$

"which, since only z has any dimensions, can be separated using the method of the Celebrated Johann Bernoulli given in the *Actis Lips.*" ("*Actis Lips.*" refers to *Acta Eruditorum*, published in Leipzig.) But Euler has another way, and it turns out to be a very important method. A little algebra transforms the previous equation into

$$dz + \frac{2z\,dt}{t-1} + \frac{dt}{tt-t} = 0.$$

Euler points out that the first term is just dz, and that in the second term, z is multiplied by $\frac{2\,dt}{t-1}$. He integrates this factor of the second term to get $2\int \frac{dt}{t-1}$. He prefers to leave it in this form for now, rather than evaluating the integral. He knows that the differential of $c^{2\int dt/(t-1)}$ will be

$$\frac{2\,dt}{t-1} c^{2\int dt/(t-1)}.$$

To take advantage of this, he multiplies his differential equation by $c^{2\int dt/(t-1)}$ to get

$$c^{2\int dt/(t-1)}\,dz + \frac{2c^{2\int dt/(t-1)} z\,dt}{t-1} + \frac{c^{2\int dt/(t-1)}\,dt}{tt-t} = 0.$$

The insight here is that the first two terms integrate to give $c^{2\int dt/(t-1)}z$, so, when he integrates the whole equation, he gets

$$c^{2\int dt/(t-1)}z + \int \frac{c^{2\int dt/(t-1)}\,dt}{tt-t} = a$$

where a is the constant of integration.

Now he starts to evaluate some integrals. First, $c^{2\int dt/(t-1)} = c^{2l(t-1)} = (t-1)^2$, ignoring a constant of integration (and recall that Euler uses l for the natural logarithm). Consequently

$$(t-1)^2 z + \int \frac{(t-1)\,dt}{t} = a$$

and, integrating again, we conclude that

$$(t-1)^2 z + t - lt = a.$$

Here, Euler has been brilliant, but he does not seem to recognize the importance of what he has discovered. His idea of multiplying by $c^{2\int dt/(t-1)}$ so that two terms of the equation can be merged together by a reverse application of the product rule is essentially discovering the technique of solving linear differential equations by using an integrating factor. It is one of the most familiar techniques in differential equations, far more useful than anything else Euler mentions in this paper. Yet he buries it in the middle of an example!

The reader may have noticed that, despite the distraction involved in discovering integrating factors, Euler still hasn't finished his example. He proceeds as before, substituting $dv = z\,dt$, $x = c^v$ and $y = c^{vt}$, then introducing a constant of integration b to get

$$lx = \frac{bx - ax + y - by - yly + ylx}{y - x}.$$

Exponentiating and substituting $b - a = f$ and $1 - b = g$ he arrives at

$$fx + gy = yly - xlx$$

as the solution to the equation he started to solve several paragraphs earlier. Euler notes that in the special case $f = 0$ and $g = 0$, this gives the solution

$$y^y = x^x.$$

Euler had promised to solve three types of equations. He describes his third type as "those in which one or the other of the indeterminates in the separate terms hold the same number of dimensions." This means that the differential equation is homogeneous in some variable x and its differentials dx and ddx. Here, he distinguishes two cases, depending on whether or not dx is taken to be constant.

If dx is constant, then ddx is zero. One equation of this form is

$$Px^m\,dy^{m+2} + Qx^{m-b}\,dx^b\,dy^{m+2-b} = dx^m\,ddy.$$

This is homogeneous in x and dx and is of degree m. P and Q are taken to be functions of y. Euler fails to mention it, but there could be more than two terms on the left and the equation would still satisfy his homogeneity requirements.

As before, Euler makes the now familiar substitution $x = c^v$ so that $dx = c^v\,dv$ and $ddx = c^v(ddv + dv^2) = 0$, because dx is taken to be constant. After a few steps, he substitutes $dv = z\,dy$ and soon gets

$$P\,dy + Qz^b\,dy = -z^{m+1}\,dy - z^{m-1}\,dz$$

which is of the first degree, and Euler sees his way safely to his solution.

In the other case, dx is not held constant, so ddx may not be zero, and the equation might take the form

$$Px^m\,dy^{m+1} + Qx^{m-h}\,dx^h\,dy^{m-h+1} = dx^{m-1}\,ddx$$

which is homogeneous of degree m in the variables x, dx and ddx.

Yet again the substitution $x = c^v$, setting $dv = z\,dy$ and taking $ddy = 0$ leads to a first degree equation

$$P\,dy + Qz^h\,dy = z^{m+1}\,dy + z^{m-1}\,dz.$$

Euler wraps up this paper with three paragraphs of closing remarks. He begins:

> From this I think it will be understood how differential equations of the second degree may be treated as one or another of the three kinds. Indeed, I easily submit that it is very rare to arrive at equations in which one or another of the indeterminates is absent; yet I think that nobody will be prepared to assail the usefulness of this discovery. It may happen that new ways will be found for the resolution of such equations. I recall that at one time the solution of a physical problem led to the solution of the equation $y^2\,ddy = x\,dx\,dy$. Neither I nor anybody with whom I corresponded had been able to reduce that equation in any way. Now, though, the reduction of that equation from the second to the first degree follows easily, as one may see from §10.

In this paragraph, Euler discusses a bit of why it was necessary in certain steps that one or the other of the differentials be held constant.

Finally, Euler suggests that the same methods that reduce second degree equations to first degree equations ought also to reduce third degree to second degree, and, in general, degree n equations to equations of degree $n - 1$.

References

[E] Euler, Leonhard, "On Differential Equations of the Second Order," a translation of E 10 by Florian Cajori, in *A Source Book in Mathematics*, David Eugene Smith, ed., pp. 638–643, Dover, New York, 1929.

Interlude: 1729–1731

World events

Princess Sophia Augusta Frederika of Anhalt-Zerbst was born on April 21, 1729. She later married Grand Duke Peter Fedorovich, heir to the throne of Russia and she adopted the name Catherine. Peter became Peter III, and, on his death in 1762, she became Catherine II "the Great," Empress of All the Russias. She was Empress for all of Euler's second St. Petersburg years, thus, in a sense, Euler's boss's boss, and she supported the Academy generously.

In 1730, Czar Peter II of Russia died and was succeeded by Anna Ivanova, known as Anna I. Under Anna, the Academy enjoyed greater support from the government than it had under Peter II.

Benjamin Banneker, son of a slave and author of a popular American almanac, was born on November 9, 1731. Banneker later surveyed the District of Columbia.

Brook Taylor (1685–1731) died in London in December 1731. Taylor invented integration by parts and the Taylor series expansion. Euler seems to have been the first one to use the term "Taylor series."

In Euler's life

In 1729, the young Leonhard Euler sent his first letter to Christian Goldbach. Their correspondence would continue until Goldbach's death in 1764. Almost 200 of their letters survive. Goldbach served as a kind of mentor for Euler, and it was he who introduced Euler to number theory through the works of Fermat.

In 1730, due in part to the more generous support of the Academy under Anna I, Euler left his part-time position in the Navy and began to work full-time at the Academy. Two key figures left the Academy, though, frustrated by the lack of support and by censorship of some publications of the Academy that endorsed the heliocentric theories of the solar system, propounded by Copernicus, over the old geocentric models favored by the church. These two were physicist Georg Bilfinger who went to Tuebingen, and Jacob Hermann, who took the chair of ethics at Basel. Hermann was Euler's mother's second cousin and a good friend of Euler.

In 1731, Euler's salary was frozen at 400 rubles, while four others, Gmelin, Krafft, Mueller and Weitbrecht, got increases to 400 rubles. This was a key event in the long-running friction between Euler and the Secretary of the Academy, Johann Schumacher. Later in his career, Euler became far more adept at salary negotiations.

Euler's other work

In 1729, Euler wrote two works on tautochrones that appear in the *Opera Omnia* in volumes about mechanics. They are E-12, De innumerabilibus curvis tautochronis in vacuo, "On infinitely many tautochrone curves in a vacuum," and E-13, Curva tautochrona in fluido resistentiam faciente secundum quadrata celeritatum, "Tautochronal curves in a fluid making resistance proportional to the square of the speed."

In 1730 he wrote E-8, a paper on laminar curves titled Solutio problematis de invenienda curva, quam format lamina utcunque elastica in singulis punctis a potentiis quibuscunque sollicitata, "Solution to the problem of finding the curve that is formed by an elastic lamina subject to force at a single point." It, too, is classified as a paper on mechanics.

In 1731, he wrote his first book, E-33, *Tentamen novae theoriae musicae*, "Essay on a new theory of music," in which he applies logarithms for the first time to the theory of musical scales. It is a very important milestone in the physics of music.

That same year he wrote another paper on tautochrones, E-24, Solutio singularis circa tautochronism, "A singular solution about tautochronism," and another paper on mechanics, E-22, De communicatione motus in collisione corporum, "On the communication of motion in colliding bodies."

Euler's mathematics

Euler wrote only three mathematics papers in these three years. In 1729 he wrote E-19, De progressionibus transcendentibus, seu quarum termini generales algebraice dari nequeunt, "On transcendental progressions, or those for which the general term cannot be given algebraically." This paper is based on his first letter to Goldbach, and in it he discovers what we now call the gamma function and explores many of its properties.

Then in 1730 he wrote up a problem that Johann Bernoulli had given him as a homework assignment. It became E-9, De linea brevissima in superficie quacunque duo quaelibet puncta jungente, "On the shortest curve joining two given points on a surface."

The following year, 1731, he wrote E-20, De summatione innumerabilium progressionum, "On the summation of infinitely many progressions," a kind of companion to E-19 in which Euler begins his study of the harmonic series and makes steps towards discovering γ, the Euler-Mascheroni constant.

5

E-19: De progressionibus transcendentibus seu quarum termini generales algebraice dari nequeunt*

On transcendental progressions, or those for which the general term cannot be given algebraically

This wonderful paper, "On transcendental progressions" and its sequel, "On the summation of innumerably many progressions," are the first fruits of the rich correspondence between Euler and Christian Goldbach, Euler's boss and sometime mentor in Russia. Euler began exchanging letters with Goldbach on October 13/24,[1] 1729. Their relationship lasted until Euler's final letter from Berlin in April 1764, the year Goldbach died. Of the 196 surviving letters between Euler and Goldbach, the first one is particularly significant.

Euler begins the correspondence rather abruptly, "Most Celebrated Sir: I have been thinking about the laws by which a series may be interpolated .... The most Celebrated Bernoulli suggested that I write to you."

The letter goes on to discuss Euler's first definition of what we now call the gamma function. In passing, Euler mentions a fact that depends on knowing that $e^{\pi i} = -1$. He then uses a process similar to the interpolation of the factorial progression that led to the gamma function to give the partial sums of the harmonic series.

These two achievements, extension of the factorial progression to discover the gamma function and interpolation of the partial sums of the harmonic series, became the basis of two of Euler's early papers. The first of these, E-19, "On Transcendental Progressions," led to further development of the gamma function, including the idea of fractional derivatives. The second, E-20, "On the summation of innumerable progressions," provided a rapidly converging approximation for the sum of the reciprocals of the squares, which, in turn, led four years later to one of Euler's most famous discoveries, that

$$\zeta(2) = \frac{\pi^2}{6}.$$

The letter

Before looking at the paper itself, we look at the letter Euler sent to Goldbach. Euler begins by proclaiming, with little preamble, that the general term of the "series" 1, 2, 6, 24, 120, etc. is given by

$$\frac{1 \cdot 2^m}{1+m} \cdot \frac{2^{1-m} \cdot 3^m}{2+m} \cdot \frac{3^{1-m} \cdot 4^m}{3+m} \cdot \frac{4^{1-m} \cdot 5^m}{4+m} \text{ etc.} \qquad (1)$$

Comm. Acad. Sci. Imp. Petropol. 5 (1730/1) 1738, 36–57; *Opera Omnia* I.14, 1–24

[1] In the 1700's, Russia still used the Julian calendar, while most of the rest of the world used the Gregorian calendar. Hence, in Russia, the date was October 13, but in the rest of the world, they called it October 24. Russia did not change to the Gregorian calendar until after the Revolution in 1914.

This is a remarkable infinite product. The modern eye immediately wants to do some cancellation among factors in the numerator to "reduce" this to

$$\frac{1}{1+m}\cdot\frac{2}{2+m}\cdot\frac{3}{3+m}\cdot\frac{4}{4+m}\text{ etc.}$$

Then, we see that the value of m "shifts" the denominator, and when m is a positive integer, each of the factors in the denominator exactly cancels with a factor in the numerator, leaving $m!$.

So, why didn't Euler use the simpler form? It turns out that Euler's form converges, albeit slowly, to $m!$ without any rearrangement of factors. The modern-looking form converges to zero. Even at this early date (Euler was only 22) he was aware of issues of convergence, though it would not become a central issue in mathematics for more than a hundred years and Euler did not do much to advance our understanding of convergence.

Euler then substitutes $m = \frac{1}{2}$ and tells us "The term with exponent $\frac{1}{2}$ is equal to this:

$$\frac{1}{2}\sqrt{\sqrt{-1}\cdot l-1},$$

which is equal to the side of the square equal to the circle with diameter = 1." Here Euler uses l to denote the natural logarithm but does not use the famous constant we now call π. The mathematical community had yet to adopt a standard symbol for the ratio of the circumference of a circle to its diameter. In part this was because people still regarded π as a ratio, and still maintained a distinction between ratios and numbers. In part it was because Euler had not yet led the way in demonstrating the value of a standardized system of notation.

That Euler's radical equals $\frac{\sqrt{-\pi}}{2}$ follows easily from knowing $e^{\pi i} = -1$ by taking logs of both sides, but the fact was not widely known at the time, and our modern notation using π and i would have been quite foreign. It is not quite clear why Euler describes $\frac{\sqrt{\pi}}{2}$ instead of $\frac{\sqrt{-\pi}}{2}$. In the paper, though not in the letter, Euler goes into more detail, substituting $\frac{1}{2}$ into formula (1) to get

$$\sqrt{\frac{2\cdot 4}{3\cdot 3}\cdot\frac{4\cdot 6}{5\cdot 5}\cdot\frac{6\cdot 8}{7\cdot 7}\cdot\frac{8\cdot 10}{9\cdot 9}\cdot\text{etc.}}$$

which he compares to Wallis's formula

$$\frac{4}{\pi}=\frac{3\cdot 3\cdot 5\cdot 5\cdot 7\cdot 7\cdots}{2\cdot 4\cdot 4\cdot 6\cdot 6\cdot 8\cdots}.$$

Then the value $\frac{\sqrt{\pi}}{2}$ follows easily.

Euler continues the letter with some remarks about partial sums of the harmonic progression. These ideas will lead to the next article, E-20. Details for that will have to wait.

Later letters

Six weeks later, Goldbach replies from Moscow with a letter dated December 1/12, 1729, in which he repeats Euler's form of the gamma function and mentions the sum of the series $1+1\cdot 2+1\cdot 2\cdot 3+1\cdot 2\cdot 3\cdot 4+\cdots$, and also admits that he doesn't know what "hyperbolic logarithms" are. As a final note, he writes "P.S. A note to you is that Fermat has observed that all numbers with the formula $2^{2^x}+1$, that is 3, 5, 17, etc. are primes, but he himself was not able to prove this, and, as far as I know, nobody since him has proved it, either." But this was a spark for another fuse, and we will return to this thread in E-26, "Observations on some theorems of Fermat."

Euler replies to Goldbach from St. Petersburg on January 8/19, 1730 with an outline of some of the integral methods he was using to analyze the form for the gamma function, steps he would detail in his paper E-19. He closes with a note "I have been able to discover nothing about the observation noticed by Fermat."

In Goldbach's reply from Moscow, written five months later on May 22/31, 1730, he echoes one of Euler's integral forms for the gamma function, and forwards a couple of other of Fermat's observations on number theory.

After this fourth letter, Euler and Goldbach write each other at a more rapid pace, but they do not return to these topics. It is clear from these letters that the mathematical insights and initiative were Euler's, but that Goldbach was a willing, if inept, partner in the correspondence.

E-19: On Transcendental Progressions

Euler began to publish the fruits of this correspondence with Goldbach in Volume V of the *Commentarii academiae scientarum Petropolitanae*, for the years 1730/31 and finally printed in 1738.

Euler wrote five of the sixteen mathematical papers in this volume. Others were by Friedrich Christoff Mayer (4), Johann Bernoulli, Daniel Bernoulli (5) and George Wolffgang Krafft. Later volumes of the *Commentarii* would contain as many as 16 papers by Euler, but this was still very early in his career. Though prolific, he is not yet dominant.

Euler's 21-page paper De progressionibus transcendentibus seu quarum termini generales algebraice dari nequeunt, "On transcendental progressions, that is, those whose general terms cannot be given algebraically," presents his results on the gamma function as outlined in the letters to Goldbach. This important paper has been translated by Stacy Langton and is available at his website, `home.sandiego.edu/~langton`. In all quotations here from E-19 we follow his excellent translation.

Euler frames the prototypical problem for this paper as finding the general term of the progression

$$1 + 1 \cdot 2 + 1 \cdot 2 \cdot 3 + 1 \cdot 2 \cdot 3 \cdot 4 + \text{etc.}$$

This is an odd way to pose the question, for, in this paper, he never actually considers the *sum* of this progression. He just wants to find a formula for the nth term.

That problem stated, Euler gives the form for the gamma function as above. After showing that the formula (1) yields the values 2 and 6 for $m = 2$ and 3, respectively, he applies Wallis' result to evaluate formula 1 for $m = \frac{1}{2}$, getting "that the term of index $\frac{1}{2}$ is equal to the square root of the circle whose diameter $= 1$." When Euler speaks of "the circle," he means the area of the circle.

Euler pauses for some commentary. He writes (we use Langton's translation):

> 3. I had previously supposed that the general term of the series 1, 2, 6, 24, etc. could be given, if not algebraically, at least exponentially. But after I had seen that some intermediate terms depended on the quadrature of a circle, I recognized that neither algebraic nor exponential quantities were suitable for expressing it. For the general term of that progression must thus include not only algebraic quantities but also those depending on the quadrature of a circle, and perhaps even on other quadratures; thus it could not be represented either by any algebraic formula, or by an exponential.

COMMENTARII ACADEMIAE SCIENTIARVM IMPERIALIS

PETROPOLITANAE.

TOMVS V.

AD ANNOS clɔ lɔcc xxx. et clɔ lɔcc xxxi.

PETROPOLI,

TYPIS ACADEMIAE.

clɔ lɔ cc xxxviii.

DE PROGRESSIONIBVS TRANSCENDENTIBVS, SEV QVARVM TERMINI GENERALES ALGEBRAICE DARI NEQVEVNT.

Auct. L. Eulero.

I.

Cum nuper occaſione eorum, quae Cel. *Goldbach* de ſeriebus cum Societate communicauerit, in expreſſionem quandam generalem inquirerem, quae huius Progreſſionis $1 + 1 . 2 + 1 . 2 . 3 + 1 . 2 . 3 . 4 +$ etc. terminos omnes daret; incidi conſiderans, quod ea in infinitum continuata tandem cum geometrica confundatur in ſequentem expreſſionem, $\frac{1 . 2^n}{1+n} . \frac{2^{1-n} 3^n}{2+n} . \frac{3^{1-n} 4^n}{3+n} . \frac{4^{1-n} 5^n}{4+n} .$ etc. quae dictae progreſſionis terminum ordine n exponit. Ea quidem in nullo caſu abrumpitur, neque ſi n eſt numerus integer neque ſi fractus, ſed ad quemuis terminum inueniendum tantummodo approximationes ſuppeditat, niſi excipiantur caſus $n = 0$, et $n = 1$, quibus ea actu abit in 1. Ponatur $n = 2$, habebitur $\frac{2.2}{1.3} . \frac{3.3}{2.4} . \frac{4.4}{3.5} . \frac{5.5}{4.6}$ etc. = termino ſecundo 2. Si $n = 3$ habebitur $\frac{2.2.2}{1.1.4} . \frac{3.3.3}{2.2.5} . \frac{4.4.4}{3.3.6} . \frac{5.5.5}{4.4.7}$ etc. = termino tertio 6.

§. 2. Quanquam autem haec expreſſio nullum vſum habere videatur in inuentione terminorum

4. Since I knew, however, of formulas involving differential quantities, which indeed in some cases could be integrated, and then produce algebraic quantities, in other cases however could not, and then represent quantities depending on the quadratures of curves, it occurred to me that perhaps there were formulas of this kind, suitable for expressing the general terms of the aforementioned progression, as well as of similar ones. Indeed, progressions whose general terms cannot be expressed algebraically, I call *transcendental*; just as everything in Geometry which surpasses the powers of ordinary Algebra is commonly called transcendental.

5. I therefore asked myself in what way differential formulas would be best suited to express the general terms of progressions. Now a general term is a formula involving not only constant quantities, but also some other non-constant quantity, say n, which gives the order or index of the terms; thus, if the third term is wanted, one should put 3 in place of n. But a differential formula must contain some variable quantity. It would not make sense, of course, to take n for this quantity; n is not the variable of integration, but after the formula has been integrated, or after its integration has been indicated, then n should serve to express the formation of the progression. Thus, a differential formula has to contain some variable quantity x, which however after integration must be set equal to some other quantity related to the progression, and this is how we get the particular term of index n.

Here we see that Euler classifies quantities into two types, algebraic and transcendental. Among the transcendental functions, sometimes he singles out exponential quantities. Because he suspects that the quantities he seeks are transcendental, he will try to express them using quadratures, that is integrals, of the form $\int p\,dx$. The problem will be to find the appropriate functions p.

In section 8, about a third of the way through this paper, and without revealing what made him think of it, Euler considers

$$\int x^e\,dx\,(1-x)^n$$

"and let it be integrated so that it becomes $= 0$ if $x = 0$, and then by setting $x = 1$, let it give the term of order n of the resulting progression." That means he wants the definite integral from 0 to 1. Both e and n are constants, with values as yet unspecified.

Apply the binomial theorem to $(1-x)^n$:

$$(1-x)^n = 1 - \frac{n}{1}x + \frac{n(n-1)}{1\cdot 2}x^2 - \frac{n(n-1)(n-2)}{1\cdot 2\cdot 3}x^3 + \text{etc.}$$

Multiply this series by $x^e\,dx$ and integrate to get

$$\int x^e\,dx(1-x)^n = \frac{x^{e+1}}{e+1} - \frac{nx^{e+2}}{1\cdot(e+2)} + \frac{n(n-1)x^{e+3}}{1\cdot 2\cdot(e+3)} - \frac{n(n-1)(n-2)x^{e+4}}{1\cdot 2\cdot 3\cdot(e+4)} + \text{etc.}$$

Integrating this from 0 to 1 gives

$$\frac{1}{e+1} - \frac{n}{1\cdot(e+2)} + \frac{n(n-1)}{1\cdot 2\cdot(e+3)} - \frac{n(n-1)(n-2)}{1\cdot 2\cdot 3\cdot(e+4)} + \text{etc.}$$

The resulting series can be used to find the nth term of the progression Euler seeks. When $n = 0$, the term is $\frac{1}{e+1}$. If $n = 1$ it is

$$\frac{1}{(e+1)(e+2)},$$

when $n = 2$,

$$\frac{2}{(e+1)(e+2)(e+3)}.$$

The fractions continue to add nicely, if tediously, and when $n = 3$, the term is

$$\frac{6}{(e+1)(e+2)(e+3)(e+4)}.$$

Euler says, "the rule which these terms follow is obvious," but he tells us what it is anyway, namely

$$\int x^e\,dx\,(1-x)^n = \frac{1\cdot 2\cdot 3\cdot 4\cdot\cdots\cdot n}{(e+1)(e+2)\cdots(e+n+1)}$$

when n is a positive integer. However, if n is not an integer, the result is better left as the series

$$\frac{1}{e+1} - \frac{n}{1\cdot(e+2)} + \frac{n(n-1)}{1\cdot 2\cdot(e+3)} - \frac{n(n-1)(n-2)}{1\cdot 2\cdot 3\cdot(e+4)} + \text{etc.},$$

which will not terminate because the numerators never have zero as a factor. Euler multiplies both sides of the last integral by $e+n+1$ to get a form he will find more useful:

$$(e+n+1)\int x^e\,dx\,(1-x)^n = \frac{1\cdot 2\cdot 3\cdot 4\cdot\cdots\cdot n}{(e+1)(e+2)\cdots(e+n)}.$$

Euler briefly works through an example in the case $e = 2$. He finds that the general term is

$$\frac{2}{(n+1)(n+2)},$$

the reciprocal of a triangular number. We leave details to the reader.

Moving to Section 10, we consider the case where e is a rational number, $e = \frac{f}{g}$. Substituting and simplifying a bit, he transforms the last formula into

$$\frac{f+(n+1)g}{g}\int x^{f/g}\,dx\,(1-x)^n = \frac{1\cdot 2\cdot 3\cdot\cdots\cdot n}{(f+g)(f+2g)(f+3g)\cdots(f+ng)}g^n,$$

which he divides by g^n, to get

$$\frac{f+(n+1)g}{g^{n+1}}\int x^{f/g}\,dx\,(1-x)^n = \frac{1\cdot 2\cdot 3\cdot\cdots\cdot n}{(f+g)(f+2g)(f+3g)\cdots(f+ng)}. \tag{2}$$

Euler recognizes this as an intimidating formula, so he reassures us with an example, $f = 1$, $g = 2$, whose nth term is

$$\frac{2\cdot 4\cdot 6\cdot 8\cdot\cdots\cdot 2n}{3\cdot 5\cdot 7\cdot 9\cdot\cdots\cdot(2n+1)}$$

and the progression itself, which Euler prefers to write as a series, will be

$$\frac{2}{3} + \frac{2\cdot 4}{3\cdot 5} + \frac{2\cdot 4\cdot 6}{3\cdot 5\cdot 7} + \text{etc.}$$

The integral formula (2) tells us that the term of index n is

$$\frac{2n+3}{2} \int dx\,(1-x)^n \sqrt{x}.$$

In the case $n = \frac{1}{2}$, this becomes

$$2 \int dx\, \sqrt{x - xx}.$$

This is a familiar integral, giving the area of a circle of diameter 1, the value we now write $\frac{\pi}{4}$. Euler could have used it to provide another proof of the Wallis formula, but he does not avail himself of the opportunity. The proof would have involved some bold manipulations on infinite products. Perhaps Euler senses the risk and so chooses not to pursue it.

Now, Euler wants to look at the progression of binomial coefficients, which he again writes as a sum

$$1 + \frac{r}{1} + \frac{r(r-1)}{1\cdot 2} + \frac{r(r-1)(r-2)}{1\cdot 2\cdot 3} + \text{etc.}$$

with general term given by

$$\frac{r(r-1)(r-2)\cdots(r-n+2)}{1\cdot 2\cdot 3\cdots(n-1)}.$$

Euler recognizes this as being related to the form

$$\frac{(f+g)(f+2g)\cdots(f+ng)}{1\cdot 2\cdot 3\cdot\cdots\cdot n},$$

the reciprocal of the form he had been working with earlier, with the values $g = -1$ and $f = r+1$.

In Section 14 (this is only Section 11), Euler will show that

$$r(r-1)\cdots 1 = \int dx(-lx)^r.$$

Just this once, Euler doesn't mind doing things out of order, so he uses this result now to say that the general term of his progression of binomial coefficients will be given by

$$\frac{r(r-1)(r-2)\cdots(r-n+2)}{1\cdot 2\cdot 3\cdots(n-1)} = \frac{\int dx(-lx)^r}{\int dx(-lx)^{n-1} \int dx(-lx)^{r-n+1}}.$$

If $r = 2$, the general term is

$$\frac{2}{\int dx(-lx)^{n-1} \int dx(-lx)^{3-n}}$$

and this gives the progression 1, 2, 1, 0, 0, 0 etc., which Euler chooses to write as a sequence rather than as a sum. Let A be the area of the circle of diameter $= 1$, that is in modern notation, $A = \frac{\pi}{4}$, and let $n = \frac{3}{2}$. Euler knows that $\int dx(-lx)^{1/2} = \sqrt{A}$ and $\int dx(-lx)^{3/2} = \frac{3}{2}\sqrt{A}$, so the term with index $n = \frac{3}{2}$ works out to be $\frac{4}{3A}$, which Euler tells us is approximately $\frac{5}{3}$, but which we would evaluate as $\frac{16}{3\pi} \approx 1.69765$.

Euler returns to the factorial numbers $1 \cdot 2 \cdot 3 \cdot \cdots \cdot n$, given by his form

$$\frac{1 \cdot 2 \cdot 3 \cdots n}{(f+g)(f+2g)\cdots(f+ng)}$$

in the case $f = 1$, $g = 0$. Because he showed in Section 10 that

$$\frac{f+(n+1)g}{g^{n+1}} \int x^{f/g}\, dx\, (1-x)^n = \frac{1 \cdot 2 \cdot 3 \cdot \cdots \cdot n}{(f+g)(f+2g)(f+3g)\cdots(f+ng)}$$

he wants to evaluate the integral

$$\int \frac{x^{1/0} dx\, (1-x)^n}{0^{n+1}}.$$

This presents a problem, with that $x^{1/0}$ in the numerator of the integrand and 0^{n+1} in the denominator! The mathematical world had not yet settled on the rules for dealing with such indeterminate forms, so Euler hunts for a substitution $u(x)$ such that $u(0) = 0$ and $u(1) = 1$ so that he can make a change of variables that will transform the integral into one he can evaluate. He substitutes $x^{g/(f+g)}$ for x and thus $\frac{g}{f+g} x^{-f/(g+f)}$ for dx. (In keeping the variable named x instead of introducing a new variable name, Euler is changing the variable, rather than changing variables.) This gives

$$\frac{f+(n+1)g}{g^{n+1}} \int \frac{g}{f+g} dx \left(1 - x^{g/(f+g)}\right)^n.$$

Again, $f = 1$, $g = 0$, so the integral is

$$\int \frac{dx(1-x^0)^n}{0^n}.$$

Euler thinks he knows how to deal with this. He evaluates $\frac{1-x^0}{0}$ by examining $\frac{1-x^z}{z}$ when z vanishes. "By a known rule,"—l'Hôpital's—he differentiates numerator and denominator with respect to z to get $-x^z\, lx$ and substitutes $z = 0$ to conclude

$$\frac{1-x^0}{0} = -l\, x.$$

This tells him that

$$\int \frac{dx\, (1-x^0)^n}{0^n} = \int dx\, (-l\, x)^n.$$

For positive integer values of n, this integral thus becomes $1 \cdot 2 \cdot 3 \cdot \cdots \cdot n$, and the integral is defined for fractional values of n as well. This is the result in Section 14 that Euler had promised back in Section 11. Euler checks his work in the case $n = 3$ by evaluating $\int dx\, (-l\, x)^3$ and getting 6.

Today we would perhaps prove that $n! = \int_0^1 (-\ln x)^n\, dx$ using integration by parts and mathematical induction rather than trying to forge Euler's use of indeterminate forms into something that would meet today's standards.

Euler turns to terms of the form $(f+g)(f+2g)\cdots(f+ng)$. Using his earlier formulas

$$\frac{f+(n+1)g}{g^{n+1}} \int x^{f/g}\, dx\, (1-x)^n = \frac{1 \cdot 2 \cdot 3 \cdot \cdots \cdot n}{(f+g)(f+2g)(f+3g)\cdots(f+ng)}$$

and

$$r(r-1)\cdots 1 = \int dx(-lx)^r$$

it follows that

$$(f+g)(f+2g)\cdots(f+ng) = \frac{g^{n+1}\int dx(-\ln x)^n}{(f+(n+1)g)\int x^{f/g}dx(1-x)^n}.$$

Now, Euler uses this form to examine some other progressions. He lets $g = 2$, $f = -1$ to generate the progression

$$1, \quad 1\cdot 3, \quad 1\cdot 3\cdot 5, \quad 1\cdot 3\cdot 5\cdot 7 \text{ etc.}$$

and finds that the term of index $n = \frac{1}{2}$ has value $= \sqrt{\frac{2}{p}}$, where p is the circumference of the circle of diameter 1. (We would write this $\sqrt{\frac{2}{\pi}}$, of course.) Euler gives no reason for switching from describing such quantities related to π in terms of the circumference of a circle rather than the area, as he has in earlier sections. He now takes $g = 1$ and $f = \frac{nn-n}{2}$ to get the progression

$$1, \quad 2\cdot 3, \quad 4\cdot 5\cdot 6, \quad 7\cdot 8\cdot 9\cdot 10 \text{ etc.},$$

and gets a general term

$$\frac{2\int dx\,(-l\,x)^n}{(nn+n+2)\int x^{(nn-n)/2}\,dx\,(1-x)^n} = \frac{2\int dx\,(-l\,x)^n}{(nn+n+2)\int dx\,(x^{(n-1)/2}-x^{(n+1)/2})^n}.$$

This is a trickier move than it may seem at first glance because it makes f depend on n. Indeed, n was supposed to be the index of a progression that depends on f and on g, or, actually on $\frac{f}{g}$. Now, though, he is looking at what is essentially a progression with two indices, $\frac{f}{g}$ and n, and looking at a sequence of terms that go along a diagonal of that progression, where f depends on n.

We're getting near the end of this paper, so, as often happens, Euler's examples start to get a bit cumbersome and repetitious. Euler notes that if he uses his main formula

$$\frac{f+(n+1)g}{g^{n+1}}\int x^{f/g}\,dx\,(1-x)^n = \frac{1\cdot 2\cdot 3\cdot\cdots\cdot n}{(f+g)(f+2g)(f+3g)\cdots(f+ng)}$$

and takes $f = n$, $g = 1$, he gets numbers of the form

$$\frac{1\cdot 2\cdot 3\cdot\cdots\cdot n}{(n+1)(n+2)(n+3)\cdots 2n},$$

numbers we now recognize as having being reciprocals of the central binomial coefficients $\binom{2n}{n}$. Euler interpolates this progression for all values $\frac{p}{2}$, first explicitly for $p = 1, 3, 5$ and then in general for p odd. He gets a form that involves a cube root as well as two integrals:

$$\sqrt[3]{1\cdot 2\cdot\cdots\cdot p\cdot\left(\frac{2p+3}{3}\right)(p+1)\int dx\,(x-xx)^{p/3}\int dx\,(xx-x^3)^{p/3}}.$$

He sets $f = 2n$, $g = 1$ to get progressions we would now say have the form

$$\frac{(n!)^3}{(3n)!},$$

and interpolates them for indices of the form $\frac{p}{3}$, yielding a form involving a fourth root and a product of two integrals. Euler shows that the pattern of complexity continues for $f = 4n$, $g = 1$ and writes "One could proceed further in the same manner to more complex progressions, by taking more complex general terms, but I will not pursue these things any further."

Instead, Euler closes with some philosophical remarks. In the last sections of the paper, Euler writes

> To round off this discussion, let me add something which certainly is more curious than useful. It is known that $d^n x$ denotes the differential of x of order n and if p denotes any function of x and dx is taken to be constant then ... the ratio of $d^n p$ to $d^x n$ can be expressed algebraically .... We now ask, if n is a fractional number, what the value of that ratio should be.

So, Euler discovers Fractional Derivatives as a toss-in at the end of this paper!

Despite the ultimate importance of the gamma function, Euler does relatively little to follow up on what he began here in E-19. He includes some of these results in his *Calculus differentialis*, published in 1755, where he devotes almost 30 pages of chapter 17 to the interpolation of series. Late in life, in E-652, De termino generali serierum hypergeometricarum, "On the general term of hypergeometric series" and E-661 Variae considerationes circa series hypergeometricae, "Several considerations about hypergeometric series" papers presented in 1776 but published posthumously, he generalizes the factorial function as $a(a+b)(a+2b)(a+3b)\cdots(a+(n-1)b)$. He interpolates that with

$$a^n \cdot \frac{a^{1-n}(a+b)^n}{a+nb} \cdot \frac{(a+b)^{1-n}(a+2b)^n}{a+(n+1)b} \cdot \frac{(a+2b)^{1-n}(a+3b)^n}{a+(n+2)b} \cdot \text{etc.}$$

In light of the ultimate importance of the gamma function, one might wonder why Euler did not do more with it. After all, we see it in a variety of mathematical contexts: complex variables, probability, Laplace transforms, fractional derivatives, among others. Why, then, did Euler stop where he did?

The answer lies in history. All these applications of the gamma function arose in the 19th and 20th centuries. Euler lived in the 18th century, and he could only solve problems from his own times. If his answers help us solve later problems, then that is our good luck, but Euler could not have anticipated how we would use his discoveries. To Euler, the issue was to interpolate the progression of factorial numbers. He stopped where he did because he had solved that problem.

6

E-9: De linea brevissima in superficie quacunque duo quaelibet puncta iungente*

On the shortest curve on a surface that joins any two given points

Even Euler had to do homework. Sometime before 1728, his mentor Johann Bernoulli at the University of Basel assigned him the task of finding "the shortest line between two given points on a surface." The manuscript that Euler submitted to Bernoulli is in the archives in Moscow (though that seems to be a copy, as it is not in Euler's handwriting). Euler then published a paper based on the homework in the *Commentarii* of 1728.

The manuscript and the paper together lay the analytical foundations for the calculus of variations. There had been earlier results along these lines. Some people trace its origins at least to Dido's solution of the isoperimetric problem as recounted in Virgil's *Aeneid*: (Lines 505–508 of Book I in the John Dryden translation, lines 365–368 of the original Virgil[1])

At last they landed, where from far your eyes
May view the turrets of new Carthage rise;
There bought a space of ground, which (Byrsa call'd,
From the bull's hide) they first inclos'd, and wall'd.

Here, the story goes, Dido fled Tyre to found Carthage. She purchased as much land as she could enclose with a bull's hide. She was clever, though, and cut the hide into strips and claimed enough land to found the city by enclosing it with the strips instead of with the hide itself. The so-called "Dido problem" is to show that the largest possible area that can be enclosed by a curve of a given length is circular. Because Carthage was a seaside city, Dido's solution would have been a circular arc rather than a circle. Whoever gave the problem its name seems to have neglected this detail.

A bit closer to Euler's time, in 1697, the same Johann Bernoulli who would later be Euler's mentor posed the "brachistochrone problem":

Given two points A and B in a vertical plane, what is the curve traced out by a point acted on only by gravity, which starts at A and reaches B in the shortest time.

Johann Bernoulli's own solution was a masterful contribution to the subject that was not yet called "Calculus of Variations." There was something *ad hoc* about Bernoulli's solution, though, so

**Comm. Acad. Sci. Imp. Petropol.* 3 (1728) 1732, 110–124; *Opera Omnia* I.25, 1–12

[1]Devenere locos, ubi nunc ingentia cernis
moenia surgentemque novae Karthaginis arcem,
mercatique solum, facti de nomine Byrsam,
taurino quantum possent circumdare tergo

Reportedly, Euler memorized the entire twelve books of the *Aeneid*, almost 10,000 lines and could still recite it in his 70s.

Euler's work might provide the first theory of the calculus of variations. As such, it should perhaps be considered as an alternative to Tom Banchoff's nomination for "The Best Homework Ever?"[2]

Both the manuscript and the paper are reprinted in the *Opera Omnia*, I.25. The manuscript takes eight pages and consists of 20 paragraphs. The paper itself is 12 pages, 36 paragraphs. The two are distinct but quite similar. Euler does not "recycle" any paragraphs from the manuscript, but he keeps the same notation in both places. He works his examples and explains his derivations in detail in the manuscript. The paper is more general and covers a little more material.

The 20th century Greek mathematician Constantin Carathéodory, in his capacity as editor of Volume I.25 of the *Opera Omnia* on calculus of variations, writes of the 1728 paper, "Beginning in paragraph 14, this work reads like a worksheet, and one watches Euler's discoveries as he makes them." It is a thrill to see the Master at work.

Euler begins with a statement of the problem and the two easy special cases. If the surface is a plane, then the shortest line between two given points is a straight line, and if the surface is a sphere, then the shortest is a segment of a great circle. Euler uses the word "linea" to mean either a curve or a straight line, and we will follow his usage here.

He then tells us that for other surfaces "convex or concave or mixed of these," no general solution is known, and "Johann Bernoulli posed this question to me, of finding a universal means of finding the equation of that shortest line determined on whatever surface might be proposed."

If the surface happens to be convex, then Euler describes what he calls a "mechanical solution." He can solve the problem by stretching a string between the two given points. Euler notes that this method gives "chords on the convex parts," and gives only the curve, not an equation describing the curve. He does not seem to notice that the solution thus found may be a local minimum and not a global minimum.

This kind of mechanical solution to a problem seems unusual for Euler. He does not indicate whether it is his own idea or he got it from somebody else. In any case, the example indicates that the concept of a valid argument is evolving at the time, and that Euler is willing to use any analytical tool to solve a problem.

In his discussion, Euler goes on to use what is then a new system of three-dimensional coordinates, a system that he attributes to Jakob Hermann. Hermann held the Chair of Mathematics at the Petersburg Academy when Euler arrived there. His system is illustrated rather unhelpfully in Figure 1. Euler describes the position of a point M relative to a point A. Along a line AP, he finds the point

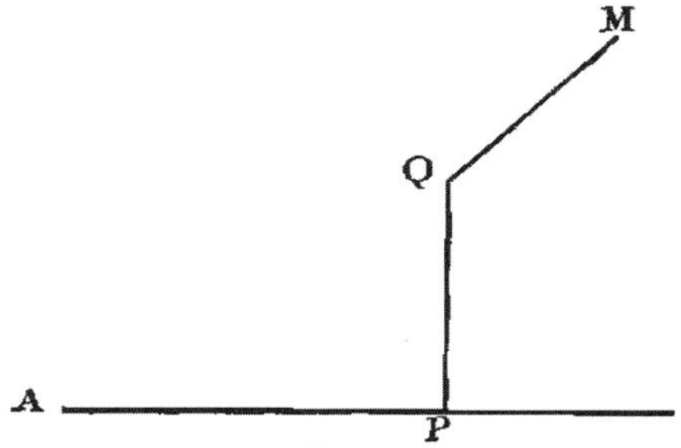

Fig. 1

[2]Former MAA President Tom Banchoff has nominated a homework solution by Cassidy Curtis as the "Best Homework Ever?" See `http://www.brown.edu/Administration/Brown_Alumni_Magazine/97/12-96/features/homework.html`

P so that M is in the plane perpendicular to the line AP at the point P. Then, he finds the point Q so that the line segment PQ is perpendicular to the segment QM. He does not specify what line to choose for AP, nor in what plane the points A, P and Q should lie. Rather, these are left to be chosen for the convenience of the problem. Finally, he denotes $t = AP$, $x = PQ$ and $y = QM$.

This system seems quite alien to the modern eye, accustomed to coordinate axes and 3-d graphics. It took over a hundred years for mathematical practice to evolve from Hermann's ideas to the ones familiar today.

All Euler's choices of segments and variables seem arbitrary, but he will describe other points in the problem relative to the same line containing A and P and relative to the same plane containing A, P and Q.

Because this is a problem about arc length, he lets $AM = a$ and reminds us that $a^2 = t^2+x^2+y^2$.

Now a quarter of the way through the article, Euler notes that "an equation will define the surface" and that an intersection of equations defines a curve on the surface. He will be interested in two kinds of such intersections. He takes sections by holding one of the three variables constant, usually the variable t. This will give him a curve in a plane. He also gives a solution in one case by giving an equation in terms of t and x.

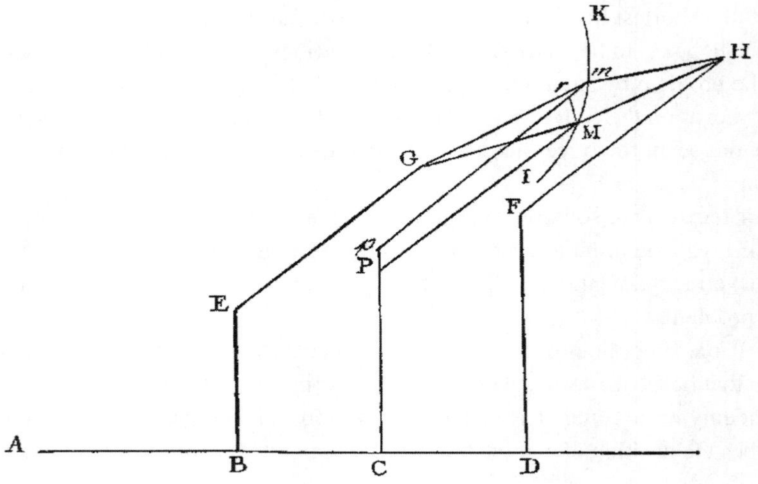

Fig. 2

Now, Euler moves to Figure 2. The reader will recognize that the coordinate system introduced in Figure 1 is being used here. The points G, I, M, m, K and H are all on the surface, and the other points, with the exception of r, are part of the coordinate system. The awkwardly named curve $IMmK$ is the curve on the surface determined by its intersection with the plane perpendicular to AC and passing through C. Euler wants to find necessary conditions on the point M that minimize the length $GM+MH$, and he informs us that to do this he will use the "method of maxima and minima." To this end, Euler tells us that if M minimizes $GM + MH$ and if m is a point on the curve IMK near M, then $GM + MH = Gm + mH$.

To a modern eye, this seems exactly wrong. We want to say that if M is the minimum, then $GM + MN < Gm + mH$. Euler, though, is thinking of differentials. The quantity $(GM + MH) - (Gm + mH)$ is a differential of the arc length at the point M, and if M is a minimum, then that differential will be zero. With this interpretation Euler's equation is correct.

Having told us his plan, Euler assigns variables. $BC = CD = a$, $BE = b$, $EC = c$, $DE = f$, $FH = g$, $CP = x$ and $PM = y$. Then $Cp = x + dx$ and $pm = y + dy$. Now, Euler calculates.

$$GM = \sqrt{a^2 + (x-b)^2 + (y-c)^2},$$
$$GM^2 = (PM - GE)^2 + (CP - BE)^2 + BC^2$$

and likewise

$$HM = \sqrt{a^2 + (f-x)^2 + (g-y)}.$$

(Euler switches inconsistently between HM and MH.) Adding

$$GM + MH = \sqrt{a^2 + (x-b)^2 + (y-c)^2} + \sqrt{a^2 + (f-x)^2 + (g-y)^2}.$$

Differentiate (actually, take differentials) and set equal to zero to get

$$\frac{(x-b)\,dx + (y-c)\,dy}{\sqrt{a^2 + (x-b)^2 + (y-c)^2}} = \frac{(f-x)\,dx + (g-y)\,dy}{\sqrt{a^2 + (f-x)^2 + (g-y)^2}}.$$

Euler often mistook necessary conditions for sufficient conditions and he does so here. He claims that the point M is determined by this equation when in fact it is only true that any point M that minimizes the length $GM + MH$ must satisfy this condition.

Now we reach section 14, the point at which Carathéodory says we can see Euler thinking.

The curve IK is determined by the surface and the point C and can be given in terms of the coordinates x and y. So we can differentiate the equation of the curve IK and get some differential equation $P\,dx = Q\,dy$. Euler also gives this in a second form as ratios: $dx : dy = Q : P$. Substituting these in the equation above produces

$$\frac{(x-b)Q + (y-c)P}{\sqrt{a^2 + (x-b)^2 + (y-c)^2}} = \frac{(f-x)Q + (g-y)P}{\sqrt{a^2 + (f-x)^2 + (g-y)^2}}$$

which is, as Euler says, "without differential quantities."

Making the string of substitutions $BC = CD = a = dt$, $DF = f = x + dx$, $FH = g = y + dy$. $BE = b = x - dx + ddx$ and $EG = c = y - dy + ddy$ gives

$$\frac{Q(dx - ddx) + P(dy - ddy)}{\sqrt{dt^2 + (dx - ddx)^2 + (dy - ddy)^2}} = \frac{Q\,dx + P\,dy}{\sqrt{dt^2 + dx^2 + dy^2}}.$$

But, by an application of the same Method of Maxima and Minima used earlier, this means that the differential of the quantity on the right,

$$\frac{Q\,dx + P\,dy}{\sqrt{dt^2 + dx^2 + dy^2}},$$

is zero. So, Euler holds P, Q and dt constant, in order to stay on the same section curve of the surface, and applies the quotient rule to get,

$$\frac{Q\,ddx + P\,ddy}{Q\,dx + P\,dy} = \frac{dx\,ddx + dy\,ddy}{dt^2 + dx^2 + dy^2}. \tag{1}$$

Moreover, Euler can describe the surface by a differential equation

$$P\,dx = Q\,dy + R\,dt. \tag{2}$$

When the surface is constrained to the curve *IK*, then $dt = 0$ so the equation reduces to $P\,dx = Q\,dy$.

Now, Euler's analysis for the general case concludes that the shortest line on a surface must satisfy the two differential equations (1.1) and (1.2).

Euler applies his analysis to three kinds of surfaces:

1. Cylinders, not necessarily circular cylinders
2. Circular cones
3. Surfaces of revolution.

Euler illustrates the case of the cylinder with Figure 3. In this perspective, Hermann's coordinate technique doesn't seem quite so bizarre. He assigns variables $AQ = t$, $QP = x$ and $PM = y$. Note that the variables are the same, but the points are labeled slightly differently than they were in Figure 2. Euler explains that the differential equation for this surface is $P\,dx = Q\,dy$ and does not depend on t. He further notes that if the cylinder is an ordinary circular cylinder in which the curve *BHC* is a circle with *A* as its center, then the differential equation is $x\,dx = -y\,dy$. These relatively simple forms allow Euler to proceed with his analysis.

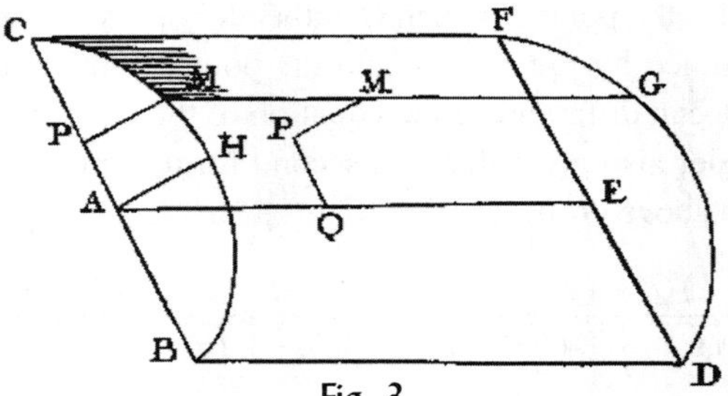

Fig. 3

He substitutes $P : Q = dy : dx$ into equation (1.1) to get

$$\frac{dx\,ddx + dy\,ddy}{dx^2 + dy^2} = \frac{dx\,ddx + dy\,ddy}{dt^2 + dx^2 + dy^2}. \tag{3}$$

Integrating (1.3) gives

$$\sqrt{dx^2 + dy^2} = m\sqrt{dt^2 + dx^2 + dy^2}$$

where m is a constant of integration. He rewrites this as

$$dx^2 + dy^2 = nn\,dt^2,$$

substituting nn for

$$\frac{1}{1 - m^2}.$$

In doing this, he is implicitly assuming that the latter quantity is non-negative. This is equivalent to assuming that $m < 1$, an assumption that he does not justify.

Integrating the square root of this last equation gives $nt = \int \sqrt{dx^2 + dy^2} + C$, from which Euler concludes "t is always proportional to an arc in a fixed section." Euler recognizes that this may not be perfectly clear, so he explains a bit.

Euler looks at the integral equation $nt = \int \sqrt{dx^2 + dy^2} + C$ a little more closely and tries to interpret the value of n. In the case $n = 0$, this is describing the shortest arc connecting two points on the same line of the cylinder parallel to the axis AE. In the case $n = 1$, this describes an arc connecting two points on the same section. Then the shortest curve traces that section.

Euler now specializes to the case of a circular cylinder, so he can substitute $x\,dx = -y\,dy$ and $xx + yy = aa$. Then

$$nt + b = \int \sqrt{dx^2 + dy^2}.$$

Differentiating both sides and squaring gives

$$n^2dt^2 = dx^2 + dy^2.$$

Now, substituting for y gives

$$n\,dt = \frac{a\,dx}{\sqrt{a^2 - x^2}} \quad \text{or} \quad nt = \int \frac{a\,dx}{\sqrt{a^2 - x^2}}$$

which Euler recognizes as a sine. Euler concludes that the shortest path on a cylinder traces a piece of a sine curve as its projection onto a plane containing the t-axis.

Euler's next example is a cone, as illustrated in Figure 4. There, $AQ = t$, $QP = x$ and $PM = y$. He notes that all transverse sections are similar. That is to say, the curve HMG is similar to the curve DEC. Euler further notes, "If in the equation, nt, nx, ny are put in the place of t, x, y, then the equation is not changed. This property is that of homogeneous equations, in which t, x and y all make the same dimension." It is interesting that Euler knew this geometric similarity property of homogeneous equations and that he uses the term "dimension" in this context rather than "degree."

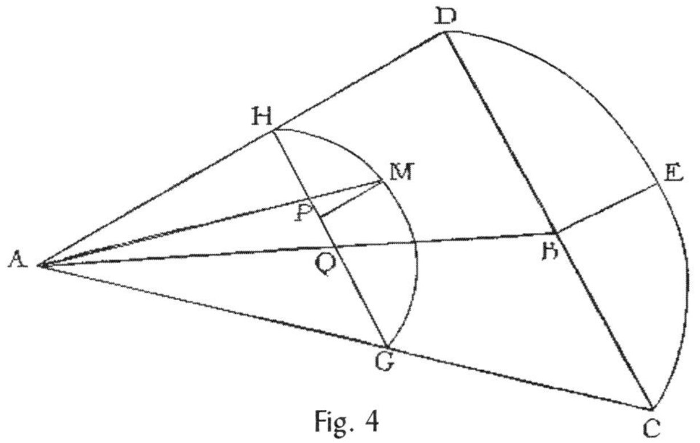

Fig. 4

From here, Euler's analysis proceeds much as before. Because the surface is a cone, he knows the equation of the surface is of the form $t \cdot F(x, y)$, and so he is able to make substitutions as before.

Euler's final example is a surface of revolution, as illustrated in Figure 5. He begins by telling us how this example is different in nature from the first two. Both the cylinder and the cone could be flattened out into a plane. The shortest path could then be found on the plane and the solution carried back to the surface itself. The special properties of a surface of revolution give Euler enough tools to work his magic.

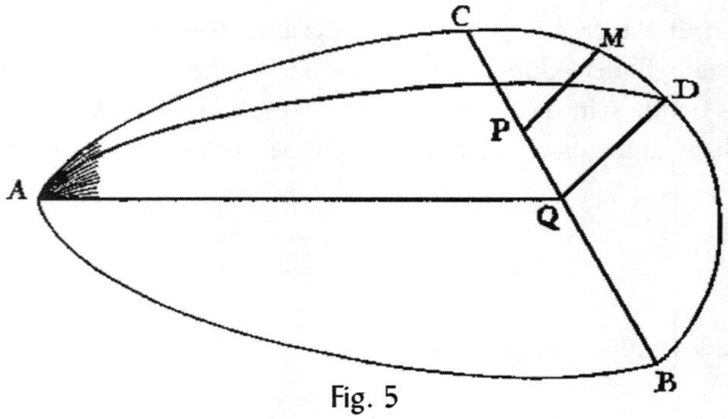

Fig. 5

Euler assigns variables $AQ = t$, $QP = x$ and $PM = y$. Because we have a surface of revolution "if I hold t constant, or $dt = 0$, then it will be the equation of a circle $xx + yy =$ Const. or $x\,dx = -y\,dy$. From this, we see that the equation for a solid of revolution is $xx + yy = T$, where T denotes some function of that variable t, and of constants." From this, Euler concludes that the differential equation describing the surface will be $x\,dx = -y\,dy + R\,dt$, "in which R depends only on t and on constants."

We see from his choice of words that Euler does not yet clearly distinguish between the solid and its surface. Moreover, he still regards the function T as being a function of t and of constants.

To continue his calculation, Euler goes back to the fundamental equation (1.1)

$$\frac{Q\,ddx + P\,ddy}{Q\,dx + P\,dy} = \frac{dx\,ddx + dy\,ddy}{dt^2 + dx^2 + dy^2}.$$

He makes his substitutions to get

$$\frac{x\,ddy - y\,ddx}{x\,dy - y\,dx} = \frac{dx\,ddx + dy\,ddy}{dt^2 + dx^2 + dy^2}$$

and then integrates to get

$$l(x\,dy - y\,dx) = l\sqrt{dt^2 + dx^2 + dy^2} + l\,a,$$

where we see that Euler again uses the symbol "l" to denote what we now call the natural logarithm function. Finally, he exponentiates both sides of this equation to get

$$x\,dy - y\,dx = a\sqrt{dt^2 + dx^2 + dy^2} \tag{4}$$

and "this, together with the natural equation of the surface expressed as $x\,dx = -y\,dy + R\,dt$, will determine the shortest line."

Though Euler now has a general solution for the shortest line on a surface of revolution, there is more to be harvested from his efforts. He notes that the value of a depends on the particular points on the surface and considers what happens if $a = 0$. In this case, equation (1.4) gives $x\,dy = y\,dx$, and so $y = nx$. "And so it is known that the periphery of the curve rotated about the axis represents the shortest line between its endpoints." Euler notes that the shortest line between two points on a sphere is the arc of a great circle, confirming what he already knew when he started the paper.

Euler's final paragraph reads:

> The Celebrated Johann Bernoulli proposed this question to me and urged me to write up my solution and to investigate these three kinds of surfaces which lead to solutions that are integrable equations. I wanted to include the solutions to these questions because they followed so easily from what I had done earlier.

In other words, he says "Once I knew what I was doing, this homework was pretty easy."

7

E-20: De summatione innumerabilium progressionum*

On the summation of innumerably many progressions

We return to Euler's first letter to Goldbach, the letter that inspired his work on the gamma function in E-19 and that would ignite his interest in number theory in E-26. In that same letter, Euler makes a remark about the partial sums of the harmonic progression.

Let us remind the reader that the harmonic progression is the sequence of reciprocals of the positive integers. It begins

$$1, \frac{1}{2}, \frac{1}{3}, \frac{1}{4}, \frac{1}{5}, \ldots.$$

Its progression of partial sums would begin

$$1, \frac{3}{2}, \frac{11}{6}, \frac{25}{12}, \frac{137}{60}, \ldots.$$

It is this progression of partial sums that Euler wants to interpolate. He tells Goldbach, without any substantiation, that "I have found that the general term, whose index is $-\frac{1}{2}$ is $-2l2$. The term whose exponent is $\frac{1}{2}$ is $2 - 2l2$" where "$l\,2$ signifies the hyperbolic logarithm of two, which is $= 0{,}69314738056$." Note that in Euler's day, as in much of the world even now, they used commas as decimal points.

Euler's claim is rather cryptic, and it is not just the translation from the Latin. When he uses the term "index," he means the upper limit of a summation. He is claiming that he can make sense of summations when the upper limit is a fraction, and that, in particular,

$$\sum_{n=1}^{-1/2} \frac{1}{n} = -2\ln 2 \quad \text{and that} \quad \sum_{n=1}^{1/2} \frac{1}{n} = 2 - 2\ln 2.$$

In the same volume of the *Commentarii* as his article about the gamma function, Euler published De summatione innumerabilium progressionum, "On the summation of innumerably many progressions." This is a little-known paper, lacking a treasure like the gamma function or a gem like fractional derivatives, but we find it to be a key paper, based on the same very general ideas as "On transcendental progressions" and leading to some of Euler's very most important work, the evaluation of $\zeta(2)$ and then to the discovery of the product-sum formula for the zeta function. Moreover, in later papers where Euler expands on some of these ideas, he discovers the important numerical analysis methods of Euler-Maclaurin series.

Euler begins, "In the previous article, the author has treated a method for finding the general terms of transcendental progressions." He asks us to look at the integral

$$\int_0^1 \frac{1 - x^n}{1 - x} dx,$$

Comm. Acad. Sci. Imp. Petropol. 5 (1730/31) 1738, 91–105; *Opera Omnia* I.14, 25–41

DE SVMMATIONE INNVMERABILIVM PROGRESSIONVM.

Auct. L. Eulero.

1.

QVae in praecedente differtatione de progreffionibus transcendentibus earumque terminis generalibus tradidi, multo latius patent, quam videri poffent; et inter alia quam plurima, ad quae accommodari poffunt, eximius earum poteft effe vfus in inueniendis fummis innumerabilium progreffionum. Quemadmodum enim in fuperiore differtatione innumerae progreffiones ad terminos generales funt reuocatae, quae communem algebram transcendunt; ita hic eandem methodum accommodabo ad terminos fummatorios inueniendos progreffionum, ad quas indefinite fummandas communis algebra non fufficit.

§. 2. Progreffio quaepiam fummari dicitur indefinite, fi detur formula numerum indefinitum n continens, quae exponat fummam tot terminorum illius progreffionis, quot n comprehendit vnitates, ita vt fi ponatur v. gr. $n = 10$ ea formula exhibeat fummam decem terminorum a primo numeratorum. Formula haec vocatur terminus fummatorius illius progreffionis, atque eft fimul

M 2 ter-

or, as Euler says

$$“\int \frac{1-x^n}{1-x}dx,$$

and take it $= 0$ if $x = 0$, and then set $x = 1$.” This peculiar wording is Euler's description of an integral from 0 to 1. It amounts to taking a particular antiderivative, the one chosen so that the constant of integration makes the value of the antiderivative equal to zero at the left-hand endpoint.

Euler continues, “Then for various values of n, this forms the progression 1, $1+\frac{1}{2}$, $1+\frac{1}{2}+\frac{1}{3}$, $1+\frac{1}{2}+\frac{1}{3}+\frac{1}{4}$, etc. This series forms the sums of the harmonic progression 1, $\frac{1}{2}$, $\frac{1}{3}$, $\frac{1}{4}$, $\frac{1}{5}$, etc.”

Euler notes that this definition of the partial sums of the harmonic series is well defined even for non-integer values of n, and he considers the case $n = \frac{1}{2}$. To evaluate this series, he finds it is necessary to integrate

$$\frac{1-\sqrt{x}}{1-x} \quad \text{or} \quad \frac{1}{1+\sqrt{x}}$$

from 0 to 1 to get $2 - 2\ln 2$. This justifies his unsubstantiated remark in his first letter to Goldbach. He wants to make sure we understand that this allows him to define all partial sums when the denominator of the index is 2, so he tells us explicitly:

> Since, in general, the term of order $n+1$ exceeds the term of order n by the fraction $\frac{1}{n+1}$, the term of order $1\frac{1}{2}$ will be $= 2\frac{2}{3} - 2l2$, and the term of order $2\frac{1}{2}$ will be $2+\frac{2}{3}+\frac{2}{5}-2l2$, etc. The interpolated series therefore will be

$\frac{1}{2}$	1	$1\frac{1}{2}$	2	$2\frac{1}{2}$	etc.
$2-2l2$	1	$2+\frac{2}{3}-2l2$	$1+\frac{1}{2}$	$2+\frac{2}{3}+\frac{2}{5}-2l2$	

Euler gives us a hint of how he means to generalize this idea of connecting sums with integrals by considering the form

> $\int \frac{1-P^n}{1-P}\,dx$ where P denotes whatever function of x. Integrate this and set it $= 0$ when $x = 0$, and then put $x = 1$. But you could set $x = k$. Then $\int \frac{1-P^n}{1-P}\,dx$ gives this progression
>
> $$k,\ k+\int P\,dx,\ k+\int P\,dx+\int P^2\,dx, \text{ etc.}$$
>
> This progression, if each term is subtracted from the next, produces
>
> $$k,\ \int P\,dx,\ \int P^2\,dx,\ \int P^3\,dx, \text{ etc.}$$
>
> whose general term is $\int P^{n-1}\,dx$. The sum of these terms is given by the formula $\int \frac{1-P^n}{1-P}\,dx$. If $P = x^\alpha : a^\alpha$, the progression will be
>
> $$k,\ \frac{k^{\alpha+1}}{(\alpha+1)a^\alpha},\ \frac{k^{2\alpha+1}}{(2\alpha+1)a^{2\alpha}}, \text{ etc.}$$

and the sum of the terms is

$$\int \frac{a^{n\alpha} - x^{n\alpha}}{(a^{\alpha} - x^{\alpha})a^{n\alpha-\alpha}}\, dx."$$

The use of a ratio to define the function P here is a little peculiar. Euler does not seem to make a habit of it. Note that in the previous section Euler had merely described P as "whatever function of x." This all seems to be a sign that Euler had not yet clarified his ideas of what functions should be.

These calculations do not really seem to be leading anywhere, but Euler sees a pattern in this, and he is about to exploit it. We follow in Euler's words:

> Thus are found the sums of the terms of all progressions whose terms are fractions of which the numerators form a geometric progression and the denominators form an arithmetic one. Thus it is easy to sum the progression
>
> $$\frac{b}{c},\ \frac{b^{i+1}}{c+e},\ \frac{b^{2i+1}}{c+2e},\ \frac{b^{3i+1}}{c+3e},\ \text{etc.}$$
>
> whose general term is
>
> $$\frac{b^{(n-1)i+1}}{c+(n-1)e}.$$
>
> This compared to that,[2] then
>
> $$\alpha = \frac{e}{c} \quad \text{and} \quad \frac{ck^{((n-1)e/c)+1}}{a^{(n-1)e/c}} = b^{(n-1)i+1}$$
>
> and
>
> $$a = \left(\frac{ck^{((n-1)e/c)+1}}{b^{(n-1)i+1}}\right)^{c/((n-1)e)} = \left(\frac{ck}{b}\right)^{c/((n-1)e)} \frac{k}{b^{ci:e}}.$$

Sometimes, Euler seems to get carried away with calculations. He wraps up this section by telling us

> ... the sum of this series is
>
> $$\int \frac{b^{(ne-nci)/c} - c^{ne/c}x^{ne/c}}{b^{(n-1)(e-ci)/c}(b^{(e-ci)/c} - c^{e/c}x^{e/c})}\, dx,$$
>
> which ought to be integrated by setting it equal to 0 if $x = 0$, then putting $x = \frac{b}{c}$.

---

[2]This literal translation of the Latin is confusing. Latin has a richer vocabulary of pronouns than English. In the Latin, it is clear that "this" is the progression

$$\frac{b}{c} + \frac{b^{i+1}}{c+e} + \frac{b^{2i+1}}{c+2e} + \text{etc.}$$

and he is comparing it to "that" progression k,

$$\frac{k^{\alpha+1}}{(\alpha+1)\alpha^{\alpha}},\ \frac{k^{2\alpha+1}}{(2\alpha+1)\alpha^{2\alpha}},\ \text{etc.}$$

Euler, ever happy to treat infinity as a number, tells us to sum this infinite progression by setting $n = \infty$. He writes "To sum

$$\frac{b}{c} + \frac{b^{i+1}}{c+e} + \frac{b^{2i+1}}{c+2e} + \text{etc.},$$

let A be the sum of the first n terms. Augment n by 1 and A will be augmented by the term of order $n+1$, which is

$$\frac{b^{ni+1}}{cn+e}."$$

Now, Euler applies calculus to series in very interesting ways. These are the most useful and important ideas in this paper. He writes:

> Since n and A are considered fluent quantities, where n is almost infinitely greater than 1, there will be differentials dn and dA, the augments relating them being 1 and
>
> $$\frac{b^{ni+1}}{cn+e}.$$
>
> And so it produces the equation
>
> $$dA = \frac{b^{nI+1}\,dn}{c+ne}.$$
>
> The integral of this will give an equation between the sum A and the number of terms n.

Euler's choice of the word "fluent" in this passage is an interesting one. In Newton's version of calculus, a fluent was any relationship between variables. Leibniz did not use the word. Euler learned calculus from Johann Bernoulli who had worked very closely with Leibniz, so it is surprising to see Euler using Newton's word. Perhaps Euler was beginning to see the limitations of the Leibniz vocabulary of "curves" and "formulas," and was already starting to form his own ideas of functions, ideas that would take another twenty years or so to mature.

Euler sets $l(c+ne) = z$. (Remember l is the natural logarithm.) Then

$$\frac{e\,dn}{c+ne} = dz$$

and $c + ne = g^z$ (where g is the number whose logarithm is 1, the number we usually denote by e today.)

As we have noted, Euler has not yet standardized e as the basis for the natural logarithms, so he still uses whatever symbol is convenient. Recall that in E-10 Euler denoted this constant as c. Euler first uses e for this purpose in 1743 in E-61.

Let us return to the calculation. A little algebra gives

$$n = \frac{g^z - c}{e}, \quad b^{ni+1} = b^{(g^z - ci + e)/e} = b^{(e-ci)/e} b^{g^z i/e}$$

and consequently $dA = b^{(e-ci)/e} b^{g^z i/e}\,dz$.

Now we consider the case $i = 0$, that is the case of constant numerators, so the series has the form

$$\frac{b}{c} + \frac{b}{c+e} + \frac{b}{c+2e} + \text{etc.}$$

Then $dA = \frac{b}{e}\,dz$ and $A = \frac{b}{e}(z + lc) = \frac{b}{e}l(C + (c + ne))$. The constant C is not yet determined. But if m is another index, then $B = \frac{b}{e}lC(c + me)$, so

$$B - A = \frac{b}{e}l\frac{c + me}{c + ne} = \frac{b}{e}l\frac{m}{n}$$

when m and n are infinite.

This calculation itself is not terribly significant, but some of its details are interesting, for example the use of two different "infinite numbers" m and n. Moreover, this constant C, which "is not yet determined" is related to Euler's constant, γ, but Euler does not delve into its properties in this paper.

Take $i = 0$ so that the progression is

$$\frac{b}{c}, \frac{b}{c + e}, \frac{b}{c + 2e}, \frac{b}{c + 3e}, \text{ etc.}$$

Then the general term is

$$\frac{b}{c + (n - 1)e}.$$

The sum of these general terms is

$$\int \frac{b^{ne/c} - c^{ne/c}x^{ne/c}}{b^{(n-1)e/c}(b^{e/c} - c^{e/c}x^{e/c})}\,dx.$$

Euler also gives us the sum of another progression of the same form:

$$\frac{b}{c}, \frac{b}{c + f}, \frac{b}{c + 2f}, \frac{b}{c + 3f}, \text{ etc.}$$

with general term

$$\frac{b}{c + (n - 1)f}.$$

The sum is given by an integral analogous to the one above, changing e to f as necessary.

Euler is unfettered by modern concerns about conditional convergence and divergent series. Even though he intends to consider them as infinite series, Euler fearlessly adds these two progressions and gets

$$\frac{2b}{c}, \frac{2bc(e + f)}{(c + e)(c + f)}, \frac{2bc + 2b(e + f)}{(c + 2e)(c + 2f)}, \text{ etc.,}$$

with general term given by

$$\frac{2bc + (n - 1)b(e + f)}{(c + (n - 1)e)(c + (n - 1)f)}.$$

Euler also gives the sum of these terms with an extraordinarily complicated integral, which we omit here.

In the same way, but more generally, multiply the first progression by p and the second one by q and obtain a progression whose general term is

$$\frac{pb}{c(n-1)e}+\frac{qb}{c(n-1)f}=\frac{(p+q)bc+(n-1)b(pf+qe)}{(c+(n-1)e)(c+(n-1)f)}.$$

Euler gives us another integral for the sum of the first n terms of this progression, but, again, we omit that integral.

The problem Euler has solved here is not the one he had originally posed. He can add two harmonic sequences to get a quadratic sequence, and then he can construct a definite integral that sums the sequence. What he had hoped to do, however, was to start with the quadratic sequence and sum that. Euler is able to adapt his solution to the other problem using the technique we now call "partial fractions."

Suppose the general term of a sequence is given by $\frac{\alpha+\beta n}{\gamma+\delta n+\varepsilon nn}$. Euler compares this to the form we saw earlier

$$\frac{(p+q)c+(n-1)(pf+qe)}{(c+(n-1)e)(c+(n-1)f)}$$

and gives formulas for c, e, f, p and q in terms of α, β, γ, δ, and ε. A typical example of such formulas is

$$e=\frac{\delta+2\varepsilon+\sqrt{\delta d-4\delta}}{2\sqrt{\gamma+\delta+\varepsilon}}.$$

Apparently, Euler did not expect his readers to know the techniques of partial fractions, or perhaps he thought that his readers would not be satisfied without seeing explicit formulas. Just knowing how to do the calculations might not have been enough.

In a short remark, Euler tells us that the same technique of partial fractions will work for "denominators of higher dimension," that is for polynomials of higher degree, like $\gamma+\delta n+\varepsilon n^2+\zeta n^3+\eta n^4+\text{etc.}$, as long as the polynomial does not have multiple roots. This is good, but not great, because the series Euler is *really* interested in is the one with denominators given by n^2, because that would solve the well-known Basel problem, the sum of the reciprocals of the squares. (We will learn a good deal more about the Basel problem when we get to E-41.) For now, suffice it to say that finding the exact value of the series $1+\frac{1}{4}+\frac{1}{9}+\frac{1}{16}+\frac{1}{25}+\text{etc.}$ was the best-known problem of the time. Solving it would be as important as solving Fermat's Last Theorem was in 1993 or solving the Riemann Hypothesis would be today. Euler has an idea, though, as we will see.

"We now treat a method which does not exclude these cases." That is, it works even if the polynomial in the denominator of the formula has multiple roots.

This method will be of continuing importance in Euler's later work. We will see it in E-25 and E-28, among others.

Start with the simple progression

$$\frac{1}{a},\ \frac{1}{a+b},\ \frac{1}{a+2b},\ \text{etc.}$$

with general term

$$\frac{1}{a+(n-1)b}.$$

As before, the sum of n such terms will be

$$\int \frac{1 - a^{nb/a}x^{nb/a}}{1 - a^{b/a}x^{b/a}}\, dx.$$

Substitute $ax = y$ and this becomes

$$\int \frac{1 - y^{nb/a}}{1 - y^{b/a}} \cdot \frac{dy}{a},$$

"in which integral, one should set $y = 1$." Remember from earlier in the article how Euler treats definite integrals.

So, we have

$$\frac{1}{a} + \frac{1}{a+b} + \frac{1}{a+2b} + \cdots + \frac{1}{a+(n-1)b} = \int \frac{1 - y^{nb/a}}{1 - y^{b/a}} \cdot \frac{dy}{a},$$

if we interpret the integral correctly.

Now comes the big trick. Multiply both sides by $y^{\alpha}\, dy$ and integrate again. Euler, alas, is working in notational generality that obscures what is really happening here, so perhaps it is best to digress from his exposition and see how his calculations work for the example that is really at the top of his mind, the great Basel Problem,

$$1 + \frac{1}{4} + \frac{1}{9} + \frac{1}{16} + \cdots + \frac{1}{(n-1)^2}.$$

Recall that

$$1 + x + x^2 + x^3 + \cdots + x^{n-1} = \frac{1 - x^n}{1 - x}.$$

Integrating gives

$$x + \frac{x^2}{2} + \frac{x^3}{3} + \frac{x^4}{4} + \cdots + \frac{x^n}{n} = \int \frac{1 - x^n}{1 - x}\, dx$$

and taking the definite integral from 0 to 1 gives the nth partial sum of the harmonic series. Recall that just integrating again didn't take Euler any closer to the solution of the Basel Problem, but, if he divides by x before he integrates, he gets

$$1 + \frac{x}{2} + \frac{x^2}{3} + \frac{x^3}{4} + \cdots + \frac{x^{n-1}}{n} = \frac{1}{x} \int \frac{1 - x^n}{1 - x}\, dx$$

and then

$$x + \frac{x}{4} + \frac{x^2}{9} + \frac{x^3}{16} + \cdots + \frac{x^n}{n^2} = \int \frac{dx}{x} \int \frac{1 - x^n}{1 - x}\, dx.$$

Now, a correct interpretation of the integrals amounts to substituting $x = 1$, and this gives an integral formula for the nth partial sum in the Basel Problem.

This is what is really going on in Euler's substitutions for x and with his multiplication by $y^{\alpha}\, dy$. He is doing the problem in a generality that will pay off later.

Let's get back to how Euler did it. Euler spends a few paragraphs working through the integral form that gives sums of terms of

$$\frac{a}{(a+(n-1)b)(\beta a+(n-1)b)}.$$

Any quadratic denominator, even those with multiple roots, can be represented in this form.

Euler again gives explicit formulas for the conversions we would now do using partial fractions. He tells us how to convert between progressions with general term

$$\frac{a}{a+(n-1)b+\frac{(n-1)(n-2)}{2}c}$$

and those with general term

$$\frac{a}{(a+(n-1)b)(\beta a+(n-1)b)}.$$

Euler shows how the "multiply and integrate again" process can be extended to cubic and quartic denominators, and tells us that "in this way, any progression can be summed if the numerator is a constant and the denominator consists of any algebraic progression whatsoever."

Although this generalization should not surprise us, it is interesting to note Euler's use of the term "algebraic progression." One of the themes of Euler's work will be the gradual evolution over his whole career of a fairly modern idea of the concept of a "function." The idea of a progression is closely related. It is clear from his wording that, at this time in his life, a progression is *algebraic* if it is a product of arithmetic progressions.

> If the sum of this progression continued to infinity is desired, then set $n =$ infinity. This substitution made into the expression
>
> $$\int \frac{dy}{a}\cdot\frac{1-y^{nb/a}}{1-y^{b/a}}$$
>
> is transmuted into:
>
> $$\int \frac{dy}{a(1-y^{b/a})}.$$
>
> This is because when y is less than 1, then the term $y^{nb/a}$ vanishes, and when $y = 1$, the expression $1 - y^{nb/a}$ vanishes.

The progression to which Euler refers here is

$$\frac{1}{a},\ \frac{1}{a+b},\ \frac{1}{a+2b},\ \text{etc.}$$

with general term

$$\frac{1}{a+(n-1)b},$$

and that sum diverges. Although Euler's ultimate results will be true, some of his calculations, like this one, are not sound.

Today, Euler is sometimes criticized for his use of infinity as a number and for his treatment of, or more precisely his failure to treat, issues of convergence and divergence. It is true that Euler did not yet have the theorems of analysis that we have today, and he did not understand the subtleties of conditional and absolute convergence. However, passages like this one suggest that Euler was wary of, if naïve about, some issues of convergence and divergence. He used infinite numbers simply because he did not have the tools of limits.

This is not to absolve Euler of all error. We will see, particularly in E-72, Euler's sometimes outrageous abuse of divergent and conditionally convergent series. To his credit, though, he usually gets correct results.

In fact, in this article, Euler is guilty of such an abuse in his first treatment of quadratic denominators using partial fractions. When he adds two general harmonic series, (a general harmonic series is one with a constant numerator and a linear progression in the denominator) to get a quadratic series, then sets $n =$ infinity, he has to decompose the convergent quadratic series into a conditionally convergent sum of terms that are harmonic, and then rearrange those terms.

Though Euler was probably unaware of the problem, his second treatment by multiple integrals does not commit such errors of conditional convergence.

Unfortunately, these beautiful integrals that give the sums and partial sums of series are, in general, difficult or impossible to evaluate exactly. Euler hasn't evaluated any of them since very early in the paper when he was finding the partial sum of the harmonic progression of index $\frac{1}{2}$. He would really like to get a value for the answer to the Basel Problem. If an exact answer can't be found by integration, then for now he will be satisfied with an approximation.

Let $b = a$ so that $\frac{b}{a} = 1$. Then

$$\int \frac{dy}{a(1-y)} = A - \frac{1}{a} l(1-y).$$

This integral describes the infinite sum of the terms of the progression $\frac{1}{a}, \frac{1}{2a}, \frac{1}{3a}$, etc., for he has already taken n to be an infinite number.

In this equation, Euler does explicitly include a constant of integration, A, and, as always, he uses l to denote the natural logarithm function. Of course, he means the integral to be a definite integral, so he gives it his usual treatment:

"When I set $y = 0$, the total integral ought to be 0, so $A = 0$ and so

$$\int \frac{dy}{a(1-y)} = -\frac{1}{a} l(1-y).$$

If this is multiplied by $y^{\alpha-2}dy$, it gives

$$-\frac{y^{\alpha-2}dy}{a} l(1-y)."$$

When Euler takes $\alpha = 1$ and $a = 1$, he describes the Basel problem. To evaluate this integral, put $1 - y = z$ so that $y = 1 - z$ and this gives

$$\frac{(1-z)^{\alpha-2}\,dz}{a} l(z)$$
$$= \left(1 - \frac{\alpha-2}{1}z + \frac{(\alpha-2)(\alpha-3)}{1\cdot 2}z^2 - \frac{(\alpha-2)(\alpha-3)(\alpha-4)}{1\cdot 2\cdot 3}z^3 + \cdots\right)\frac{dz}{a} l(z).$$

This step is less formidable than it might look. Euler has reversed the order of the equality, and the expression inside the parentheses is just the binomial expansion of $(1-z)^{\alpha-2}$. Euler now sets out to integrate the expression on the right. Because it is true that

$$\int z^{\eta}\,dz\,lz = C - \frac{z^{\eta+1}}{(\eta+1)^2} + \frac{z^{\eta+1}lz}{\eta+1},$$

the integral of that series will be

$$\frac{1}{a}\left(C - z + z\,lz + \frac{\alpha-2}{1\cdot 4}z^2 - \frac{\alpha-2}{1\cdot 2}z^2\,lz - \frac{(\alpha-2)(\alpha-3)}{1\cdot 2\cdot 9}z^3 + \frac{(\alpha-2)(\alpha-3)}{1\cdot 2\cdot 3}z^3\,lz + \text{etc.}\right).$$

This integral, if we make $y = 0$ and $z = 1$, ought to be 0. Because of this,

$$C = 1 - \frac{\alpha-2}{1\cdot 4} + \frac{(\alpha-2)(\alpha-3)}{1\cdot 2\cdot 9} - \frac{(\alpha-2)(\alpha-3)(\alpha-4)}{1\cdot 2\cdot 3\cdot 16} + \text{etc.}$$

It is clear from this integral that if α is an integer greater than 1, then the number of terms will be finite and the sum of the progression will be well defined. And even if the number of terms is infinite, the sum of the series is given by another infinite series which always converges faster than the given series and so is useful in determining the sum.

Here, Euler demonstrates that he is well aware of some of the issues of convergence of series, or at least rate of convergence. One might say that, at this point in his career, he has an "I know it when I see it" understanding of the topics.

Euler is going to split the infinite series into two parts, one that he can sum exactly and the other that converges rapidly. In later papers, this technique will evolve into the method we now call Euler-Maclaurin series. After first reminding us that he is taking $b = a$, he notes that the infinite sum that gives

$$\int \frac{-y^{\alpha-2}}{a}\,dy\,l(1-y)$$

is

$$\frac{1}{\alpha a} + \frac{1}{2(\alpha+1)a} + \frac{1}{3(\alpha+2)a} + \frac{1}{4(\alpha+3)a} + \text{etc.}$$

Euler now gives confusing directions "This sum being had, if in that integral is put $y = 1$, but make $y = 1 - z, \ldots$." This, in Euler's parlance, means that the upper bound of integration is to be 1, and that he is now going to make the substitution $y = 1 - z$. He gets the following formidable expression for the integral:

$$\frac{1}{a}\left\{\begin{array}{l} 1 - \dfrac{\alpha-2}{1\cdot 4} + \dfrac{(\alpha-2)(\alpha-3)}{1\cdot 2\cdot 9} - \text{etc.} - z + \dfrac{\alpha-2}{1\cdot 4}z^2 - \dfrac{(\alpha-2)(\alpha-3)}{1\cdot 2\cdot 9} + \text{etc.} \\ \qquad + z\,lz - \dfrac{\alpha-2}{1\cdot 2}z^2\,lz + \dfrac{(\alpha-2)(\alpha-3)}{1\cdot 2\cdot 3}z^3\,lz - \text{etc.} \end{array}\right\}.$$

Let $y = 1$, so that $z = 1 - y = 0$, and the sum of the series

$$\frac{1}{\alpha a} + \frac{1}{2(\alpha+1)a} + \frac{1}{3(\alpha+2)a} + \frac{1}{4(\alpha+3)a} + \text{etc.}$$

equals the sum of the series

$$\frac{1}{a} - \frac{\alpha-2}{1\cdot 4\cdot a} + \frac{(\alpha-2)(\alpha-3)}{1\cdot 2\cdot 9\cdot a} - \text{etc.}$$

This follows by eliminating all terms in the previous expression that contain z, including (without comment) those containing $z^n\, lz$. Euler now multiplies both sides by a to get two equal series

$$\frac{1}{\alpha} + \frac{1}{2(\alpha+1)} + \frac{1}{3(\alpha+2)} + \text{etc.}$$

and

$$1 - \frac{\alpha-2}{1\cdot 4} + \frac{(\alpha-2)(\alpha-3)}{1\cdot 2\cdot 9} - \text{etc.}$$

When $\alpha = 1$, the first of these series is the Basel Problem. Euler is about to show us how the second series can be manipulated to converge very rapidly.

"And so I have come upon a way to find a strongly converging series the sum of which is equal to the proposed series."

Euler will be manipulating the integral $\int -y^{\alpha-2}\, dy\, l(1-y)$.

By the usual Taylor series expansion

$$-l(1-y) = y + \frac{y^2}{2} + \frac{y^3}{3} + \frac{y^4}{4} + \text{etc.},$$

multiplying by $-y^{\alpha-2}\, dy$ and integrating gives

$$\int -y^{\alpha-2}\, dy\, l(1-y) = \frac{y^\alpha}{\alpha} + \frac{y^{\alpha+1}}{2(\alpha+1)} + \frac{y^{\alpha+2}}{3(\alpha+2)} + \text{etc.}$$

Finally, Euler reveals his hand and sets $\alpha = 1$. Because he has already taken $a = 1$, the big series above becomes

$$\begin{aligned} &1 + \frac{1}{4} + \frac{1}{9} + \frac{1}{16} + \text{etc.} \\ &- z - \frac{1}{4}z^2 - \frac{1}{9}z^3 - \frac{1}{16}z^4 - \text{etc.} \\ &+ z\, lz + \frac{1}{2}z^2\, lz + \frac{1}{3}z^3\, lz + \text{etc.} \end{aligned}$$

and this all equals

$$\frac{y}{1} + \frac{yy}{4} + \frac{y^3}{9} + \text{etc.}$$

Euler notes that

$$z + \frac{1}{2}zz + \frac{1}{3}z^3 + \text{etc.} = -l(1-z) = -ly$$

and, finding this multiplied by lz in the third line of the series above, he substitutes and rearranges to get

$$1 + \frac{1}{4} + \frac{1}{9} + \frac{1}{16} + \frac{1}{25} + \text{etc.} = \frac{y+z}{1} + \frac{y^2+z^2}{4} + \frac{y^3+z^3}{9} + \frac{y^4+z^4}{16} + \text{etc.} + ly\, lz.$$

Since this is true whenever $y + z = 1$, Euler can take any values for y and z satisfying this condition. He chooses the values for which the series on the right converges most rapidly, that is $y = z = \frac{1}{2}$. This gives

$$1 + \frac{1}{4} + \frac{1}{9} + \frac{1}{16} + \text{etc.} = 1 + \frac{1}{8} + \frac{1}{36} + \frac{1}{128} + \frac{1}{400} + \text{etc.} + \square l2.$$

Here, the pattern of the denominators is $n^2 2^{n-1}$, and Euler has made up the notation $\square l2$ to mean what we would write as $(\ln 2)^2$.

Euler knows logarithms very accurately, and he gives $\square l2$ as 0.480453, only six decimal places, though he must know many more. Euler estimates the sum of the other part of the series as 1.164481, and so he concludes that

$$1 + \frac{1}{4} + \frac{1}{9} + \frac{1}{16} + \cdots + \text{etc. is approximately } 1.644924$$

which is, of course, correct to six decimal places. Euler notes proudly that to get this accuracy by summing the series of reciprocals of squares directly would take over a thousand terms.

Euler closes by observing that this method works whenever the denominators are an algebraic progression. He writes that, done this way, it requires that the numerators be constants, but that it would not be difficult to extend the method to any progression where the numerators, also, were algebraic progressions.

This concludes one of Euler's lesser known papers, but one that plays a key role in much of his subsequent mathematics. He develops and extends these tools on the interpolation of progressions. They will eventually grow to be a major part of Chapter 17 of his great *Calculus Differentialis*, published in 1755. He sharpens his skills at using calculus to manipulate series, particularly the trick of multiplying by an appropriate power of the variable and then integrating again. He does a clever manipulation of a series to accelerate dramatically its rate of convergence. And finally he gets an accurate estimate of the value that will be the solution to the Basel Problem. This work in 1730/31, based on the letters with Goldbach of 1729 and 1730, would provide valuable clues about how to solve the Basel Problem. He will eventually give his solution in E-41, written in 1734/35 and published in 1740.

This paper had a number of other sequels. The next year, he reiterates the main results of E-20 in E-25, Methodus generalis summandi progressiones, "General methods for summing progressions," and ends that paper with a remark that the method did *not* work on two other progressions:

$$1 + \frac{1}{3} + \frac{1}{7} + \frac{1}{15} + \cdots + \frac{1}{2^n - 1} \quad \text{and} \quad \frac{1}{3} + \frac{1}{7} + \frac{1}{8} + \frac{1}{15} + \frac{1}{24} + \frac{1}{26} + \text{etc.}$$

The denominators in the second series are one less than powers of integers. Subsequent investigations of this second series in E-72 "Variae observationes circa series infinitas" led Euler to discover the Product-Sum formula for the Riemann zeta function, the formula that adorns the cover of William Dunham's book *Euler, the Master of Us All*. Other papers that followed immediately and directly from E-20 include those in which Euler discovers and evaluates what we now call the Euler-Mascheroni constant. Euler wrote on this subject throughout his life.

Interlude: 1732

World events

George Washington (1732–1799) was born on February 22, 1732. Washington led the American armies in the American Revolution and served as the first President of the United States from 1789 to 1797.

The composer Franz Joseph Haydn (1732–1809) was born on March 31 of this year.

In Euler's life

In January 1732, Euler and Goldbach exchanged four letters, but then did not write again until October 1735.

Euler's other work

In 1732, Euler wrote his first paper on astronomy, E-14, as well as his first paper on number theory, E-26 (see below). The astronomy paper is titled Solutio problematis astronomici ex datis tribus stellae fixae altitudinibus et temporum differentiis invenire elevationem poli et declinationem stellae, "Solution to problems of astronomy: given the altitudes and time differences for three fixed stars, to find the elevation of the pole and the declination of the star."

He wrote yet another paper on tautochrones, E-21, Quomodo data quacunque curva inveniri oporteat aliam quae cum data quodammodo iuncta ad tautochronismum producendum sit idonea, "How to find a tautochrone curve that joins a given point to another given curve."

Euler's mathematics

Euler's first paper in number theory was E-25, Observationes de theoremate quodam Fermatiano, aliisque ad numeros primos spectantibus, "Observations on a theorem of Fermat and others on looking at prime numbers." In this paper, Euler shows that F_5 is not a prime number.

He continued his studies of the harmonic series and related series in E-25, Methodus generalis summandi progressiones, "A general method for summing progressions."

He began to develop general methods in the calculus of variations in E-27, Problematis isoperimetrici in latissimo sensu accepti solutio generalis, "A general solution to isoperimetric problems in the widest accepted sense."

8

E-25: Methodus generalis summandi progressiones*

General methods for summing progressions

Euler must have been pleased with his results in E-20, applying tools of calculus to the summing of series, for he returns to the subject in this paper and will return to these methods several more times. He opens by writing, "I have just last year put forth a method for summing innumerable progressions," referring to E-20. He proposes to extend those results. In doing so, Euler discovers what we now call the Euler-Maclaurin summation formula, which gives a relation between the integral of a function and the sum of its values at equally spaced points.

Euler begins his analysis by defining his variables. "If the general term of exponent n is given by t, and if the sum of terms up to that term t is s, then

$$t = \frac{ds}{dn} - \frac{dds}{1 \cdot 2 \cdot dn^2} + \frac{d^3s}{1 \cdot 2 \cdot 3 \cdot dn^3} - \frac{d^4s}{1 \cdot 2 \cdot 3 \cdot 4 \cdot dn^4} + \text{etc."} \tag{1}$$

There are several things about this that may not be entirely clear. First, Euler is using the word "exponent" where we would use the word "index." Second, Euler does not tell us why this formula might be true. He does not really explain it until 1755 in his differential calculus textbook, *Institutiones calculi differentialis*. David Pengelley [P] gives a nice account of this episode and relates it to earlier work by Jakob Bernoulli. Like Euler, we won't try to explain it here.

Euler jumps straight to equation (3) below, but we will try to fill in a little bit.

Solving formula (1) for $\frac{ds}{dn}$ gives

$$\frac{ds}{dn} = t + \frac{dds}{1 \cdot 2 \cdot dn^2} - \frac{d^3s}{1 \cdot 2 \cdot 3 \cdot dn^3} + \frac{d^4s}{1 \cdot 2 \cdot 3 \cdot 4 \cdot dn^4} - \text{etc.} \tag{2}$$

Then integrating with respect to n and substituting for $\frac{ds}{dn}$ as given in equation (2) yields

$$s = \int t\,dn + \alpha t + \frac{\beta\,dt}{dn} + \frac{\gamma\,d^2t}{dn^2} + \frac{\delta\,d^3t}{dn^3} + \text{etc.} \tag{3}$$

where

$$\begin{aligned}
\alpha &= \frac{1}{2} \\
\beta &= \frac{\alpha}{2} - \frac{1}{6} \\
\gamma &= \frac{\beta}{2} - \frac{\alpha}{6} + \frac{1}{24} \\
\delta &= \frac{\gamma}{2} - \frac{\beta}{6} + \frac{\alpha}{24} - \frac{1}{120} \\
\varepsilon &= \frac{\delta}{2} - \frac{\gamma}{6} + \frac{\beta}{24} - \frac{\alpha}{120} + \frac{1}{720}.
\end{aligned}$$

**Comm. Acad. Sci. Imp. Petropol.* 6 (1732/3) 1738, 68–97; *Opera Omnia* I.14, 42–72

This is Euler's first statement of the Euler-Maculaurin formula. He apparently felt a bit of remorse in expecting his reader to see why this inversion formula is valid. He gives more details when he presents this same material again in E-47, (see [C], p. 434). We get a complete explanation in 1755.[1]

Euler tells us that in this he is following Jakob Bernoulli's *Ars Conjectandi*. Indeed in the *Institutiones calculi differentialis* Euler himself would acknowledge Bernoulli's work. He would multiply these coefficients $\alpha, \beta, \gamma, \delta$, etc. by their corresponding factorial numbers, 2, 6, 24, 120, etc. and give the results the name "Bernoulli numbers." Havil [H] gives a wonderful account of their importance in a variety of contexts, especially in describing the properties of γ, the so-called Euler-Mascheroni constant. (Note that this gamma is different from the gamma in Euler's list of coefficients.)

Continuing, Euler gets by substitution

$$s = \int t\,dn + \frac{t}{2} + \frac{dt}{12\,dn} - \frac{d^3t}{720\,dn^3} + \frac{d^5t}{30240\,dn^5} - \text{etc.}$$

Knowing this may have been obscure, Euler does an example. Take $t = n^2 + 2n$. Then $dt = 2n\,dn + 2\,dn$, $ddt = 2\,dn^2$ and $d^3t = 0$, and all higher differentials are zero as well. Applying his formula, Euler tells us that

$$s = \int (n^2 + 2n)\,dn + \frac{n^2 + 2n}{2} + \frac{2n + 2}{12} = \frac{n^3}{3} + \frac{3n^2}{2} + \frac{7n}{6} = \frac{2n^3 + 9n^2 + 7n}{6}.$$

"Where is the constant of integration?" the reader may ask. Euler doesn't bother to mention that he chooses his constant of integration so that the formula works for some known value. Because the formula is correct for $n = 1$, he knows he has chosen the correct constant.

In modern terms, Euler is telling us that

$$\sum_{k=1}^{n} (k^2 + 2k) = \frac{2n^3 + 9n^2 + 7n}{6}.$$

In this formula, some readers may notice a hint of the origins of γ, the Euler-Mascheroni constant. Here, Euler expresses the sum of a series as the integral of that series, with some correction terms. In this case, the series is given by the values of a polynomial, and the correction terms are polynomials determined by the derivatives of the first polynomial. We will see this in E-43 when Euler compares $\sum \frac{1}{k}$ and $\int \frac{1}{n}\,dn$ and finds that the correction terms have a limit that has come to be denoted γ.

This method can be applied to sums and differences of series of polynomials as well as to their derivatives and integrals.

Now we consider geometric series. Euler writes the general form of a finite geometric series as

$$x^a + x^{a+b} + x^{a+2b} + x^{a+3b} + \cdots + x^{a+(n-1)b}.$$

He sets this sum equal to s, and goes through the common algebraic manipulations, subtracting x^a, adding x^{a+nb} and dividing by x^b to get

$$s = \frac{s - x^a + x^{a+nb}}{x^b}.$$

[1] E-212, *Institutiones calculi differentialis*, Part II Chapter 5 sections 132 to 138, reprinted in the *Opera Omnia* Series I Volume 10.

Solving for s gives a form of the formula for the sum of a finite geometric series:

$$s = \frac{x^a - x^{a+nb}}{1 - x^b}.$$

"If x is a fraction less than 1 and n an infinitely large number, then $x^{a+nb} = 0$ and

$$s = \frac{x^a}{1 - x^b}."$$

We turn from infinite numbers to infinitely small quantities. Let ω be an infinitely small quantity, and let $x = 1 - \omega$. Then $x^a = 1 - a\omega$, etc., and the formula for the sum of a geometric series reduces to

$$s = \frac{nb\omega}{b\omega} = n.$$

This will be useful later.

Now we look at sums of the form

$$s = x^a + 2x^{a+b} + 3x^{a+2b} + \cdots + nx^{a+(n-1)b}.$$

Rather than use calculus techniques along the lines developed in E-20, Euler does the more familiar algebraic manipulations on this sum. He subtracts x^a and adds $(n+1)x^{a+nb}$, divides by x^b, and subtracts s to get a geometric series that he can sum using formulas developed earlier. It looks like this:

$$\frac{s - x^a + (n+1)x^{a+nb}}{x^b} - s = x^a + x^{a+b} + x^{a+2b} + \cdots + x^{a+(n-1)b} = \frac{x^a - x^{a+nb}}{1 - x^b}.$$

Solving for s gives

$$s = \frac{x^a - x^{a+nb}}{(1 - x^b)^2} - \frac{nx^{a+nb}}{1 - x^b}.$$

For $x < 1$ and $n = \infty$ the sum of the series is

$$\frac{x^a}{(1 - x^b)^2}.$$

Note that in 1732, people had not entirely clarified their ideas on order relations. When Euler writes $x < 1$, he is comparing the size of x with the size of 1, the idea we would now express using absolute value signs. In Euler's day, though, the symbol "<" compared the sizes of quantities, not the order in which they appear on the number line. It would not be fair to say that Euler was wrong when he wrote $x < 1$ instead of $|x| < 1$, in part because he was using the symbol "<" correctly, following the definition the symbol had at the time, and in part because the absolute value signs would not be invented for another hundred years.[2]

[2] According to Jeff Miller's "Earliest Use of Various Mathematical Symbols" web page [M]:

> Karl Weierstrass (1815–1897) used | | in an 1841 essay "Zur Theorie der Potenzreihen," in which the symbol appears on page 67. He also used the symbol in 1859 in "Neuer Beweis des Fundamentalsatzes der Algebra," in which the symbol appears on page 252. This latter essay was submitted to the Berlin Academy of Sciences on December 12, 1859.

Euler's formula for the sum does not work if $x = 1$ to sum the progression $1+2+3+4+\cdots+n$, as he notes, because "the numerator and denominator will vanish." Euler overcomes this with a little calculation using infinitesimals. Again, let ω be an infinitely small quantity, and let $x = 1-\omega$. Then, as before

$$1 - x^b = b\omega.$$

Somehow Euler knows when to ignore terms containing ω^2 and when not to. He does not ignore them in his next step when he writes

$$x^a = 1 - a\omega + \frac{a(a-1)\omega^2}{2}$$

or when he finds similar expressions for x^{a+nb} and for $x^{a+(n+1)b}$. Substituting these expressions into his formula for s, he gets

$$s = \frac{(n^2b^2 + nb^2)\omega^2}{2b^2\omega^2} = \frac{n(n+1)}{2}.$$

It is reassuring that Euler's method leads to what he knows to be a correct result. He could also have arrived at this result using the ideas we saw at the beginning of this article.

These techniques could be generalized to sum any progression where terms are formed by multiplying terms of a geometric series by polynomials. This nicely complements the results of E-20, where Euler summed progressions with geometric sequences in the numerators and polynomials in the denominators.

We turn to the sum

$$s = x + 2x^2 + 3x^3 + 4x^4 + \cdots + nx^n.$$

Euler applies the same tools of calculus that were so effective in E-20, and that we will see again in E-28. He divides the sum by x, multiplies by dx, and integrates, to get a geometric series, and then uses his formula for summing a geometric series. He gets

$$\int \frac{s\,dx}{x} = x + x^2 + x^3 + \cdots + x^n = \frac{x - x^{n+1}}{1-x}.$$

Now he differentiates this to get

$$\frac{s\,dx}{x} = \frac{dx - (n+1)x^n\,dx + nx^{n+1}\,dx}{(1-x)^2}$$

then solves for s to get

$$s = \frac{x - (n+1)x^{n+1} + nx^{n+2}}{(1-x)^2}.$$

This is a special case of a formula Euler found a few pages earlier, taking $a = b = 1$.

Euler extends this to progressions where the coefficients form an arithmetic sequence and the variables form a geometric one, that is, sums of the form

$$s = ax^\alpha + (a+b)x^{\alpha+\beta} + (a+2b)x^{\alpha+2\beta} + \cdots + (a+(n-1)b)x^{\alpha+(n-1)\beta}$$

though he stops his calculations at the form

$$\frac{\beta}{b}\int x^{(\beta a-\alpha b-b/b)} s\,dx = \frac{x^{\alpha\beta/b} - x^{(\alpha\beta+nb\beta)/b}}{1-x^2}$$

and expects the well-motivated reader to differentiate it and solve for s as above.

Euler makes a peculiar notational change, using arithmetic sequences of the form $an+b$, where, up to now he has been denoting them $a+bn$. With this new notation, he turns to progressions with quadratic coefficients, where the general term is given by

$$(an+b)(cn+e)x^{\alpha+(n-1)\beta}.$$

After a dense page of the same kinds of calculations, we find s hidden, this time inside a double integral, and the reader is again left to rescue s with an armory of differentiation and algebra.

Euler assures us that he could use this technique to sum any progression of this kind with polynomial coefficients, but one gets the impression he doesn't really want to do it.

Now we turn to progressions more closely related to the ones we saw in E-20. We look at progressions of quotients with geometric numerators and linear denominators. The general term looks like

$$\frac{x^{\alpha+(n-1)\beta}}{an+b}.$$

The problem here is to find s if

$$s = \frac{x^{\alpha}}{a+b} + \cdots + \frac{x^{\alpha+(n-1)\beta}}{an+b}.$$

Though he has already solved this problem in E-20, he gets the same answer with a different kind of calculation. We sketch the new way to do it because he uses this technique for much of the rest of this paper.

First, multiply the equation by px^{π} and then differentiate to get

$$px^{\pi}\,ds + p\pi x^{\pi-1}s\,dx = \frac{p(\alpha+\pi)x^{\alpha+\pi-1}\,dx}{a+b} + \cdots + \frac{p(\alpha+n\beta-\beta+\pi)x^{\alpha+(n-1)\beta+\pi-1}\,dx}{an+b}.$$

In this formula, p and π are constants that will be given values in the next step. (Here π is not the familiar constant related to circles.) Euler wants to change variables, so he lets

$$p\alpha + pn\beta - p\beta + p\pi = an+b.$$

This makes

$$p = \frac{a}{\beta} \quad\text{and}\quad \pi = \beta - \alpha + \frac{b\beta}{a}.$$

After we make these substitutions, we see that this factor px^{π} was chosen to make the integration work out nicely. The step resembles one of the key steps we saw in E-10. We substitute, integrate and solve for s to get

$$s = \frac{\beta}{a}x^{(a\alpha-a\beta-b\beta)/a}\int x^{(a\beta+b\beta-a)/a}\,dx\,\frac{1-x^{n\beta}}{1-x^{\beta}}.$$

This is the same result, but in a different form with different constants, as Euler found in E-20.

Euler uses the same technique, multiplying again by px^π to sum progressions of terms of the form

$$\frac{x^n}{(an+b)(cn+e)}.$$

This time, though, he has to go through the process twice, first using the same values for p and π as he used earlier in this paper, and then again with

$$p = c \quad \text{and} \quad \pi = 1 - \frac{b}{a} - \frac{e}{c}.$$

This process fills two pages.

Euler skips a few steps in summing progressions of the forms

$$\frac{x^n}{(an+b)(cn+e)(fn+g)} \quad \text{and} \quad \frac{x^n}{(an+b)(cn+e)(fn+g)(hn+k)}.$$

The same methods work, but both the calculations and the answers get more and more outrageous.

Moreover, there are some difficulties if the denominators have repeated factors because the denominators of solutions involve expressions like $(ae - bc)(ag - bf)(ak - bh)$. Such expressions become zero if there are any repeated factors or multiple roots.

Euler continues in this vein summing progressions of these and similar forms and, as Voltaire was later to complain, filling many pages with calculations. Frankly, it is hard to get excited about most of these examples. The only variety seems to be picking exactly the right values of p and π so that when he multiplies by px^π the integrals work out nicely. Otherwise, it seems like sixteen pages of the same thing over and over again with only small variations.

It seems as if Euler would never get to the end of this article. Eventually, though, he explains that these methods allow him to sum any progression of the form $\frac{AP}{BQ}x^{\alpha n+\beta}$ where A and B are any algebraic progressions, and P and Q are arithmetic progressions. He needs P and Q separately so that he knows at least one linear factor of the numerator and of the denominator.

At the very end, Euler gives us two progressions that he would like to be able to sum, but for which he can nor make these methods work. The first is

$$1 + \frac{1}{3} + \frac{1}{7} + \frac{1}{15} + \cdots + \frac{1}{2^n - 1}.$$

Euler adds that the methods do not work for a second progression,

$$\frac{1}{3} + \frac{1}{7} + \frac{1}{8} + \frac{1}{15} + \frac{1}{24} + \frac{1}{26} + \text{etc.}$$

Here, the terms are of the form

$$\frac{1}{a^\alpha - 1},$$

where a and α are integers greater than 2. That means the denominators are one less than non-trivial powers of integers. Euler concedes that Goldbach had used other methods to show the progression sums to 1. Euler will extend Goldbach's results in spectacular ways in E-72, written in 1737, published in 1744.

References

[C] Chabert, Jean-Luc, et al. *A History of Algorithms: From the Pebble to the Microchip*, translated by Chris Weeks, Springer, New York, 1999.

[H] Havil, Julian, *Gamma: Exploring Euler's Constant*, Princeton University Press, 2003.

[M] Miller, Jeff, *Earliest Use of Various Mathematical Symbols*, `http://members.aol.com/jeff570/mathsym.html`, revision of December 5, 2004.

[P] Pengelley, David, "Dances between Continuous and Discrete: Euler's Summation Formula," Presented at Euler E2K+2 Conference, Rumford, Maine, 2002. Online at `http://www.math.nmsu.edu/~davidp/euler2k2.pdf`.

9

E-26: Observationes de theoremate quodam Fermatiano aliisque ad numeros primos spectantibus*

Observations on theorems that Fermat and others have looked at about prime numbers

Where's the number theory? Many of Euler's most important and best-known contributions to mathematics are in number theory. Of about 800 published works, 96 of them, well over ten percent, are number theory, and the subject fills four of the 29 volumes of his mathematical works. André Weil devotes half of his book on the history of number theory [W] to the work of Euler. Yet, none of Euler's first 25 works have dealt with number theory. Why did that happen?

Mathematicians in the 1600's had not taken number theory very seriously. The foremost contributions were by the Great Amateur, Pierre de Fermat, and most of his work was published only posthumously. Little import was attached to the work at the time he was doing it. Mersenne did a bit of number theory that still bears his name. Pell did some, but apparently not on the equation that is named after him. Descartes discovered a pair of amicable numbers, but he apparently did this only because Fermat had found a pair, and the two had a very competitive relationship. If the other great mathematicians of the century, Newton, Leibniz, Bernoulli and Huygens, did any number theory, it is virtually forgotten. Thus, until 1732, Euler had little incentive to pursue questions of number theory.

When Euler writes this paper in 1732, he is fairly well known, but he is not yet regarded as "great." Real respect and fame will not come until he solves the Basel problem (E-41) in 1734 and publishes his *Mechanica* (E-15 and E-16) in 1736. Before these accomplishments, Euler simply was not famous enough to risk his reputation pursuing what would have been regarded as minor problems in amateur mathematics. Thus, as we have seen, Euler has pursued problems in mainstream mathematics and physics, problems that are "safe," established and recognized by others as important.

Two things happened, one well known and one not well known. First, Euler began to correspond with Christian Goldbach. We have seen two fruits of Euler's first letter to Goldbach, E-19 and E-20. In a letter from Moscow dated December 1, 1729, replying to Euler's first letter, Goldbach adds a note "P. S. Have you noticed the observation of Fermat that all numbers of this form $2^{2^{x-1}}+1$, that is 3, 5, 17, etc., are prime numbers, but he did not dare to claim he could demonstrate it, nor, as far as I know, has anyone else been able to prove it." Euler, in his next letter to Goldbach, wrote that he, also, was unable to find a proof.

It was probably not wise of Goldbach to bring this up, for it could only distract the young Euler from the important work he was doing establishing his own career. Moreover, Euler would have little chance of solving these problems, and even if he did succeed, results in number theory counted for little. Nonetheless, Euler did succeed, and it did not detract from his other work. One might say that Goldbach was very lucky to mentor Euler rather than some lesser genius.

Comm. Acad. Sci. Imp. Petropol. 6 (1732/3) 1738, 103–107; *Opera Omnia* I.2, 1–5

A second, much less well-known event occurred in 1730. At the time, a great philosophical debate on the foundations of science was raging through Europe [B, C]. On one side were the Newtonians, who believed that the way to understand the world was to perform experiments and to base theory on the resulting observations. Then they tested their theories with more experiments and further refined them, if necessary. The philosophical shortcoming of this method is that it describes what happens without explaining *why* it happens. The other side, the Wolffians, combined the ideas of Descartes and Leibniz, and were championed by Christian Wolff. They believed that the world was based on principles set by God and that, if we could understand God's design, we could predict not only how but also why the world works as it does. Then they would do experiments only to confirm that their logic had been correct. The name Christian Wolff now lives mostly in footnotes, but then he was an important philosopher and scientist. We mentioned him and his prodigious output in the Preface. In 1730, he published his *Elementa matheseos universae*, or *Elements of the exact science of the universe*, (the word *matheseos* translates better as "exact science" than the more limited scope of "mathematics"). In that work, Wolff included number theory, giving the subject a validation it had not enjoyed for some time.

So, with this context in mind, it makes sense that none of Euler's first 25 works involved number theory, and it also makes sense that he would now take up the subject. He will pursue the topic with great success throughout his life, and we will get to see his work in E-26, and later in E-29, E-36, E-54, E-98 and E-158.

Euler begins by showing that if a prime number has the form $a^n + 1$, then n must be a power of 2. He uses a rather strange narrative style to do this, rather than the more familiar theorem-proof format he will use later in his career. He writes:

> It is known that every quantity $a^n + 1$ always has divisors if n is an odd number or if n is divisible by an odd number other than 1. For a^{2m+1} can be divided by $a+1$, and $a^{p(2m+1)}+1$ can be divided by a^p+1, no matter what number is substituted in the place of a. On the other hand, if n is a number that is divisible by no odd number except 1, it follows that n is a power of 2, and no number can be assigned as a divisor of $a^n + 1$. For which reason, if there are prime numbers of the form $a^n + 1$, they all must be included in the form $a^{2^m} + 1$.

Euler moves on to examine the possible values of a. Obviously, if a is odd, then the form $a^n + 1$ is divisible by 2. If a is even and $n = 2$, then $a^2 + 1$ could be prime or composite. Euler dwells on the composites in giving several examples:

> Next, if a denotes an even number, it happens in countless events that the form produces a composite number. Thus, the formula $a^2 + 1$ can be divided by 5 when $a = 5b \pm 3$, and $30^2 + 1$ can be divided by 17, and $50^2 + 1$ by 41. In a similar way, $10^4 + 1$ has a divisor 73, and $6^8 + 1$ is divisible by 257.

Thus, Euler demonstrates that if n is not a power of 2, then $a^n + 1$ is not prime; that if a is odd, it is not prime; and if a is even, then it need not be prime. All of this gives Euler no reason to expect that $2^{2^m} + 1$ must be prime. He writes

> A table of prime numbers which extends beyond 100,000 does not detect any case in which the form $2^{2^m} + 1$ has any divisor. On the strength of this and other reasons, Fermat was led to announce that he did not doubt $2^{2^m} + 1$ always to be a prime number and proposed to Wallis and other English mathematicians to prove this distinguished theorem. He confessed that he himself did not have a proof of this, but nonetheless asserted that it was true.

Euler agrees that the first four numbers of this form, 5, 17, 257 and 65537, are, indeed prime. He does not include the case $n = 0$, which produces the prime number 3, though Goldbach's letter used $x - 1$ where Euler has been using m, so Goldbach would have included the prime 3 on his list. Euler writes, though,

> But I don't know how it followed at once that $2^{2^5} + 1$ was a prime number. Namely I observed some days ago that the number could possibly be divided by 641, and it will be revealed at once that this is so. For it is that $2^{2^5} + 1 = 2^{32} + 1 = 4294967297$.

Euler did not use commas to separate long numbers.

He notes that this leaves us with no formula to generate an arbitrarily large prime number, but does not explain how he came upon this factorization.

Euler goes on to discuss numbers of the form $2^n - 1$. Prime numbers of this form are now called Mersenne primes, though Euler does not mention Mersenne. He notes that if n is not prime, then any number of the form $a^n - 1$ will have divisors, so for a number of Mersenne's form to be prime, n must be prime. However, this is not sufficient. Euler gives as examples $2^{11} - 1 = 2047 = 23 \cdot 89$, and $2^{23} - 1$, which is divisible by 47. Euler here mentions Christian Wolff's interest in the subject and writes

> I see that the Celebrated Wolff did not notice this in his *Elementa matheseos universae*, where he investigated perfect numbers and listed 2047 among the primes.

Euler notes that whenever $2^n - 1$ is prime, then the number $2^{n-1}(2^n - 1)$ is a perfect number. This means that the proper divisors of the number sum to the number itself. This fact was known to Euclid, but Euler says that it makes the study of numbers of the form $2^n - 1$ interesting. For $2^n - 1$ to be prime, we have seen that n must be prime, but that the converse is not true; if n is prime, then $2^n - 1$ may not be prime. Euler suggests "it would be worthwhile if we could study those cases when $2^n - 1$ is not a prime number, even though n is such" and he gives some results, without proving them.

For example, if $n = 4m - 1$, and if $8m - 1$ is prime, then $2^n - 1$ is always divisible by $8m - 1$. This excludes the cases where n is 11, 23, 83, 131, 179, 191, 239, etc. It was the value $n = 11$ that had fooled Christian Wolff two years earlier.

There are other cases, though, where n is prime and not of this form, yet $2^n - 1$ is composite. Euler notes that if $n = 37$, then $2^n - 1$ is divisible by 223, and provides other divisors for the cases $n = 43$, 29 and 73. Euler claims

> Yet I dare to claim that aside from the cases noted, all of the prime numbers less than 50, and perhaps less than 100, make $2^{n-1}(2^n - 1)$ a perfect number, whence the eleven values 1, 2, 3, 5, 7, 13, 17, 19, 31, 41, 47 of n yield perfect numbers.

The numbers 41 and 47 do not belong on this list, but Euler will not correct his error until 1750 after his colleague in Petersburg, Christian Winsheim, notices the errors in 1749.

Euler seems to think that he has given the main results of this paper; that Fermat's formula does not always generate prime numbers and that Wolff's list of perfect numbers has the mistake in it. Now, he offers some clues about how he came upon these results. "I have deduced these observations from theorems which are not inelegant, proofs of which I do not have, but I am nonetheless truly most certain of their truth. The theorem is this: $a^n - b^n$ can always be divided by $n+1$ if n is a prime number and if a and b cannot be divided by it. I believe the proof of this to be difficult because it is not true unless $n + 1$ is a prime number."

This is, on closer look, a form of the Fermat theorem. Now we state the theorem as follows: if p is prime and p does not divide a, then $a^{p-1} \equiv 1 \pmod{p}$. Euler does not yet seem to know that Fermat knew this theorem, and he does not yet offer a proof. As we will see, he will prove it several times, first just a few years later in 1736 in E-54, then again in 1747 in E-134, in 1758 in E262 and in 1760 in E-271.

Euler writes "This leads one to discover other theorems no less elegant, which I suppose to be rather valuable, and which I have been unable to prove absolutely, and which are therefore appended here to be contemplated." Then, Euler gives as theorems six conjectures in number theory.

Theorem 1. *If n is a prime number, every $n-1$ power divided by n leaves either nothing or 1 as a remainder.*

This is a slightly different form of Fermat's theorem, also accounting for the case where a is a multiple of the prime.

Theorem 2. *Letting n be a prime number to a power with exponent $n^{m-1}(n-1)$ divided by n^m always leaves a remainder of* 0 *or* 1.

This is a special case of the more general Euler-Fermat theorem, and it will take decades for Euler to get it all sorted out. We will peek into Euler's future and untangle it a bit. In 1758, in E-271, Euler will define a function he describes as "the multitude of numbers less than n and prime to n." Gauss will later denote this function $\varphi(n)$. Euler will prove several properties of $\varphi(n)$. If n is prime, then $\varphi(n) = n - 1$. If n is a power of a prime, say $n = p^m$, then $\varphi(p^m) = (p-1)p^{m-1}$. If m and n are relatively prime, then $\varphi(mn) = \varphi(m)\,\varphi(n)$.

Then, Euler will prove that, in modern notation, if a and n are relatively prime, then $a^{\phi(n)} \equiv 1 \pmod{n}$.

In the case n is prime, this gives the Fermat theorem in the form $a^{p-1} \equiv 1 \pmod{p}$.

In the case that n is a prime power, say $n = p^m$, this gives that $a^{p^{m-1}(p-1)} \equiv 1 \pmod{p^m}$. This is the result claimed by Euler, but with the variables renamed and in the case when a and n are relatively prime.

Theorem 3. *Let m, n, p, q, etc. be unequal prime numbers and let A be the smallest number divisible by all of them. Consider $m-1$, $n-1$, $p-1$, $q-1$, etc. In this case I say that every power with exponent A, when a^A is divided by mnpq etc., leaves either* 0 *or* 1 *as a remainder, unless a can be divided by some one of the numbers m, n, p, q, etc.*

This is yet another special case of the Euler-Fermat theorem, in the case the exponent factors into prime factors that do not repeat, so called square-free numbers.

Euler closes with three conjectures, he calls them theorems, that seem unrelated to the Euler-Fermat theorem.

Theorem 4. *Denoting by $2n+1$ a prime number, 3^n+1 can be divided by $2n+1$ if either $n = 6p+2$ or $n = 6p+3$, and $3^n - 1$ can be divided by $2n+1$ if either $n = 6p$ or $n = 6p-1$.*

Theorem 5. *$3^n + 2^n$ can be divided by $2^n + 1$ if $n = 12p+3$ or $12p+5$ or $12p+6$ or $12p+8$. Also $3^n - 2^n$ can be divided by $2n+1$ if $n = 12p$ or $12p+2$ or $12p+9$ or $12p+11$.*

Theorem 6. *In the same conditions as* $3^n + 2^n$*, then* $6^n + 1$ *will be divisible by* $2n + 1$*, and* $6^n - 1$ *will divide it under the same conditions as* $3^n - 2^n$.

We see that this is a peculiar paper, and it is not typical of Euler's other work in number theory. It starts with proofs without theorems and ends with theorems without proofs. Already Euler knows most of the Euler-Fermat theorem, though its complete synthesis is almost 30 years away. It is quite a first step into the field of number theory, refuting an important conjecture of Fermat and extending another. It contains the germ of one of the most important results in number theory, the Euler-Fermat theorem.

Still, it shows immaturity in its structure, and it is clearly rooted in the 17th century tradition of number theory of claiming results without giving their proofs. Fermat's habit of doing this tantalized and annoyed mathematicians for almost 300 years.

Most importantly, this short paper opens Euler's eyes to number theory, a field in which he will make some of his most important and exciting discoveries.

References

[B] Boss, Valentin, *Newton in Russia*, Harvard University Press, Cambridge, 1972.

[C] Calinger, Ronald, "The Newtonian-Wolffian Confrontation in the St. Petersburg Academy of Sciences, Cahiers d'histoire mondiale 11 (1968) 417–436.

[W] Weil, André, *Number Theory: An approach through history From Hammurapi to Legendre*, Birkhäuser, Boston, 1983.

10

E-27: Problematis isoperimetrici in latissimo sensu accepti solutio generalis*

An account of the solution of isoperimetric problems in the broadest sense

Often when an article appeared in the *Commentarii* of the Petersburg Academy, somebody would write a short summary of the article. All of the summaries for an entire volume appear at the beginning of the volume, after the Table of Contents and before the first article begins. The style of the summaries is usually more formal and narrative than the style of the articles themselves. That raises the question, who wrote the summaries? Would the author write his own summary, or would the editor write it? Experts disagree.

The summary of this article contains a small bit of evidence suggesting it was written by an editor and not by Euler himself. The summary opens with a narrative describing how difficult it is to find "among innumerable quantities and innumerable curves satisfying some condition, one that possesses a certain property of maximum or minimum."

We proceed to a description of the elementary problem to cut a straight line into two segments so that the product of the two is maximized, and then ask the same question with three segments instead of two. Surely Euler would have seen that these examples do not fit in a paper on the calculus of variations, where answers are curves and not numerical values.

We are told that there are three kinds of optimization problems that the methods in this paper will help solve. First, from among "absolutely all curves," we are to find the one possessing some kind of maximum or minimum. The Brachistochrone Problem is cited as an example. Such problems will be solved, we are told, by considering two consecutive elements on the desired curve. We will learn what that means when we get to the paper itself.

The second kind of case is to find from among curves possessing some quality A, that one which maximizes or minimizes some quality B. An example is "to find, from among all closed curves of a given length, the one that encloses the maximum space," the classical isoperimetric problem. Such problems, we are told, "cannot be solved unless three curve elements can be drawn into the calculation."

The third kind requires that the curves satisfy two constraints, A and B, then be maximized with respect to a property C. Such problems require four curve elements. More constraints make it all progressively more difficult. This paper will discuss problems of the first two types. Some of those were solved earlier by Bernoulli, Taylor and Hermann, but this paper will answer more general questions and use more general techniques.

All of this is in the "Summarium," and except for the simple example, it is repeated in the first three sections of the text of the paper itself.

Euler gives us a quick review of the family of results that he intends to generalize in this paper. He reminds us that a cycloid has the property that no other curve with the same endpoints will allow a body to descend from the upper point to the lower in a shorter amount of time. A catenary is the

***Comm. Acad. Sci. Imp. Petropol.* 6 (1732/3) 1738, 123–155; *Opera Omnia* I.25, 13–40

curve of a given length between two points with the lowest center of gravity. A sector of a circle has the largest area of all curves joining A and B with that same length (Figure 1).

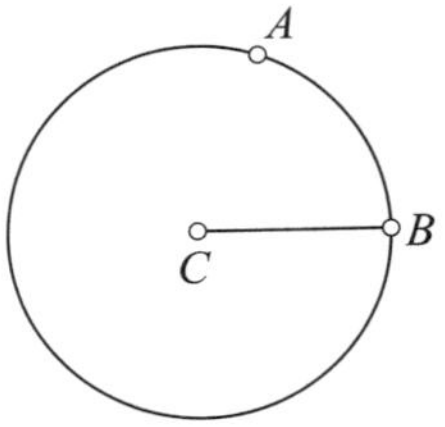

Fig. 1

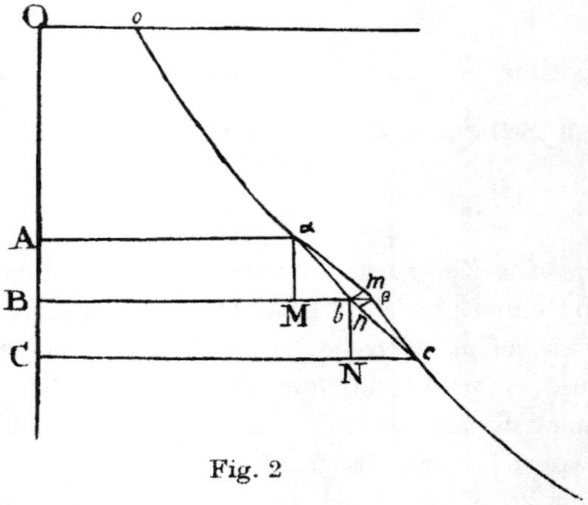

Fig. 2

Now, Euler begins his analysis of the first class of optimization problems, those that require only two curve elements for the analysis. For this, he refers us to Figure 2. Exactly this figure will appear four years later in E-56, where he returns to and improves on the results of this paper.

The general problem here is to determine from among all curves the one which has a given property of maximum or minimum. He begins with an axis *OA*, and two equal segments on that axis *AB* and *BC*. If *oa* is the desired curve, then these two equal segments have corresponding curve elements ab and bc, and the elements along the abscissa bM and cN. Now, he assigns variables, $s =$ arc length of oa, $x = OA$ and $y = Aa$.

Note that these conventions about axes are more similar to Jakob Hermann's system that we saw in E-9 than they are to what we usually see today. The x axis goes down, the y axis goes across, and the y coordinate is measured as a distance from the x axis rather than as a distance along the y axis.

To be consistent with the notation he has chosen, Euler further declares $AB = BC = dx$, $bM = dy$ and $ab = ds$, from which it follows that $cN = dy + ddy$ and $bc = ds + dds$.

Rather than try to develop a new system of notation, Euler gives us a dictionary to translate the lengths of segments if the curve were to pass through β instead of b. The lengths would be

$$a\beta \text{ instead of } ab$$
$$\beta c \text{ instead of } bc$$
$$M\beta \text{ instead of } Mb$$
$$cN - b\beta \text{ instead of } cN.$$

We need a few more points. Point m is on $a\beta$ so that mb is perpendicular to $a\beta$, and point n is on bc so that $n\beta$ is perpendicular to bc.

Now, if the curve is optimal (a word Euler never used), then two other curve elements $a\beta$ and βc will make a curve still having that optimal property. If the curve passes through β instead of b, then Euler tells us how much each segment in his diagram either increases or decreases. For example, ab grows by βm and bc decreases by bn.

There are two pairs of similar triangles here, the pair βbm and baM and the pair βbn and cbN. This gives

$$\beta m = \frac{bM \cdot b\beta}{ab} \quad \text{and} \quad bn = \frac{cN \cdot b\beta}{bc}.$$

Substitute this into the dictionary of changes as follows:

$$\text{in place of } ab \text{ put } ab + \frac{bM \cdot b\beta}{ab}$$
$$\text{in place of } bc \text{ put } bc - \frac{cN \cdot b\beta}{bc}$$
$$\text{for } bM \text{ put } bM + b\beta$$
$$\text{for } cN \text{ put } cN - b\beta$$

and, he reminds us, keep the elements AB, BC unchanged. Then Bb grows by the element $b\beta$ and the arc oab grows by the particle βm, that is by $\frac{bM \cdot b\beta}{ab}$. For some reason, Euler uses the word "particle" for this last length.

Euler is ready to apply the principle of maximum or minimum. At a maximum or a minimum, the differential must be zero. He explains it differently here than he had in E-9, the step we described at the time as seeming "exactly wrong." He tells us that if the elements ab, bc are on the curve that has the maximum or minimum property, and if these elements are close to the elements $a\beta$, βc, and if we evaluate the property via ab, bc and along $a\beta$, βc and subtract, then the difference ought to be zero. It is worth comparing this to Fig. 2 in E-9.

Euler is stuck for a good notation now. He can either speak of a general property A, to be maximized, or he can take a particular example. He does not know how to define a problem in calculus of variations as we might find in a textbook today, to find the function y of x that minimizes $\int_{x_1}^{x_2} f(x, y, y')\, dx$. So, for want of notation, Euler proceeds to an example. Suppose that we know the curve from o to a that minimizes $\int x^n\, dx$, and we want to extend the curve to c. Then for ab, bc to be elements of the minimizing curve, then $OA^n \cdot ab + OB^n \cdot bc$ ought to be equal to $OA^n \cdot a\beta + OB^n \cdot \beta c$. That implies, by a substitution and a bit of algebra that Euler leaves to us,

$$\frac{OA^n \cdot bM \cdot b\beta}{ab} = \frac{OB^n \cdot cN \cdot b\beta}{bc},$$

or, canceling, that

$$\frac{OA^n \cdot bM}{ab} = \frac{OB^n \cdot cN}{bc}.$$

In this last equation, the differential of the left-hand side is zero, so the left-hand side is constant. Call it a^n. By substituting the differentials for the elements, again a task Euler leaves to us, we get the differential equation

$$x^n\, dy = a^n\, ds.$$

Euler regards this as having solved the problem, even without seeking a solution to the differential equation itself. He generalizes a bit, saying that if $\int P\, dx$ is to be minimized, where P is any kind of function of x, then it will produce the solution

$$P\,dy = A\,ds,$$

where A is a homogeneous constant of P. It is not clear what makes a constant "homogeneous."

Euler adds that if P is a function of y instead of x, then switching the coordinates will produce the equation

$$P\,dx = A\,ds.$$

We move on to some more examples. Euler asks us to minimize $\int x^m y^n\,ds$ and gets for a solution

$$\frac{xy\,dx\,ddy}{ds^2} + my\,dy - nx\,dx = 0.$$

This is a second degree equation that he says can be reduced to the first degree using the techniques of E-10, "but it is complicated, and it appears that the variables cannot be separated."

Maximizing or minimizing $\int x^m s^n\,ds$ gives the solution $x\,dx\,ddy + m\,ds^2\,dy = 0$, an equation that does not involve n.

To finish his discussion of problems of the first kind, Euler maximizes or minimizes

$$\int \frac{ds^m\,dy^n}{dx^{m+n-1}}$$

and more generally

$$\int \frac{P\,ds^m\,dy^n}{dx^{m+n-1}}.$$

We turn to problems of the second kind, "to find from among all curves oa with a quality A, that one that has a property B to a maximum or minimum degree." For such problems, he asks us to consider three curve elements instead of the two elements that sufficed for problems of the first kind.

As before, Euler uses axis OA (Figure 3). It has three equal elements, AB, BC, CD, and corresponding curve elements ab, bc, cd. Draw next Aa, Bb, Cc, Dd and then perpendicular to them

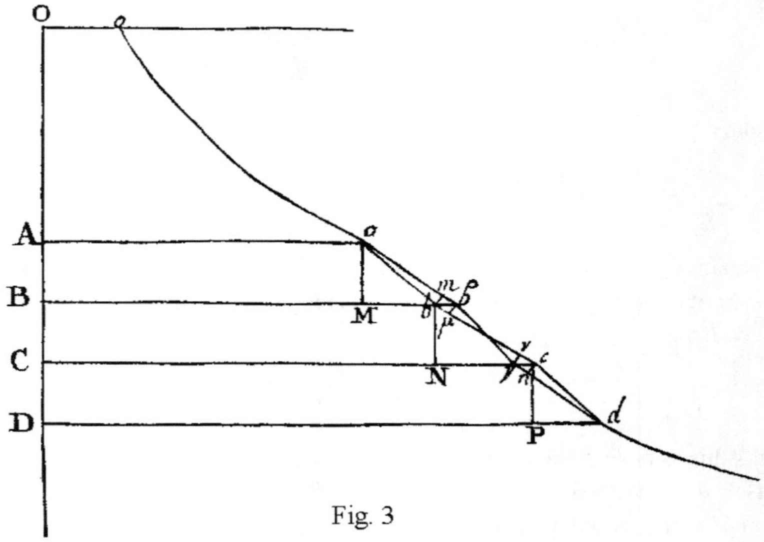

Fig. 3

aM, bN, cP, as before. Assigning variables, Euler asks us to "Call Oa, x; Aa, y and oa, s, and then

$$AB = BC = CD = dx,\ bM = dy \quad \text{and} \quad ab = ds,$$

and then it follows that

$$cN = dy + ddy,\ bc = ds + dds,\ dP = dy + 2ddy + d^3y \quad \text{and} \quad cd = ds + 2dds + d^3s."$$

As before, we compare the curve having elements *ab*, *bc*, *cd* to a nearby curve having elements $a\beta$, $\beta\gamma$, γd. The figure is a bit hard to read in this detail. Euler wants to compare the value of the function P on the original curve, *oabcd*, with its value on a nearby alternative curve, $oa\beta\chi d$. In changing curves, we are told that *ab* grows by βm, *bc* is diminished by $b\mu + c\nu$, *cd* grows by γn, bM grows by $b\beta$, cN decreases by $b\beta + c\gamma$, and dP grows by $c\gamma$. By similar triangles, these increments and decrements are

$$\beta m = \frac{bM \cdot b\beta}{ab}, \quad b\mu = \frac{cN \cdot b\beta}{bc}, \quad c\nu = \frac{cN \cdot c\nu}{bc}, \quad \gamma n = \frac{dP \cdot c\gamma}{cd}.$$

We spend a few lines explaining why, if there were two properties to be satisfied, an analysis using three curve elements would not suffice. The two properties themselves would then be enough to determine both intermediate points β and γ, and then the curve itself would be determined.

This said, he continues with his analysis of the problem in which the curve satisfies a single property, which he says can be put in the form

$$P \cdot b\beta - (P + dP) \cdot c\gamma = 0.$$

He gave a few reasons for this formula in the previous paragraphs, but even without looking closely, it makes sense that there ought to be a function like P telling us, if we change b to β, how much we have to change c in order to preserve the property.

Though Euler can be fairly general about the conditions that the curves must satisfy, he still can't handle much generality in the property to be optimized, so he proceeds to an example. Again, he uses $\int y^n\,dx$. As before, but this time with three curve elements instead of two, the differential must be zero at a maximum or minimum, so

$$Aa^n + Bb^n + Cc^n = Aa^n + B\beta^n + C\gamma^n,$$

hence,

$$B\beta^{n-1} \cdot b\beta - C\gamma^{n-1} \cdot c\gamma = 0.$$

This is not immediately obvious, so we will provide a few steps that Euler chose to omit. Subtracting, we get

$$Bb^n + Cc^n = B\beta^n + C\gamma^n.$$

Substituting $Bb = B\beta + \beta b$ and $Cc = C\gamma - c\gamma$, and expanding, gives

$$B\beta^n + n \cdot B\beta^{n-1} \cdot b\beta + \frac{n(n-1)}{2} B\beta^{n-2} \cdot b\beta^2 + \cdots + C\gamma^n - n \cdot C\gamma^{n-1} \cdot c\gamma$$
$$+ \frac{n(n-1)}{2} C\gamma^{n-2} \cdot c\gamma^2 - \cdots = B\beta^n + C\gamma^n.$$

Subtracting, then neglecting all higher powers of $b\beta$ and $c\gamma$ gives the result he claims.

Again, without explaining why, Euler tells us that the quantity by which P changes is y^{n-1}. This, too, is tricky, but it follows similarly from solving $B\beta^{n-1} \cdot b\beta - C\gamma^{n-1} \cdot c\gamma = 0$ for $c\gamma$ and substituting into $P \cdot b\beta - (P + dP) \cdot c\gamma = 0$, then solving for dP.

This gets us to section 13 of the 39 sections in this paper. Euler continues in this vein, considering integrals to be optimized of a variety of forms, $\int x^m s^n\, ds$, $\int y^n\, ds$, $\int ds^m\, dy^n$, $\int x^n\, ds$, among others. Eventually, he realizes that the forms he considers can be generalized and he considers first $\int T\, dx$ where $dT = M\, dy + N\, dx$, then $\int T\, ds$ where $dT = L\, ds + M\, dy + N\, dx$. In each case, the challenge in the analysis is to find that function P that measures how much the integral changes as the curve is perturbed. Eventually, in section 25, Euler realizes that his results need some organizing, and he gives a table giving the form of P depending on whether T in the integral to be optimized is integrated with respect to ds, dx or dy, and depending on how the differential dT depends on $M\, dy$, $N\, dx$ and $L\, ds$. His table only describes 12 of the 48 possible combinations. He also describes three cases that look like

$$\int \frac{T\, ds^m\, dy^n}{dx^{m+n-1}},$$

for various forms of dT.

A typical line in the table is case III, which tells us that if we are to optimize $\int T\, ds$ and if $dT = N\, dx$, then T depends only on x and not on y, y' or s. The function P has the form $P = d \cdot Tq$, where $q = \frac{dy}{ds}$.

Euler turns to a familiar example, the Brachistochrone Problem. He asks us to minimize the time of descent given by the integral $\int \frac{ds}{\sqrt{x}}$. Here, $T = \frac{1}{\sqrt{x}}$ and the integrand involves ds, so the third line of the table tells us that $P = d \cdot Tq$, which is to be set equal to 0. This makes $P = d \cdot \frac{q}{\sqrt{x}} = 0$, so $\frac{q}{\sqrt{x}}$ is constant. Call that constant $\frac{1}{\sqrt{a}}$, from which we get that $dy\sqrt{a} = ds\sqrt{x}$, so $a\, ds^2 - a\, dx^2 = x\, ds^2$. That makes

$$ds = \frac{dx\sqrt{a}}{\sqrt{a - x}}$$

and $s = C - 2\sqrt{a(a-x)}$, which Euler recognizes as the shape of a cycloid. Thus, Euler confirms his new technique by using it to get a known answer to a familiar problem.

He turns to a second example, to find a curve oa which rotated around the axis Oo, normal to oA, will generate a solid that gives minimum resistance when moved through a fluid in the direction of the Oo axis. Citing other sources, he says that the integral to be minimized is

$$\int \frac{x\, dx^3}{ds^2}.$$

This corresponds to case XIII in his table, with $T = x$, $m = -2$ and $n = 0$. The table tells us that the function P which must remain zero has the form $P = d \cdot T\, ds^{m-2} dy^{n-1}(m\, dy^2 + n\, ds^2)$. Substituting gives

$$P = d \cdot -\frac{2x\, dy}{ds^4} = 0$$

(where this d-dot notation means we are to take the differential of the expression beyond the dot). Euler claims this leads to the differential equation $x\, dx^3 dy = a\, ds^4$, which will determine the curve that generates the solid of minimum resistance. This again was a known result, one Euler had found in his work on the masting of ships several years earlier.

Euler turns to problems of the second class, in which the curve must satisfy a property A and then either maximize or minimize a property B. The same table helps in problems like this, too. The function P, found using A, should equal zero, because the value of A has to remain constant. The function P derived from B must be equal to zero as well, but for a different reason. That is a necessary condition for a curve to be a maximum or a minimum. The plan is to find P in two different ways and get a differential equation that describes the desired curve.

This said, he turns to another classic, the isoperimetric problem. Suppose $A = s = \int ds$ constrains the length of the curve, and $B = \int y\,dx$ is the area, which is to be maximized. The constraint fits Case III of Euler's table and makes $P = dq$, whereas the integral to be maximized fits Case I, in which $P = M\,dx$. Here, $M = 1$, so $P = dx$. Setting these two values of P equal to each other gives $a\,dq = dx$. (The value a sneaks in for a rather subtle reason, namely that maximizing P is the same as maximizing aP, for $a > 0$, so we are not really sure if $P = dq$ or if $P = a\,dq$. Euler is not helpful in this step.) From this, the derivation proceeds:

$$
\begin{aligned}
aq &= \frac{a\,dy}{ds} = x,\\
dy &= \frac{x\,dx}{\sqrt{a^2 - x^2}},\\
y^2 + x^2 &= a^2.
\end{aligned}
$$

Hence, the circle is the isoperimetric curve, another familiar result.

Euler gives us a second example of a problem with a constraint. He wants to find a curve oa of fixed length that will give the maximum volume when rotated about the axis Oo. To solve this, we must maximize $B = \int x^2\,dy$ subject to the constraint $A = s = \int ds$. From A he gets that $P = dq$, as before, and from B he gets that $P = 2x\,dx$. The differential equation thus becomes $a^2\,dq = 2x\,dx$. Integrating gives $a^2\,dy = x^2\,ds \pm b^2\,ds$. Euler ends the example by remarking "the nature of an elastic curve is expressed by this equation."

Continuing his exhibition of examples, Euler asks us to find the curve oa which, rotated about the axis Oo, sweeps out a volume of a given area and produces the minimum surface area. The volume constraint is given by $A = \int xx\,dy$ and the surface area is determined by $B = \int x\,ds$. The table again tells us that $P = 2x\,dx$ and that $P = d{\cdot}xq$, which leads to the equations $2x\,dx = a\,d{\cdot}xq$ and $x^2 \pm b^2 = \frac{ax\,dy}{ds}$. This reduces to

$$
dy = \frac{(x^2 + b^2)\,dx}{\sqrt{a^2x^2 - (x^2 \pm b^2)^2}}.
$$

Euler says this is a circle if $b = 0$. He also makes the remark, difficult to fathom, that this is a catenary if a is infinite and $bb = ae$. Today this remark can only be interpreted in terms of limits and the ratio of bb to a as a goes to infinity.

Euler introduces a new and more complicated kind of problem with his next example. He seeks the curve oa from among all curves of a given length that will put the center of gravity of the area under the curve as far as possible from the axis Oo. The constraint is $A = \int ds$ and the quantity to be maximized is

$$
B = \frac{\int x\,ds}{s}.
$$

When we find the form for A in the table, it tells us that $P = dq$. Likewise, the form of B implies that $P = d \cdot xq$. Thus $a\,dq = d \cdot xq$ or $aq = xq - b$. Substituting $x - a$ for x gives $xq = b$ so $x\,dy = b\,ds$, which is an equation for the catenary.

Euler could not have worked this example if s had not been one of the constraints of the problem because the form

$$\frac{\int x\,ds}{s}$$

is not found in the table. Euler substitutes $s = \int s\,dx$ to get

$$B = \int \frac{ds \int s\,dx}{s^2}$$

and develops another means of solving the problem that does not rely on treating s as a constant. He finds that if $B = \int \left(s^n \int s\,dx\right) ds$, then

$$P = c^{\int \frac{sq\,ds\,dx - n\,ds\,dq \int s\,dx}{s^2 q\,dx + s\,dq \int s\,dx}} \left(s^{n+1} q\,dx + s^n\,dq \int s\,dx\right).$$

In the given example, $n = -2$ and solving the resulting differential equation again produces a catenary.

This leads Euler to a new class of examples, to maximize or minimize properties with forms he writes like $\int T\,dx \int V\,dx$ and $\int T\,dy \int V\,ds$. Euler means us to consider the integral involving V to be part of the integrand of the integral involving T. There are nine such cases, one for each combination of dx, dy and ds in each part of the integral. He continues his table, numbering the cases XVI through XXIV, giving the value of P for each of these nine new forms.

Euler demonstrates the use of his new table with a complicated example finding the curve of a chain of variable density so that its center of gravity is as far as possible from the line Oo.

Next, we turn to the theory behind the third class of problems, satisfying two properties, A and B, and at the same time maximizing or minimizing a property C. His analysis is based on Figure 4

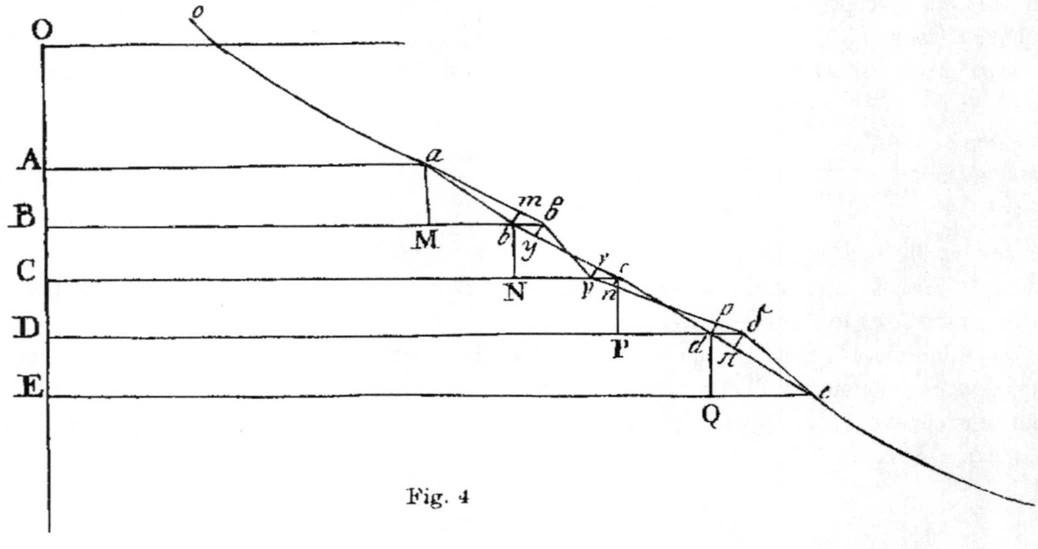

Fig. 4

and closely parallels his previous arguments. In place of the function P he found before, he derives three functions that he denotes P, p and π, that depend on the forms of properties A, B and C, and must satisfy $P + mp + n\pi = 0$, for constants m and n.

From this, Euler says it is easy to extend the techniques to other classes of problems, by which he means problems with even more constraints. For an example, Euler proposes to find a curve from among all those with a given length (property A) and enclosing a given area (property B) and which encloses a given volume when rotated about the axis Oo (property C) that will have the minimum surface area (property D). Property A is given by the formula $\int ds$ and yields $P = dq$. The second property is given by $\int y\,dx$ and makes $P = dx$. The third is expressed by $\int xx\,dy$ and makes $P = x\,dx$. The surface area property to be minimized is described by $\int x\,ds$ and leads to $P = d \cdot xq$. All of these are found using Euler's table, and he has chosen to call them all P rather than to try to find a fourth symbol to add to the collection of P, p and π that he used earlier in his analysis. Introducing constants a, b and c, he puts these four differentials together to obtain

$$a\,dq + b\,dx + 2cx\,dx + d\,xq = 0$$

so that

$$aq + bx + cx^2 + xq = f$$

which, if we substitute for q, equals

$$a\,dy + bx\,ds + cx^2\,ds + x\,dy = f\,ds.$$

Euler says of this "innumerable curves are contained in this form."

Euler offers yet another example, to find a curve with given length and enclosing a given area with the largest possible volume in its solid of revolution about the axis Oo. The three properties are then $\int ds$, $\int y\,dx$ and $\int x^2\,dy$. The tables give the corresponding values of P as dq, dx and $x\,dx$, so the required curve will satisfy the equation

$$a\,dq + b\,dx + 2x\,dx = 0, \quad \text{or} \quad aq + bx + xx = c,$$

that is, substituting for q,

$$a\,dy + bx\,ds + xx\,ds = c\,ds.$$

For his next example, Euler asks us to find, from among the curves of a given length and enclosing a given area, the curve that will allow the quickest descent. As before, the constraints give dq and dx as values of P. The criterion of quickest descent is given by $\int \frac{ds}{\sqrt{x}}$, which gives $P = d \cdot \frac{q}{\sqrt{x}}$. The calculations to find the equation of the curve proceed as follows:

$$a\,dq + dx + b\,d \cdot \frac{q}{\sqrt{x}} = 0,$$

$$aq + x + \frac{bq}{\sqrt{x}} = c,$$

so, substituting for q,

$$a\,dy + x\,ds + \frac{b\,dy}{\sqrt{x}} = c\,ds.$$

As usual, Euler regards a differential equation as a solution to the problem.

We return to questions of fluid flow for the next example. We are to fix the area and the volume of the solid of revolution, then find the curve that gives the minimum resistance to motion in a fluid in the direction of the axis of rotation. As before, the first two conditions give for P the values dx and $x\,dx$. The fluid resistance is given by

$$\int \frac{x\,dx^3}{ds^2}$$

so the third value of P is

$$d \cdot \frac{x\,dy}{ds^4}.$$

This leads, after the usual computations, to

$$ax + bxx + \frac{x\,dx^3\,dy}{ds^4} = c.$$

In this last equation, Euler notes that if $c = 0$ and if x is sufficiently small, then the term bxx may be neglected. This allows him to put the equation into the form

$$x\,ds^4 = a\,dx^3\,dy.$$

Without warning, Euler has also replaced a with its negative reciprocal. This equation gives algebraic curves of order 4:

$$y^4 - 2by^3 + 2x^2y^2 - 18bx^2y + x^4 + 27b^2x^2 = 0$$

and

$$y^4 + 6by^3 + 2x^2y^2 + 12b^2y^2 - 10bx^2y + 8b^3y - b^2x^2 + x^4 = 0.$$

These are the curves that produce the solids of minimum resistance that Euler seeks.

This is Euler's last example in this paper. He concludes with some remarks about how some of the entries are related to other entries. For example, if a form contains $\int s\,dx$, so that Case VIII would apply, it can be transformed using integration by parts to contain $sx - \int x\,ds$ so that Case III might apply instead.

This is a long and difficult paper. This paper is Euler's first effort in organizing his results on the subject into a cohesive discipline and contains a wealth of examples. Four years later, Euler explains some of these same points in a more unified and better-organized presentation in E-56. The real comprehensive presentation will have to wait until 1744 and the publication of E-65, his important book on Calculus of Variations, *Methodus inveniendi lineas curves maximi minimive proprietate gaudentes sive solutio problematis isoperimetrici latissimo sensu accepti*, "Method of finding curved lines enjoying properties of maximum or minimum, or solution of isoperimetric problems in the widest accepted sense."

Interlude: 1733

World events

On February 12, 1733, James Oglethorpe founded a British colony at Savannah, Georgia.

Then on Friday the 13th of July, a fleet of 22 Spanish treasure ships left Havana, Cuba. The next day, all but one of them was sunk off the Florida Keys in a hurricane.

French mathematician, physicist and political scientist Jean-Charles de Borda (1733–1799) was born on May 4 of this year. He devised the voting scheme known as the "Borda count."

In Euler's life

According to Clifford Truesdell [T], in 1733 Euler wrote "when Professor Daniel Bernoulli, too, went back to his native land, I was given the professorship of Higher Mathematics, and soon thereafter the directing senate ordered me to take over the Department of Geography, on which occasion my salary was increased to 1200 rubles." His skills at salary negotiation certainly had improved.

Euler's other work

All five of Euler's papers in 1733 were mathematical.

Euler's mathematics

Euler wrote five mathematical papers in 1733. The first, E-11, was Constructio aequationum quarundam differentialium, quae indeterminatarum separationem non admittunt, "Construction of differential equations that do not admit separation of variables." It is a shorter version of E-28, Specimen de constructione aequationum differentialium sine indeterminatarum separatione, "Example of the solution of a differential equation without separation of variables." E-29, De solutione problematum Diophanteorum per numeros integros, "On the solution of problems of Diophantine about integer numbers" is about the Pell equation. He also wrote E-30, an obscure paper on the theory of equations titled De formus radicum aequationum cuiusque ordinis conjectatio, "Inferences on the forms of roots of equations and their orders." His fifth paper of the year was E-31, Constructio aequationis differentialis $ax^n\,dx = dy + y^2\,dx$, "The solution of the differential equation $ax^n\,dx = dy + y^2\,dx$."

References

[T] Truesdell, Clifford, *An Idiot's Fugitive Essays on Science: Methods, Criticism, Training, Circumstances*, Springer, New York, 1984.

11

E-11: Constructio aequationum quarundam differentialium quae indeterminatarum separationem non admittunt*

Construction of differential equations which do not admit separation of variables

Because this article is a little different from the others, we begin with a bit of background about some of the leading scientific journals of the 1730s.

This paper appeared in the *Nova acta eruditorum* in 1733. In part because the journal came out monthly, and in part because it was run in a more businesslike manner than the *Commentarii*, the *Nova acta eruditorum* had a much shorter backlog than the *Commentarii*, which was supposed to be published every year or two. In fact, because of political turmoil in Russia and chronically tight budgets, the *Commentarii* and its successor series had long publication delays. In the 1730s, it typically took five years for a paper to see print in the *Commentarii*. Hence the present article, E-11, published in 1733, was probably written in 1732 or 1733, and the previous article, E-10, published in 1732, had been written by 1728. In fact, this article, E-11, was probably written shortly after E-28 and E-31, because it discusses topics that are presented in more detail there. This happens because Eneström assigned the E-numbers in the order in which manuscripts were published and that only roughly corresponds to the order in which they were actually written.

The *Commentarii* and the *Nova acta eruditorum* were different kinds of journals. Each volume of the *Commentarii, academiae scientiarum imperialis petropolitanae*, was divided into four "Classes"—Mathematica, Physica, Historica and Observationibus Astronomicae et Physicae. Mathematica included mathematics, physics and astronomy. Physica was mostly medicine and biology, with some chemistry. History always consisted of three or four essays on ancient history and archaeology by T. S. Bayer. The section on observations gave weather reports for St. Petersburg, as well as detailed measurements of eclipses and comets. Papers in the *Commentarii* were full-length scientific articles by members of the Petersburg Academy. A volume of the *Commentarii* contained the written results of one or two years of research at the Academy, which were carefully edited and published all at once, usually three to six years later. After the political turmoil in Russia in 1741, the publication delays grew to eleven years. When things settled down again, the Academy ended the *Commentarii* series and began anew with the *Novi Commentarii*. Several years later and under similar circumstances, the *Novi Commentarii* were succeeded by the *Acta*, then the *Nova Acta* of the Petersburg Academy (not to be confused with the *Nova acta eruditorum*). When the official language of the Academy switched from Latin to French, the *Nova Acta* were succeeded by the *Mémoires*, and so forth. This continued until the Russian Revolution of 1914, when the eighth series of journals ended. The name of the Academy changed, and its official language became Russian.

Meanwhile, the *Nova acta eruditorum* was the successor to the *Acta eruditorum* which had been founded and edited by Leibniz, among others, in 1682. Both series came out monthly and covered

**Nova acta eruditorum* 1733, 369–373; *Opera Omnia* I.22, 15–18

a great variety of scholarly topics in the sciences and humanities. Issues were timely enough that there could be lively debates, often conducted in the form of book reviews or "Extracts of letters" from one participant to another. Newton and Leibniz conducted part of their priority dispute for the discovery of calculus in this venue.

In the present article, Euler seems to be using the *Nova acta eruditorum* to make a kind of research announcement, probably to protect his priority on the results in E-28 and E-31 that would not actually be published for another five years. It is hard to tell whether this article had much impact. The results, as presented here, are too sketchy for anybody else to make sense of them, and Euler's correspondence only mentions this article twice, once in a letter to D. Bernoulli in 1734 and again in a letter to the Italian mathematician and astronomer G. Poleni in 1735.

In the 1730s, the word "Constructio" in the title of this article had two close mathematical meanings when it was applied to equations. First, it meant the process of finding a formula or differential equation that describes a physical or geometrical situation. Second, it meant the process of "constructing" or finding a solution to a differential equation. In this article, Euler uses the word in both senses. Also, "geometry" had a much broader meaning than it now has, encompassing all of what we now call "mathematics," while "mathematics" then included what we now might call the "exact sciences," including mathematics, physics, astronomy and optics. The people we would now call mathematicians were then usually called "Geometers," always capitalized. The word "mathematician" still held its old connotation of astrology and occult arts, so Geometers avoided using it.

The paper itself begins with a general description of constructions and differential equations. Euler tells us "Constructions which make use of geometry in determining whatever quantities are of two kinds: Those which refer to Geometric constructions, either plane or solid or linear; and those which refer to the quadrature or rectification of curves."

The classification of geometric constructions here is an ancient one. It is the same as Descartes describes in Book II of his *Geometria*. (Descartes wrote it in French under the title *La Géométrie*, but at the time most scholars read it in Latin, then the universal language of science, so here we use the Latin title, *Geometria*.) It would have been familiar to all of Euler's readers. Plane problems were those using just lines and circles in their solution. Solid problems additionally require conic sections and linear problems require more complex curves, like exponential curves or spirals. The word "linear" has quite a different meaning today.

Euler continues, "Geometry used to concern itself with algebraic equations, their roots and intersections. Now there are kinds of constructions, which can be called transcendental, which arise in solving differential equations." This seems to be Euler's first use of the word "transcendental," though Leibniz had used the word itself with almost exactly the same meaning about fifty years earlier.

Euler moves closer to the purpose of this paper and writes "In some equations, the variables cannot be separated. The first example of this was

$$dy + \frac{y^2\,dx}{x} = \frac{x\,dx}{x^2 - 1}.$$

If this equation could be solved, then the equation which gives the perimeter of an ellipse could be given as well." This is essentially a three sentence summary of Euler's paper E-28.

Euler goes on to tell us that similar methods could be used to construct the equation $ax^n\,dx = dy + y^2\,dx$, the equation he calls the Riccati equation. He gives only the very roughest idea how this might be done, telling us that it involves the differential

$$n(n+4)\,dz\,(1-z^2)^{\frac{-n-4}{2n+4}} + 2\,dz\,(1-z^2)^{\frac{-n-4}{2n+4}}\left(c^{\frac{2z\sqrt{f}}{n+2}} + c^{\frac{-2z\sqrt{f}}{n+2}}\right),$$

where f is a constant and c is the base for the natural logarithms, the number we now denote by e. Details of this have to wait until E-31 is published in 1738.

12

E-28: Specimen de constructione aequationum differentialium sine indeterminatarum separatione*

Example of the solution of a differential equation without separation of variables

In 1733, separation of variables was still the only well-known general method of solving differential equations. Euler cites a paper by Johann Bernoulli in *Comm. Acad. Sci. Imp. Petropol.* 1 that shows how some equations can be solved even if the variables do not separate. For some reason, Euler chooses not to cite his own paper, E-10, as another general means of solving differential equations. In that paper, Euler had introduced the integrating factors. Perhaps Euler did not yet recognize that his own method was quite general and would solve any first-order equation of the form we now call "linear."

This paper uses series to solve an inseparable equation involving the arc length of an ellipse. The Editors of the *Opera Omnia* might well have chosen to put this paper in the volume on Differential Equations, but they decided instead to put it as the first article in Series I volume 17 on Elliptic Integrals.

Euler notes the similarities between two equations,

$$dy + \frac{y^2\,dx}{x} = \frac{x\,dx}{x^2 - 1},$$

which is the equation he derives and studies in this paper, and $dy + y^2\,dx = x^2 dx$, the so-called Riccati equation. Though they appear to be similar, the Riccati equation can be solved using clever transformations and separation of variables, but his own equation cannot. Watson [W] gives a nice account of the early history of the Riccati equation.

Now, Euler begins his derivation. He is considering the arc length of the ellipse shown in Figure 1. He describes the figure as follows.

$AC = a$, the major axis of the ellipse

$BC = b$, the minor axis of the ellipse

AT is the tangent to the ellipse at A

CT cuts the ellipse at M

$AM = s$ is the length of the arc AM

$AT = t$

$CP = x$

Note that Euler uses the variable t twice here, once as a point and once as a length. This was a common practice in the eighteenth century and it often gets confusing.

**Comm. Acad. Sci. Imp. Petropol.* 6 (1732/3) 1738, 168–174; *Opera Omnia* I.20, 1–7

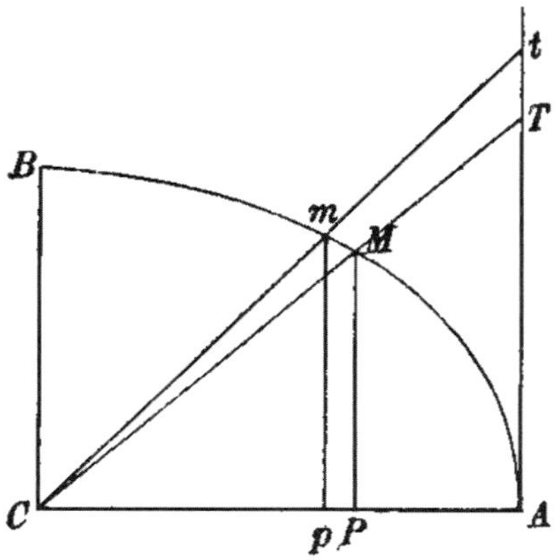

Fig. 1.

It goes without saying that tT is taken to be small, that Ct cuts the ellipse at m, that MP is perpendicular to CA at P, and that mp is perpendicular to CA at p. So, Euler is talking about the ellipse that we would describe with the formula

$$\frac{x^2}{a^2} + \frac{y^2}{b^2} = 1,$$

where he uses MP or PM where we would use y.

Euler tells us that "from the nature of an ellipse,

$$PM = \frac{b\sqrt{a^2 - x^2}}{a}\text{,''}$$

and "because of the analogy $CP : PM = CA : AT$, we have $tx = b\sqrt{a^2 - x^2}$ or

$$x = \frac{ab}{\sqrt{bb + tt}}.\text{''}$$

Adding to the arc AM the element Mm and drawing mp and Ct near to MP and CT respectively, gives

$$Mm = ds = \frac{-dx\sqrt{a^4 - (a^2 - b^2)x^2}}{a\sqrt{a^2 - x^2}}$$

and $Tt = dt$.

From the equation

$$x = \frac{ab}{\sqrt{b^2 + t^2}}$$

it follows that

$$dx = \frac{-abt\,dt}{(b^2 + t^2)^{3/2}}.$$

(Note the use of the fractional exponent in the denominator. Euler does this only rarely.) Further,

$$\sqrt{a^2 - x^2} = \frac{at}{\sqrt{b^2 + t^2}}$$

and

$$\sqrt{a^4 - (a^2 - b^2)x^2} = \frac{a\sqrt{b^4 + a^2tt}}{\sqrt{b^2 + t^2}}.$$

From this it follows that

$$ds = \frac{b\,dt\sqrt{b^4 + a^2tt}}{(bb + tt)^{3/2}}.$$

Now, Euler wants to integrate this last differential with t ranging from zero to infinity to get the arc length of the quarter ellipse. He doesn't mention that it would also be nice to be able to integrate it from zero to T to get the arc length AM. Both of these integrals are difficult, but Euler has a clever plan involving a tricky substitution. The trick seems to come from nowhere, but in the final section of this article, we get a clue as to why the substitution works. He writes, "To find this integral by means of series, I put $a^2 = (n + 1)b^2$, which produces

$$ds = \frac{b^2\,dt\sqrt{(b^2 + t^2) + nt^2}}{(b^2 + t^2)^{3/2}}$$

and the above irrational [that is, the square root in the numerator] is made a binomial whose one member is $b^2 + t^2$ and the other is the simple term nt^2. Now I resolve $\sqrt{(b^2 + t^2) + nt^2}$ by the usual way into the series

$$(b^2 + t^2)^{1/2} + \frac{Ant^2}{(b^2 + t^2)^{1/2}} + \frac{Bn^2t^4}{(b^2 + t^2)^{3/2}} + \frac{Cn^3t^6}{(b^2 + t^2)^{5/2}} + \text{etc.}$$

in which, for the sake of brevity, I write

$$A = \frac{1}{2} \qquad B = -\frac{1}{2} \cdot \frac{1}{4} \qquad C = \frac{1 \cdot 1 \cdot 3}{2 \cdot 4 \cdot 6} \qquad D = -\frac{1 \cdot 1 \cdot 3 \cdot 5}{2 \cdot 4 \cdot 6 \cdot 8}."$$

Here, Euler is using the binomial theorem for the exponent $\frac{1}{2}$, and the values A, B, C, D are the fractional binomial coefficients.

Multiplying the series by

$$\frac{b^2\,dt}{(b^2 + t^2)^{3/2}}$$

gives

$$ds = \frac{b^2\,dt}{b^2 + t^2} + \frac{Ab^2nt^2\,dt}{(b^2 + t^2)^2} + \frac{Bb^4n^2t^4\,dt}{(b^2 + t^2)^3} + \frac{Cb^6n^3t^6\,dt}{(b^2 + t^2)^4} + \text{etc.}$$

and the arc of the ellipse, s, will be the integral of this series.

Euler asks us to notice that "the integrals of all these terms can be reduced to the integral of the first term $\int \frac{bb\,dt}{bb+tt}$. In fact, $\int \frac{bb\,dt}{bb+tt}$ is the arc of the circle of radius b whose tangent is t. Because those circular arc lengths are known, I will integrate, and it follows . . . "

$$\int \frac{b^2t^2\,dt}{(b^2+t^2)^2} = \frac{1}{2}\int \frac{bb\,dt}{bb+tt} - \frac{1}{2}\frac{b^2t}{bb+tt}$$

[integrating by parts, with

$$dv = \frac{2t\,dt}{(b^2+t^2)^2} \quad \text{so that} \quad v = \frac{-1}{b^2+t^2},]$$

$$\int \frac{b^2t^4\,dt}{(b^2+t^2)^3} = \frac{1\cdot 3}{2\cdot 4}\int \frac{b^2\,dt}{bb+tt} - \frac{1\cdot 3}{2\cdot 4}\frac{b^2t}{bb+tt} - \frac{1}{4}\frac{b^2t^3}{(bb+tt)^2},$$

$$\int \frac{b^2t^6\,dt}{(b^2+t^2)^4} = \frac{1\cdot 3\cdot 5}{2\cdot 4\cdot 6}\int \frac{b^2\,dt}{bb+tt} - \frac{1\cdot 3\cdot 5}{2\cdot 4\cdot 6}\frac{b^2t}{bb+tt} - \frac{1\cdot 5}{4\cdot 6}\frac{b^2t^3}{(bb+tt)^2} - \frac{1}{6}\frac{b^2t^5}{(bb+tt)^3}$$

from which the law for the integrals of the remaining terms is apparent enough.

Euler can't get the length of an arbitrary arc of an ellipse, but he is willing to settle for a quarter of an ellipse for now. He tells us:

> If the quarter part of the ellipse *AMB* is required, then one makes t become infinite, and then all the algebraic terms in the integrals above will vanish. Then the circular arc $\int \frac{bb\,dt}{bb+tt}$, putting $t=\infty$, gives a quarter part of the periphery of a circle whose radius is b, or *BC*, the arc length we will denote with the letter e.

Here, it works out that $e = \frac{1}{4}2\pi b$.

Thus, Euler's integrations by parts do not simplify the integrals for the case of general arc length, but they do result in a considerable simplification in the case of the quarter ellipse. He continues as follows:

$$\int \frac{b^2\,dt}{bb+tt} = e$$

$$\int \frac{b^2t^2\,dt}{(bb+tt)^2} = \frac{1\cdot e}{2}$$

$$\int \frac{b^2t^4\,dt}{(bb+tt)^3} = \frac{1\cdot 3\cdot e}{2\cdot 4}$$

$$\int \frac{b^2t^6\,dt}{(bb+tt)^4} = \frac{1\cdot 3\cdot 5\cdot e}{2\cdot 4\cdot 6}$$

etc. From this, then, it follows that the quarter part of the ellipse is given by

$$AMB = e\left(1 + \frac{1}{2}An + \frac{1\cdot 3}{2\cdot 4}Bn^2 + \frac{1\cdot 3\cdot 5}{2\cdot 4\cdot 6}Cn^3 + \frac{1\cdot 3\cdot 5\cdot 7}{2\cdot 4\cdot 6\cdot 8}Dn^4 + \text{etc.}\right)$$

and substituting in the places of A, B, C, D, their actual values gives

$$AMB = e\left(1 + \frac{1\cdot n}{2\cdot 2} - \frac{1\cdot 3\cdot n^2}{2\cdot 2\cdot 4\cdot 4} + \frac{1\cdot 1\cdot 3\cdot 3\cdot 5\cdot n^3}{2\cdot 2\cdot 4\cdot 4\cdot 6\cdot 6} - \frac{1\cdot 1\cdot 3\cdot 3\cdot 5\cdot 5\cdot 7\cdot n^4}{2\cdot 2\cdot 4\cdot 4\cdot 6\cdot 6\cdot 8\cdot 8} + \text{etc.}\right).$$

If n is small enough so that $\frac{a^2-b^2}{b^2}$ (that is, n) vanishes and the ellipse is almost a circle, then this series converges rapidly; and in this case, therefore, the perimeter of the ellipse is found easily. When n is a very small quantity, that is $a = b + \omega$, and ω is infinitely small, then $n = \frac{2\omega}{b}$ (ignoring higher powers of ω) and $AMB = e\left(1 + \frac{\omega}{2b}\right)$ approximately.

When $a = 0$ then A and C are the same point, and that makes $AMB = BC = b$. This is the case when $n = -1$, and then

$$\frac{b}{e} = 1 - \frac{1}{2 \cdot 2} - \frac{1 \cdot 3}{2 \cdot 2 \cdot 4 \cdot 4} - \frac{1 \cdot 1 \cdot 3 \cdot 3 \cdot 5}{2 \cdot 2 \cdot 4 \cdot 4 \cdot 6 \cdot 6} - \text{etc.}$$

The sum of this series therefore expresses the ratio of the radius to the quarter part of the periphery of the circle. Euler uses this wording rather than the modern notation $\frac{b}{e} = \frac{2}{\pi}$ because the symbol π, as noted previously, was not yet standard notation. It is worth noting that, though this series does converge as advertised, the convergence is slow.

Euler is now going to demonstrate some of the brilliance of his n substitution earlier in the analysis. He writes:

> Now we look at that value n which appears in the series found in §5 [about a page earlier], used in the rectification of an ellipse for which the ratio of its axes is $\sqrt{(n+1)} : 1$. Now we use some of the methods I have used elsewhere [Euler means E-25] to help resolve the series. The following easy method will work. I put $n = -x^2$ and get this series:
>
> $$1 - \frac{1 \cdot x^2}{2 \cdot 2} - \frac{1 \cdot 1 \cdot 3 \cdot x^4}{2 \cdot 2 \cdot 4 \cdot 4} - \frac{1 \cdot 1 \cdot 3 \cdot 3 \cdot 5 \cdot x^6}{2 \cdot 2 \cdot 4 \cdot 4 \cdot 6 \cdot 6} - \text{etc.}$$
>
> Set this series equal to s, and then
>
> $$\frac{ds}{dx} = -\frac{1 \cdot x}{2} - \frac{1 \cdot 1 \cdot 3 \cdot x^3}{2 \cdot 2 \cdot 4} - \frac{1 \cdot 1 \cdot 3 \cdot 3 \cdot 5 \cdot x^5}{2 \cdot 2 \cdot 4 \cdot 4 \cdot 6} - \text{etc.''}$$

Note that if n is positive, there is a problem with making the substitution $n = -x^2$. We can solve the problem and make n negative by reversing the axes, because that interchanges a and b.

The calculations are very much like those in E-25, for Euler tells us to "multiply by x and differentiate again" so

$$\frac{d.x\,ds}{x\,dx} = -x - \frac{1 \cdot 1 \cdot x^3}{2 \cdot 2} - \frac{1 \cdot 1 \cdot 3 \cdot 3 \cdot x^5}{2 \cdot 2 \cdot 4 \cdot 4} - \text{etc.}$$

"Now I multiply again by dx and divide by x^3 and get." (including one more term than before)

$$\int \frac{1}{x^3} \int \frac{d.x\,ds}{x} = \frac{1}{x} - \frac{1 \cdot x}{2 \cdot 2} - \frac{1 \cdot 1 \cdot 3 \cdot x^3}{2 \cdot 2 \cdot 4 \cdot 4} - \frac{1 \cdot 1 \cdot 3 \cdot 3 \cdot 5 \cdot x^5}{2 \cdot 2 \cdot 4 \cdot 4 \cdot 6 \cdot 6} - \text{etc.}$$

The notation here is a little unfamiliar, as perhaps it was in Euler's time, but Euler's commentary makes sure that we interpret the integrals and differentials correctly. Note that the dx that was in the denominator in the previous equation disappeared on the left-hand side when he multiplied by dx. Note also that Euler is not really ignoring the constant of integration. He already knows that he wants it to be zero.

This last series is just the first one (in this section) divided by x, so it is $\frac{s}{x}$. From this, we get the equation

$$\int \frac{1}{x^3} \int \frac{d.x\,ds}{x} = \frac{s}{x}.$$

Differentiating this leads to (in a couple of steps which Euler omits)

$$x^2\,ds - sx\,dx = \int \frac{d.x\,ds}{x}.$$

Differentiating again will produce

$$x^2\,dds + x\,dx\,ds - s\,dx^2 = \frac{d.x\,ds}{x} = dds + \frac{dx\,ds}{x}.$$

Solving this differential equation would lead to the summation of the given series, which, in turn, would give the rectification of the ellipse.

Euler wrote, "Since in this equation s has dimension 1, it can be reduced by my method from Volume III of the *Commentarii* to a simpler differential equation by making the substitution $s = c^{\int p\,dx}$ where c denotes the number with log. [sic][1] equal to 1. This done, it makes $ds = c^{\int p\,dx}p\,dx$ and $dds = c^{\int p\,dx}(dp\,dx + pp\,dx^2)$ and the equation is transformed into this:

$$x^2\,dp + x^2p^2\,dx + px\,dx - dx = dp + pp\,dx + \frac{p\,dx}{x}$$

which, divided by $xx - 1$ is changed into this:

$$dp + pp\,dx + \frac{p\,dx}{x} = \frac{dx}{x-1}.$$

To simplify this, I will put $p = \frac{y}{x}$ and it becomes

$$dy + \frac{yy\,dx}{x} = \frac{x\,dx}{x^2-1}.$$

In this equation, the variables cannot be separated, nor does any other construction seem to solve it."

The article Euler mentions "from Volume III of the *Commentarii*" is E-10 and is partially translated in Smith, *A Source Book In Mathematics* [E].

Euler devotes the last section of this paper to deducing the same differential equation geometrically. Differential equations that arise from geometric constructions still command a particular respect that equations that arise analytically or algebraically, or even practically, do not yet command. He doesn't quite do this. First, he transforms this last differential equation with some substitutions. He begins:

> Now, this last equation will be deduced by a geometric construction. Denote the semiaxis AC by r, which we called a before, and r is to be considered a variable. Let q denote the quarter perimeter of the ellipse. Then
>
> $$-xx = n = \frac{r^2-b^2}{b^2} \quad \text{and} \quad x = \frac{\sqrt{b^2-r^2}}{b}.$$
>
> Let $q = es$, where $s = ec^{\int y\,dx/x}$ and $lq - le = \int \frac{y\,dx}{x}$. (Here, as usual, the symbol l denotes the natural logarithm). From that
>
> $$y = \frac{x\,dq}{q\,dx} = \frac{(r^2-b^2)\,dq}{qr\,dr}.$$

[1]Here, as in the case with π, the symbol e as the basis for the natural logarithms was not yet standard. Note also the period in "log." Euler still thinks of this as an abbreviation for "logarithm" and not as a function notation.

Here, when r is greater than b, I put in place of xx the value $-n$: Then

$$\frac{dx}{x} = \frac{dn}{2n} \quad \text{and} \quad \frac{x\,dx}{xx-1} = \frac{dn}{2(n+1)},$$

and substituting this into the other equation gives

$$2\,dy + \frac{y^2\,dn}{n} = \frac{dn}{n+1}.$$

This is the form of the equation that Euler intends to derive by constructive geometry. He is now ready to begin his construction.

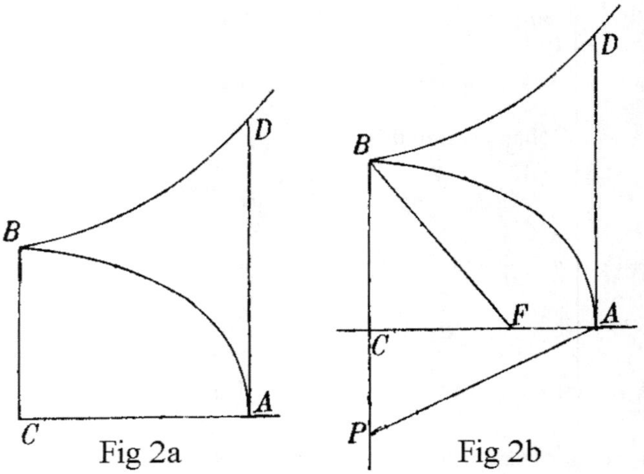

Fig 2a Fig 2b

Let BCA be a quarter of an ellipse with center C and semi-axis $BC = 1$ (Figure 2a); I put 1 in place of b, which can serve well, by homogeneity. Let the other semi-axis $AC = r$. From A is erected a normal AD of length equal to the arc length of the ellipse AB.

That point D will be on a curve BD, which is constructed in this same manner. Thus $AD = q$.

This means that the curve BD is the graph of some otherwise unknown function $f(r)$, where r the semi-axis of the ellipse, and $f(r)$ the arc length of the corresponding quarter ellipse. Clearly, $f(0) = 1$ and $f(1) = \frac{\pi}{2}$. Euler, of course, could not have used such notation because he hadn't invented it yet. He continues:

"If F is the focus of the ellipse, then $CF = \sqrt{r^2 - 1}$ (Figure 2b). To BF, draw the normal FT, with P on the line BC. Then $CP = r^2 - 1 = n$." At last we have a geometric interpretation for Euler's ingenious substitution n.

Euler notes that when $AC < BC$, the focus lies on BC and the value of n will be negative. When this happens, the points B and C ought to be exchanged.

Next is drawn the tangent DT to the curve BD at D (Figure 2c). Then $AT = \frac{q\,dr}{dq}$. Intersecting AP from T is drawn the normal straight line TG, cutting AP at the point O and cutting DA, extended, at G.

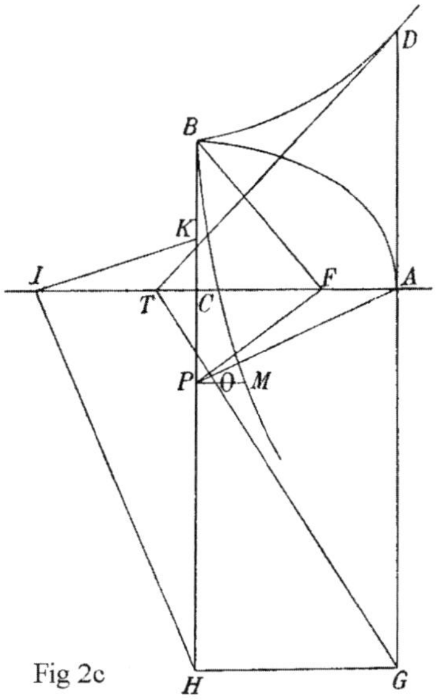

Fig 2c

Because of similar triangles PCA and TAG,

$$AG = \frac{rq\,dr}{(r^2-1)\,dq}.$$

Making $CH = AG$ and making $CI = CB = 1$, draw HI and erect its perpendicular IK. Then

$$CK = \frac{(r^2-1)\,dq}{rq\,dr} = y.$$

Now let CK be equal to PM [thus describing PM as a function of the position of P, that is as a function of $n = CP$, and indifferent to the fact that the axis for the independent variable is vertical.] Then the curve that is sought is BM. That curve has the property, calling $CP = n$ and $PM = y$, that

$$2y + \frac{y^2\,dn}{n} = \frac{dn}{n+1}.$$

Thus Euler has given a geometric derivation of the differential equation he had found earlier by using series.

References

[E] Euler, Leonhard, "On Differential Equations of the Second Order", a translation of E-10 by Florian Cajori, in *A Source Book in Mathematics*, David Eugene Smith, ed., pp. 638–643, Dover, New York, 1929.

[W] Watson, G. N., *A treatise on the theory of Bessel functions*, Cambridge University Press, London, 1922.

13

E-29: De solutione problematum Diophanteorum per numeros integros*

On the solution of problems of Diophantus about integer numbers

Euler's letters do not show when he first became interested in Diophantine equations. Probably he learned of them in the works of Fermat and of Gregory of St. Vincent. His letters show that he had read both of their works carefully.

This paper, E-29, is Euler's second on number theory, appearing in Volume 6 of the *Commentarii*, the same volume in which he factored the fifth Fermat number. This is the article in which Euler first misattributes the Pell equation to Pell, and the mistake still persists. The main result of this paper is to show how quadratic Diophantine equations can be reduced to the Pell equation. The paper is interesting as a seed from which much of Euler's number theory will evolve.

Euler tells us that solving a Diophantine problem means finding whole numbers that satisfy a formula in two or more variables. As an example, he tells us that the formula $3x^2 + 2$ is never a perfect square. That is the whole example. He does not put it into the form $3x^2 + 2 = y^2$, so that it has two or more variables, nor does he tell us at this time *why* it has no solutions. Much later in his 1770 book *Algebra*, Euler gives what we would now describe as a (mod 3) proof.

Some formulas though will have infinitely many solutions. Euler is particularly interested in the formula $ax^2 + bx + c$ and when it is a perfect square. The main point of this paper will be to show how to find more solutions once a first one is known.

Euler will return to this very topic in 1762 (E-279), De resolutione formularum quadraticarum indeterminatarum per numeros integros, "On the solution of unknowns in quadratic formulas by integer numbers." There he gives sharper and more mature proofs and derivations of these same results, though the methods are essentially the same.

Suppose that n is a solution, or as Euler writes it, that $an^2 + bn + c$ is a square number whose root is m, and suppose that another solution exists. Euler guesses that it will have the form

$$\alpha n + \beta + \gamma\sqrt{an^2 + bn + c},$$

where α, β and γ are whole numbers. He further supposes that, when this is substituted into the formula $ax^2 + bx + c$, it will give the square of some value of the form

$$\delta n + \varepsilon + \zeta\sqrt{an^2 + bn + c},$$

again with the new Greek symbols representing integers. This is an insightful guess on Euler's part.

Euler has no fear of calculations, so he makes the substitutions and matches terms. This enables him to eliminate some variables, and the whole thing reduces to $\alpha = \sqrt{a\gamma^2 + 1}$. To solve this, Euler seeks integer values of γ that will make $a\gamma^2 + 1$ a perfect square.

Comm. Acad. Sci. Petropol. 6 (1732/3) 1738, 168–174; *Opera Omnia* I.2, 6–17

Suppose now that p is such a number, and that $q = \sqrt{ap^2+1}$. Then, matching terms with the forms above, we get

$$\alpha = q, \quad \gamma = p, \quad \beta = \frac{b(q-1)}{2a}, \quad \delta = ap, \quad \varepsilon = \frac{bp}{2} \quad \text{and} \quad \zeta = q.$$

This gives the proof for the following theorem:

Theorem. *If $ax^2 + bx + c$ is a square in the case $x = n$, then it will also be a square when*

$$x = qn + \frac{bq-b}{2a} + p\sqrt{an^2+bn+c}$$

and the root of that square will be

$$apn + \frac{bp}{2} + q\sqrt{an^2+bn+c}.$$

If bp can be divided by 2, then the root of the square will be a whole number. Also, the value substituted for x will be a whole number if $bq - b$ can be divided by $2a$.

The theorem had been proved by the calculations that led to its statement.

Of course, even if the divisibility criteria are not satisfied, the new solutions generated in this way will at least be rational numbers.

Still more solutions can be generated by repeating the substitutions, as Euler demonstrates. He shows that if $x = A$, $y = E$ is one solution generated in the manner of the theorem, and if $x = B$, $y = F$ is the next solution, then another solution can be found by taking

$$x = 2qB - A + \frac{b(q-1)}{a} \quad \text{and} \quad y = 2qF - E$$

respectively. Readers who have worked with continued fractions will have a sense of déjà vu. This should not be surprising, because many people first see continued fractions when studying the formula $ap^2 + 1$.

Note that now Euler can generate a sequence of solutions for which $ax^2 + bx + c$ starting either with two consecutive solutions or starting with one solution and a solution to the corresponding Pell formula $ap^2 + 1$. He asserts that $ap^2 + 1$ can always be made a square, so the only problem might be whether or not $ax^2 + bx + c$ can be a perfect square. Euler turns to investigating this.

The simplest case is when c itself is a perfect square, say $c = dd$. Then $n = 0$, $m = d$ provides an initial solution. If p, q is a solution to the corresponding Pell equation, ap^2+1, then the sequence of values of x is given by

$$0,\ dp + \frac{b(q-1)}{2a},\ 2dpq + \frac{b(q^2-1)}{a}, \ldots,\ A,\ B,\ 2qB - A + \frac{b(q-1)}{A}$$

and the corresponding values of the roots are

$$d,\ dq + \frac{bp}{2},\ d(2q^2-1) + bpq, \ldots,\ E,\ F,\ 2qF - E.$$

As a second example, Euler considers $b = 0$, $d = 1$, a special case of his first example, because 1 is a perfect square. This gives the familiar form $ax^2 + 1$. If one non-trivial solution p, q can be found, then it can be combined with the trivial solution 0, 1, to get a sequence of solutions with the values of x being

$$0,\ p,\ 2pq,\ 4pq^2 - p, \ldots,\ A,\ B,\ 2qB - A$$

and the values of y being

$$1,\ q,\ 2q^2 - 1,\ 4q^3 - 3q, \ldots,\ E,\ F,\ 2qF - E.$$

There is still the problem of finding that first pair of values, p and q, that satisfy $ax^2 + 1 = y^2$. In typical Euler fashion, he provides a few special and easy examples before he reveals a general method.

If, for example, a has the form $a = e^2 - 1$, then there is the easy solution $p = 1, q = e$. A slightly more complicated example arises if $a = e^2 + 1$. Then we may take $p = 2e$ and $q = 2e^2 + 1$. Examples get progressively more complicated until finally we consider $a = \frac{1}{4}\alpha^2 k^2 e^{2h} \pm 2\beta e^{\mu - 1}$, in which case we may take $p = ke$ and $q = \frac{1}{2}\alpha k^2 e^{k+1} \pm 1$.

"But if a is a kind of number that cannot be reduced in any way to one of these formulas," then there is a method, according to Euler, found by Pell and Fermat that always works. Euler gives us the method, attributing it to Pell, but citing the description by Wallis in his *Algebra*.

This is the origin of one of the great misnamings in the history of mathematics. Though Wallis does mention Pell a number of times, he does *not* cite him in his discussions of the equation $ax^2 + 1 = y^2$. Euler apparently remembered this incorrectly and attributed the equation to Pell, and the name stuck. It would be more accurate historically to call the equation "Fermat's equation," but there are many other things named after Fermat, and to try to correct the mistake at this late date would only cause confusion and deprive us of a good story.

Euler describes the method with an example. Suppose $a = 31$. Then it is required to find values of p and q that make $31p^2 + 1 = q^2$, or, equivalently, $\sqrt{31p^2 + 1} = q$. Since $6 > \sqrt{31} > 5$, this makes $q > 5p$. Introduce a new variable, a so that $q = 5p + a$. This is confusing, since another a already equals 31, but this a is the first term of what will be a sequence of remainders. We know that $0 < a < p$. Substituting that into $31p^2 + 1 = q^2$ gives $6p^2 + 1 = 10ap + a^2$. Solving this for p gives

$$p = \frac{5a + \sqrt{31a^2 - 6}}{6},$$

and that allows the estimate that $p > a$, though it takes a bit of thinking to figure out exactly why. Remember that that $0 < a < p$. Now, introduce a new variable b such that $p = a + b$, and substitute again.

Continuing this process, and introducing new variables c to g, produces the following calculations.

$$5a^2 = 2ab + 6b^2 + 1 \qquad a = \frac{b + \sqrt{31b^2 + 5}}{5} \qquad a = b + c$$

$$3b^2 = 8bc + 5c^2 - 1 \qquad b = \frac{4c + \sqrt{31c^2 - 3}}{3} \qquad b = 3c + d$$

$$2c^2 = 10cd + 3d^2 + 1 \qquad c = \frac{5d + \sqrt{31d^2 + 2}}{2} \qquad c = 5d + e$$

$$\begin{array}{rclrclrcl} 3d^2 & = & 10de + 2e^2 - 1 & d & = & \dfrac{5e + \sqrt{31e^2 - 3}}{3} & d & = & 3e + f \\ 5e^2 & = & 8ef + 3f^2 + 1 & e & = & \dfrac{4f + \sqrt{31f^2 + 5}}{5} & e & = & 2f + g \\ f^2 & = & 12fg - 5g^2 + 1 & f & = & 6g + \sqrt{31g^2 + 1} & & & \end{array}$$

This last equation, $f = 6g + \sqrt{31g^2 + 1}$, is satisfied if $g = 0$ and $f = 1$. Now, back-substitution gives $e = 2$, $d = 7$, $c = 37$, $b = 118$, $a = 155$, $p = 273$, and finally $q = 1520$. And, indeed, $31 \cdot 273^2 + 1 = 2310400 = 1520^2$, as required.

Some readers will note the similarities between this algorithm and the Euclidean algorithm, and the back-substitution resembles the way the Euclidean algorithm leads to solving Diophantine equations of the form $ax + by = 1$. Euler will improve upon this five years later in E-71, "De fractionibus continuis dissertatio."

Euler admits that this is a bit of work, so he provides a table of the smallest values of p and q for each value of a from 2 to 68 that is not a perfect square.

Is this good for anything? Euler offers the fact that $\frac{p}{q}$ is a rational approximation of $\sqrt{a}$, and that further iterations give successively better approximations. He invents a peculiar notation to give us a sequence of rational approximations for $\sqrt{6}$, starting with $p = 5, q = 2$:

$$\frac{1,\ 5,\ 49,\ 485,\ 4801,\ 47525,\ 470449,\ 4656965\,\text{etc.}}{0,\ 2,\ 20,\ 198,\ 1960,\ 19402,\ 192060,\ 1901198\,\text{etc.}}.$$

The last ratio, $\frac{4656965}{1901198}$ differs from $\sqrt{6}$ by only

$$\frac{1}{2(1931198)^2\sqrt{6}},$$

that is accurate to more than 12 decimal places.

Next, he asks if any triangular numbers, that is, numbers of the form

$$\frac{x^2 + x}{2}$$

are square numbers. For that to happen, $2x^2 + 2x$ must also be a square number, and this is of the form $ax^2 + bx + d^2$, considered earlier. Using the theory he developed, he finds that the values of x that give square numbers are 0, 1, 8, 49, 288, 1681, 9800, etc., and that they give the squares of 0, 1, 6, 35, 204, 1189, 6930, etc., respectively.

Euler concludes the article with an analysis of what other polygonal numbers can be perfect squares.

14

E-30: De formis radicum aequationum cuiusque ordinis conjectatio*

Inferences on the forms of roots of equations and of their orders

Euler must have been delighted with the method for reducing the degree of differential equations that we saw in E-10. It was natural for him to try to do the same thing for polynomial equations. In this paper, he builds on the 16th century work of Cardano, Del Ferro and Bombelli, though he does not cite them, and shows how to reduce some polynomial equations to equations of lower degree. Of course, his methods cannot work all of the time, but Euler couldn't know that. Abel would not prove the "unsolvability of the quintic" for almost another hundred years.

In this paper, Euler gives some useful results and provides some hope for the doomed effort to find the roots of any polynomial. Euler's main tool will be a kind of auxiliary equation that we now call a *resolvent*. Some sixty years later Lagrange will build an extensive theory of equations based on these same ideas. It is hard to know whether Lagrange was influenced by this paper; it is not one of Euler's better known results. Whether he knew this paper or not, Lagrange carried these ideas much farther than Euler.

Euler begins with an explanation of how to solve cubic equations. For most of his readers, this was probably a review, motivating the techniques that were to come. Today, the topic is seldom taught, so modern readers may well not have seen the derivation before. We will show how Euler did it.

We want to solve a cubic equation $x^3 = ax + b$. Euler expects all of his readers to know that any cubic equation $y^3 + ay^2 + by + c = 0$ can be reduced to this form by the substitution $y = x - \frac{a}{3}$. Now, suppose that one of the roots x of the equation has the form

$$x = \sqrt[3]{A} + \sqrt[3]{B}.$$

This might seem like another of Euler's brilliant insights that a solution has a particular form. We have seen such leaps of imagination before, for example in E-10, E-19 and E-27, and we will see more of them in the future. This time, though, Euler had seen it before, perhaps in the work of Bombelli in the 1500s.

Then A and B are the roots of some quadratic equation $z^2 = \alpha z - \beta$, where $A + B = \alpha$ and $AB = \beta$. Now, knowing values for a and b, we seek values for α and β, from which we can use the quadratic formula to find A and B. Substituting $x = \sqrt[3]{A} + \sqrt[3]{B}$ into $x^3 = ax + b$ gives

$$x^3 = A + B + 3\sqrt[3]{AB}(\sqrt[3]{A} + \sqrt[3]{B}) = 3x\sqrt[3]{AB} + A + B.$$

Matching terms with $x^3 = ax + b$ gives $a = 3\sqrt[3]{AB} = 3\sqrt[3]{\beta}$ and $b = A + B = \alpha$. That makes

$$\alpha = b \quad \text{and} \quad \beta = \frac{a^3}{27}.$$

**Comm. Acad. Sci. Imp. Petropol.* 6 (1732/3) 1738, 216–231; *Opera Omnia* I.6, 1–19

So, if we use the quadratic formula to find the roots of the equation

$$z^2 = bz - \frac{a^3}{27},$$

and call those roots A and B, then one of the roots of the original cubic $x^3 = ax + b$ will be

$$x = \sqrt[3]{A} + \sqrt[3]{B}.$$

Euler knows about roots of unity. If μ and ν are cube roots of 1, equal to

$$\frac{-1+\sqrt{-3}}{2} \quad \text{and} \quad \frac{-1-\sqrt{-3}}{2}$$

respectively, then $\mu\nu = 1$ and the other roots are

$$\mu A + \nu B \quad \text{and} \quad \nu A + \mu B.$$

This completes Euler's analysis of cubic equations. He does not provide an example the way he usually does. Instead, he moves on to "biquadratics," or fourth degree equations of the form

$$x^4 = ax^2 + bx + c.$$

Again, any biquadratic can be reduced to this form by a linear substitution. Euler supposes that one of the roots has the form

$$x = \sqrt{A} + \sqrt{B} + \sqrt{C}$$

and again, he looks at the cubic equation that has as its roots A, B and C. If that equation is

$$z^3 = \alpha z^2 - \beta z + \gamma,$$

then

$$\alpha = A + B + C, \quad \beta = AB + AC + BC \quad \text{and} \quad \gamma = ABC.$$

Squaring x gives

$$x^2 = A + B + C + 2\sqrt{AB} + 2\sqrt{AC} + 2\sqrt{BC}$$

and substituting α gives

$$x^2 - \alpha = 2\sqrt{AB} + 2\sqrt{AC} + 2\sqrt{BC}.$$

Square that and substitute further to get

$$x^4 - 2\alpha x^2 + \alpha^2 = 4AB + 4AC + 4BC + 8\sqrt{ABC}(\sqrt{A} + \sqrt{B} + \sqrt{C}) = 4\beta + 8x\sqrt{\gamma}$$

or

$$x^4 = 2\alpha x^2 + 8x\sqrt{\gamma} + 4\beta - \alpha^2.$$

Now we can match coefficients with the original equation and find

$$\alpha = \frac{a}{2}, \quad \gamma = \frac{b^2}{64} \quad \text{and} \quad \beta = \frac{c}{4} + \frac{a^2}{16}.$$

This leads to the cubic equation

$$z^3 = \frac{a}{2}z^2 - \frac{4c + a^2}{16}z + \frac{b^2}{64}$$

that has three roots, A, B and C. Those combine to tell us that one of the roots of the original biquadratic is

$$x = \sqrt{A} + \sqrt{B} + \sqrt{C}.$$

The remaining three roots are

$$\sqrt{A} - \sqrt{B} - \sqrt{C}, \quad \sqrt{B} - \sqrt{A} - \sqrt{C}, \quad \text{and} \quad \sqrt{C} - \sqrt{A} - \sqrt{B}.$$

There is something just a little out of order in this solution. To solve cubic equations, Euler looks at sums of two cube roots, but to solve the biquadratic, he looked at sums of three *square* roots. Wouldn't it make a better pattern if he used three *fourth* roots? Euler apparently thinks it would, too, because he shows how to solve the biquadratic in a slightly different way, using a sum of three fourth roots.

Set $z = \sqrt{t}$. Then his cubic equation

$$z^3 = \frac{a}{2}z^2 - \frac{4c + a^2}{16}z + \frac{b^2}{64}$$

can be rewritten as

$$\left(t + \frac{4c + a^2}{16}\right)\sqrt{t} = \frac{at}{2} + \frac{b^2}{64}.$$

Squaring this and solving for t^3 gives a cubic equation

$$t^3 = \left(\frac{a^2}{8} - \frac{c}{2}\right)t^2 + \left(\frac{ab^2}{64} - \frac{cc}{16} - \frac{a^2c}{32} - \frac{a^4}{256}\right)t + \frac{b^4}{4096}.$$

This equation has three roots. If we call them E, F and G, then the root x of the original biquadratic has the form that fits the pattern

$$x = \sqrt[4]{E} + \sqrt[4]{F} + \sqrt[4]{G}.$$

To make the pattern clearer, Euler notes that to solve a quadratic equation $x^2 = a$, he can solve the first-degree equation $z = a$, and find its root, a. Then the roots of the original quadratic are related to the square root of a. In particular, $x = \sqrt{a}$ or $x = -\sqrt{a}$.

Euler asks that we call the equation in z that has degree one less than the original equation the *aequationem resolventem*, or the *resolvent equation*. Euler tells us again that the resolvent equation for the quadratic is

$$z = a.$$

For the cubic it is

$$z^2 = bz - \frac{a^3}{27},$$

and for the biquadratic it is (changing the t in its earlier form to z and rearranging the middle term a bit)

$$z^3 = \left(\frac{a^2}{8} - \frac{c}{2}\right)z^2 - \left(\frac{a^4}{256} + \frac{a^2c}{32} + \frac{c^2}{16} - \frac{ab^2}{64}\right)z + \frac{b^4}{4096}.$$

Now, Euler will try to extend this pattern to quintic, or fifth degree and higher degree equations. With our modern knowledge of Abel's results and with well over 200 years of hindsight, we know that Euler will fall short, but it will be fascinating to see how close he comes.

A quintic equation has the form

$$x^5 = ax^3 + bx^2 + cx + d$$

and its resolvent would be a fourth degree equation

$$z^4 = \alpha z^3 - \beta z^2 + \gamma z - \delta.$$

If the resolvent has roots A, B, C and D, then Euler would hope that the original quintic would have a root

$$x = \sqrt[5]{A} + \sqrt[5]{B} + \sqrt[5]{C} + \sqrt[5]{D}.$$

A general equation

$$x^n = ax^{n-2} + bx^{n-3} + cx^{n-4} + \text{etc.}$$

would have a resolvent equation

$$z^{n-1} = \alpha z^{n-2} - \beta z^{n-3} + \gamma z^{n-4} - \text{etc.}$$

which, in turn would have $n-1$ roots, A, B, C, D, etc., and one of the roots of the original equation would be

$$x = \sqrt[n]{A} + \sqrt[n]{B} + \sqrt[n]{C} + \sqrt[n]{D} + \text{etc.}$$

He recognizes that there may be problems with this. He writes, "If the given equation has more than four dimensions, it has not thus far been possible to define a resolvent equation." However, Euler knows certain cases in which it can be made to work. If the resolvent equation can be defined, and if it turns out that all terms after the first three vanish, then he can use the resolvent to construct solutions to the given equation. Euler attributes this result to de Moivre.

He works backwards. Suppose the resolvent has only three terms, so it looks like

$$z^{n-1} = \alpha z^{n-2} - \beta z^{n-3}$$

which, since roots that are zero do not matter, reduces to

$$z^2 = \alpha z - \beta.$$

If this resolvent has roots A and B, then the original equation would have a root

$$x = \sqrt[n]{A} + \sqrt[n]{B}.$$

As before,

$$\alpha = A + B \quad \text{and} \quad \beta = AB.$$

Squaring $x = \sqrt[n]{A} + \sqrt[n]{B}$ and substituting β for AB gives

$$\sqrt[n]{A^2} + \sqrt[n]{B^2} = x^2 - 2\sqrt[n]{\beta}.$$

Higher powers of x lead to

$$\sqrt[n]{A^3} + \sqrt[n]{B^3} = x^3 - 3x\sqrt[n]{\beta},$$
$$\sqrt[n]{A^4} + \sqrt[n]{B^4} = x^4 - 4x^2\sqrt[n]{\beta} + 2\sqrt[n]{\beta^2}$$

and finally, for the nth power of x,

$$\sqrt[n]{A^n} + \sqrt[n]{B^n} = x^n - nx^{n-2}\sqrt[n]{\beta} + \frac{n(n-3)}{1\cdot 2}x^{n-4}\sqrt[n]{\beta^2}$$
$$- \frac{n(n-4)(n-5)}{1\cdot 2\cdot 3}x^{n-6}\sqrt[n]{\beta^3} + \frac{n(n-5)(n-6)(n-7)}{1\cdot 2\cdot 3\cdot 4}x^{n-8}\sqrt[n]{\beta^4} - \text{etc.} = \alpha.$$

If the original equation has this form, then the resolvent will exist and have the form

$$z^{n-1} = \alpha z^{n-2} - \beta z^{n-3} \quad \text{or} \quad z^2 = \alpha z - \beta$$

with roots A and B. The original equation will have as one of its roots

$$x = \sqrt[n]{A} + \sqrt[n]{B}.$$

Moreover, its other roots, Euler tells us, will be related to the nth roots of unity. If $\mu^n = \nu^n = \mu\nu = 1$, then the other roots of the original equation will have the form

$$x = \mu\sqrt[n]{A} + \nu\sqrt[n]{B}.$$

In the case $n = 5$, the original equation of this form is

$$x^5 - 5x^3\sqrt[5]{\beta} + 5x\sqrt[5]{\beta^2} = \alpha.$$

Then the resolvent equation is $z^2 = \alpha z - \beta$, and Euler gives us all five roots of the original equation in terms of A and B, the roots of the resolvent equation, and in terms of the four complex fifth roots of unity, all of which have the form

$$\frac{-1 \pm \sqrt{5} \pm \sqrt{-10 \mp 2\sqrt{5}}}{4}.$$

Euler says that in a similar manner he can deal with seventh degree polynomials, but he gives no details.

Next, Euler turns to polynomials that have a property he calls *recoprocas*, or reciprocal. We might call them "palindromes." They have the property that, in Euler's words, "if I put $\frac{1}{y}$ for y, their form will not change." Euler does not give an example, so we will provide one. Take, for example, $y^4 + ay^3 + by^2 + ay + 1 = 0$. Substituting $\frac{1}{y}$ for y gives

$$\frac{1}{y^4} + \frac{\alpha}{y^3} + \frac{\beta}{y^2} + \frac{\alpha}{y} + 1 = 0.$$

When we multiply by y^4, we get the original polynomial. This property occurs exactly when the sequence of coefficients is a palindrome.

Moreover, reciprocal polynomials of odd degree "can always be divided by $y+1$ and the equation that results will be reciprocal in which the maximum dimension of y will be even." We note that this is closely related to a fact about numbers. If a number is a palindrome with an odd number of digits (when written in our usual base-ten notation), then it is always divisible by 11. An example is $14{,}641 = 11 \times 1331$. Euler's fact implies that we need only look at reciprocal equations of even degree.

Such an equation of degree four looks like

$$y^4 + ay^3 + by^2 + ay + 1 = 0.$$

We will look for a resolvent equation for this. Suppose it can be factored into two quadratics

$$y^2 + \alpha y + 1 = 0 \quad \text{and} \quad y^2 + \beta y + 1 = 0.$$

That makes

$$\alpha + \beta = a \quad \text{and} \quad \alpha\beta + 2 = b \quad \text{so} \quad \alpha\beta = b - 2.$$

Then α and β will be the two roots of this equation

$$u^2 - au + b = 0.$$

This shows how to factor reciprocal polynomials of degree four into two quadratics. Similarly, a reciprocal polynomial of degree six leads to a cubic equation and one of degree eight leads to a biquadratic, both of which Euler shows how to handle.

When Euler tries to find the polynomials of the form $y^2 + \zeta y + 1 = 0$ that factor reciprocal equations of degree 10, though, it leads to a fifth degree equation, and Euler can only solve fifth degree equations if their resolvent equations have that special form we saw earlier. Never one to shrink from calculations, he works through and finds the form a reciprocal polynomial of degree 10 must have for its factorization into quadratics to lead to a quintic with the required resolvent. In fact, he even does it for an arbitrary reciprocal polynomial of even degree, though its form requires 26 terms to express. We omit it. The interested reader may find it in the *Opera Omnia*, series I volume 6 on page 13, or on page 225 of the original from 1738.

We move on to another special form, reciprocal polynomials with the form $y^{2n} + py^n + 1 = 0$. Rather than consider these as quadratics in y^n, Euler looks at its quadratic factors

$$\begin{gathered} y^2 + \alpha y + 1 \\ y^2 + \beta y + 1 \\ y^2 + \gamma y + 1 \\ \text{etc.} \end{gathered}$$

Using the condition we omitted above, Euler finds that the Greek coefficients $\alpha, \beta, \gamma, \delta$ etc. must be the n roots of the equation

$$u^n - nu^{n-2} + \frac{n(n-3)}{1 \cdot 2}u^{n-4} - \frac{n(n-4)(n-5)}{1 \cdot 2 \cdot 3}u^{n-6} + \cdots \pm p = 0.$$

This has the form of equations with three terms in their resolvents that Euler discussed earlier, and so this case can be solved.

One wonders why Euler did not consider $y^{2n} + py^n + 1 = 0$ as a quadratic in y^n. It works out rather simply, but either he did not see it or he chose to do it this other way.

As before, Euler is trying to find a technique that works to find resolvents for third and fourth degree equations, hoping to extend that technique to fifth and higher degree equations. Euler has one more idea. He looks at a third degree resolvent equation $z^3 = \alpha z^2 - \beta z + \gamma$ and asks again what kinds of equations can have this as a resolvent? As before, he supposes the resolvent has roots A, B, C, and notes the familiar values for α, β, γ in terms of those roots. This time, though, he introduces another variable,

$$p = \sqrt[n]{AB} + \sqrt[n]{AC} + \sqrt[n]{BC}.$$

Euler squares x and substitutes p to get

$$\sqrt[n]{A^2} + \sqrt[n]{B^2} + \sqrt[n]{C^2} = x^2 - 2p$$

and

$$\sqrt[n]{A^2B^2} + \sqrt[n]{A^2C^2} + \sqrt[n]{B^2C^2} = p^2 - 2x\sqrt[n]{\gamma}$$

and continues with higher powers of x, up to the fifth power. It is a little surprising that the forms do not get more complicated than they do:

$$\sqrt[n]{A^5} + \sqrt[n]{B^5} + \sqrt[n]{C^5} = x^5 - 5px^3 + 5x^2\sqrt[n]{\gamma} + 5p^2x - 5p\sqrt[n]{\gamma}$$

and

$$\sqrt[n]{A^5B^5} + \sqrt[n]{A^5C^5} + \sqrt[n]{B^5C^5} = p^5 - 5p^3x\sqrt[n]{\gamma} + 5p^2\sqrt[n]{\gamma^2} + 5px^2\sqrt[n]{\gamma^2} - 5x\sqrt[n]{\gamma^3}.$$

In this tangle of symbols, Euler sees a recursive relationship, giving roots of mth powers in terms of $(m-1)$st, $(m-2)$nd and $(m-3)$rd powers:

$$\begin{aligned}\sqrt[n]{A^m} + \sqrt[n]{B^m} + \sqrt[n]{C^m} &= x\left(\sqrt[n]{A^{m-1}} + \sqrt[n]{B^{m-1}} + \sqrt[n]{C^{m-1}}\right) \\ - p\left(\sqrt[n]{A^{m-2}} + \sqrt[n]{B^{m-2}} + \sqrt[n]{C^{m-2}}\right) &+ \sqrt[n]{\gamma}\left(\sqrt[n]{A^{m-3}} + \sqrt[n]{B^{m-3}} + \sqrt[n]{C^{m-3}}\right)\end{aligned}$$

and similarly for $\sqrt[n]{A^mB^m} + \sqrt[n]{A^mC^m} + \sqrt[n]{B^mC^m}$. This relationship would look a lot simpler if Euler had a notation for the roots of the powers, but he does introduce a notation to give a recursive relation between roots of mth powers and roots of $2m$th powers.

For fourth degree equations, Euler takes three pages of calculations to find that he can get four equations with four unknowns, α, β, γ and p, and thus find his third degree resolvent equation.

This makes Euler optimistic that the technique can be extended to higher degrees. Finally, Euler considers the fourth degree resolvent equation

$$z^4 = \alpha z^3 + \beta z^2 + \gamma z + \delta$$

again with roots A, B, C, D. As before, he wants to know what kinds of equations can have this as their resolvent. He again sets $x = \sqrt[n]{A} + \sqrt[n]{B} + \sqrt[n]{C} + \sqrt[n]{D}$. This time, he introduces two new forms,

$$p = \sqrt[n]{AB} + \sqrt[n]{AC} + \sqrt[n]{AD} + \sqrt[n]{BC} + \sqrt[n]{BD} + \sqrt[n]{CD}$$

and

$$q = \sqrt[n]{ABC} + \sqrt[n]{ABD} + \sqrt[n]{ACD} + \sqrt[n]{BCD}.$$

He is confident that it will be possible to find recursive relationships involving p, q and the coefficients of the resolvent polynomial, and thus find resolvents for equations of the fifth degree and higher. He resolves to complete this at some other time.

That takes a while. In 1762, Euler will write a sequel to this paper, E-282, De resolutione aequationum cuiusvis gradus, "On the resolution of equations of whatever degree." There, he will make a better use of roots of unity, as well as do some numerical examples notably missing in this paper. Of course, he will still fall short of solving the problem. After all, it is unsolvable.

15

E-31: Constructio aequationis differentialis $ax^n\,dx = dy + y^2\,dx$*

Solution of the differential equation $ax^n\,dx = dy + y^2\,dx$

This is the paper people have been waiting for. Five years passes between the time the results in this paper were announced back in E-11 and the time those results appear in print. You can almost feel the excitement. Even the Editor's summary, which sometimes drags on for a couple of pages, reflects the thrill. The five-line summary reads:

> "There is great excitement (*maxime agitate*) among Geometers about this equation first proposed by the illustrious Count Riccati. Until now, nobody has been able to solve it, except for certain given values of n. And so it is a great accomplishment and a benefit to all by Euler who has overcome all difficulties and given a universal solution to these equations."

Euler tells us that he hopes to extend his techniques from E-28 on the arc length of the ellipse to the Riccati equation (named for Jacopo Riccati (1676–1754), not his son, Vincenzo (1707–1755), also a mathematician). Euler compares the Riccati equation

$$ax^n\,dx = dy + y^2\,dx$$

to the differential equation describing the arc length of an ellipse

$$dy + \frac{y^2\,dx}{x} = \frac{x\,dx}{x^2 - 1}$$

from E-28. A solution to this differential equation will give y, the arc length of a quadrant of an ellipse with minor axis 1, in terms of x the length of its major axis. What these equations have in common, in Euler's eyes, is that separation of variables does not seem to solve them.

Euler begins with a review of the differential equations part of E-28. He uses different notation and his summary reviews only those parts that will be useful here. He supposes that Z is a curve satisfying the differential equation and that dZ has the form $PR\,dz$. He assumes that he can write R as a series

$$R = 1 + AgQ + ABg^2Q^2 + ABCg^3Q^3 + ABCDg^4Q^4 + \text{etc.},$$

where Q is a function of z and where g is a parameter of the curve. In the case of the ellipse, g is the length of the major axis and Q is, in the notation of E-28,

$$\frac{t^2}{b^2 + t^2}.$$

---

*_Comm. Acad. Sci. Imp. Petropol._ 6 (1732/3) 1738, 124–137; _Opera Omnia_ I.22, 19–35

To express Z, Euler multiplies R by P and integrates the product to get

$$Z = \int P\,dz + \int AgPQ\,dz + \int ABg^2PQ^2\,dz + \int ABCg^3PQ^3\,dz + \text{etc.}$$

In E-28, Euler knew enough about how P and Q were related to determine that

$$\int PQ\,dz = \alpha \int P\,dz + O_1$$
$$\int PQ^2\,dz = \alpha\beta \int P\,dz + O_2$$
$$\int PQ^3\,dz = \alpha\beta\gamma \int P\,dz + O_3, \text{ etc.,}$$

where the O's represent what he calls *quantitates algebraicas*. He sets $z = h$, the lower bound of integration in each integral so that the quantities outside the integral, the O's, disappears, and is able to express Z as a product of an integral and a series.

For the Riccati equation, Euler again wants to use a series

$$R = 1 + AgQ + ABg^2Q^2 + ABCg^3Q^3 + \text{etc.}$$

but he does not yet know enough about R to make progress. Bear with him. Returning to the Riccati equation $ax^n\,dx = dy + y^2\,dx$, Euler substitutes $y = \frac{dt}{t\,dx}$, where dx is supposed to be constant. We have seen this idea of holding a differential constant in E-10. This makes

$$ax^n\,dx = \frac{ddt}{t\,dx} \quad \text{so that} \quad ax^n t\,dx^2 = ddt.$$

Euler seems to be going backwards. His first order Riccati equation is in danger of becoming a second order equation.

Now, much to the dismay of the Author's word processor, and probably his typesetter as well, Euler briefly introduces variables in the old German Fraktur alphabet as he rewrites t as a series

$$1 + \mathfrak{A}x^{n+2} + \mathfrak{B}x^{2n+4} + \mathfrak{C}x^{3n+6} + \text{etc.}$$

Then

$$ddt = (n+1)(n+2)\mathfrak{A}x^n\,dx^2 + (2n+3)(2n+4)\mathfrak{B}x^{2n+2}\,dx^2 + (3n+5)(3n+6)\mathfrak{C}^{3n+4}\,dx^2 + \text{etc.}$$

Now, from above we know that $ax^n t\,dx^2 = ddt$, so Euler writes the other side of the equation as a series as well,

$$ax^n t\,dx^2 = ax^n\,dx^2 + \mathfrak{A}ax^{2n+2}\,dx^2 + \mathfrak{B}ax^{3n+4}\,dx^2 + \text{etc.}$$

As we have seen, Euler loves to match terms in series, and it works here, too. Doing so for these last two series gives

$$\mathfrak{A} = \frac{a}{(n+1)(n+2)}, \quad \mathfrak{B} = \frac{\mathfrak{A}a}{(2n+3)(2n+4)}, \quad \mathfrak{C} = \frac{\mathfrak{B}a}{(3n+5)(3n+6)}, \text{ etc.}$$

For the sake of brevity, he eliminates the Fraktur and writes $ax^{n+2} = f$, and gets that

$$t = 1 + \frac{f}{(n+1)(n+2)} + \frac{f^2}{(n+1)(n+2)(2n+3)(2n+4)} + \frac{f^3}{(n+1)(n+2)(2n+3)(2n+4)(3n+5)(3n+6)} + \text{etc.}$$

Out of what seemed like nothing, Euler now has a series the sum of which is related to the solution of the Riccati equation.

Now, Euler returns to the idea that $dZ = PR\,dz$ and makes the assumption that

$$P = \frac{1}{(1+bz^\mu)^\nu} \quad \text{and} \quad Q = \frac{z^\mu}{1+bz^\mu}.$$

This, he hopes, will lead to a good choice of R and that, in turn, might lead to finding Z.

P and Q are well enough related that Euler can find the integrals analogous to the ones he needed in E-28. In fact, he skips the easier cases of $\int P\,dz$, $\int PQ\,dz$ and $\int PQ^2\,dz$ and dives right into

$$\int \frac{z^{\theta\mu}\,dz}{(1+bz^\mu)^{\nu+\theta}} = \frac{(\theta-1)\mu+1}{b\mu(\nu+\theta-1)} \int \frac{z^{(\theta-1)\mu}\,dz}{(1+bz^\mu)^{\nu+\theta-1}} - \frac{1}{b\mu(\nu+\theta-1)} \cdot \frac{z^{(\theta-1)\mu+1}}{(1+bz^\mu)^{\nu+\theta-1}}.$$

The reader is supposed to recognize that this means

$$\int PQ^\theta\,dz = \frac{(\theta-1)\mu+1}{b\mu(\nu-\theta-1)} \int PQ^{\theta-1}\,dz - \frac{1}{b\mu(\nu+\theta-1)} \cdot zPQ^{\theta-1}.$$

Remember, he chooses his particular antiderivative so that the rational function outside the integral is zero. Now from this, he can discern that his coefficients α, β, γ are

$$\alpha = \frac{1}{b\mu\nu}, \quad \beta = \frac{\mu+1}{b\mu(\nu+1)}, \quad \gamma = \frac{2\mu+1}{b\mu(\nu+2)}, \quad \text{etc.}$$

Let us pause to check the score. The solution to the Riccati equation is named Z. We have set $dZ = PR\,dz$. We could find P if only we knew a, b, μ and ν. We can write R as a series involving a function Q and coefficients A, B, C, etc. Because Q has almost the same form as P, and if we know one, now that we know α, β, γ, etc., we can find the other. That leaves us needing parameters a, b, μ and ν and coefficients A, B, C, etc. How close are we to a solution? Well, there are 18 sections in this paper, and we are about to begin section 8. There is still a lot to do.

Euler turns his attention to those coefficients A, B, C, which "I see ought to be of this form $\frac{1}{\sigma\tau}$." Without any other justification,

"I make $A = \dfrac{1}{\pi(\pi+\rho)}$, $B = \dfrac{1}{(\pi+2\rho)(\pi+3\rho)}$, $C = \dfrac{1}{(\pi+4\rho)(\pi+5\rho)}$, etc."

Recall that

$$R = 1 + AgQ + ABg^2Q^2 + ABCg^3Q^3 + ABCDg^4Q^4 + \text{etc.}$$

and, "for the sake of brevity," set $gQ = q^2$. This makes

$$R = 1 + \frac{q^2}{\pi(\pi+\rho)} + \frac{q^4}{\pi(\pi+\rho)(\pi+2\rho)(\pi+3\rho)} + \text{etc.}$$

Now, Euler does some series manipulations that remind us of E-20 and its sequels, especially E-28 and the elliptic arc length differential equation he was telling us about before. He is going to multiply, subtract and differentiate until something interesting happens. First, let $S = R - 1$, so that

$$S = \frac{q^2}{\pi(\pi+\rho)} + \frac{q^4}{\pi(\pi+\rho)(\pi+2\rho)(\pi+3\rho)} + \text{etc.}$$

Multiply by $\rho q^{(\pi-\rho)/\rho}$ and differentiate with respect to q to get

$$\frac{\rho d(q^{(\pi-\rho)/\rho}S)}{dq} = \frac{q^{\pi/\rho}}{\pi} + \frac{q^{(\pi+2\rho)/\rho}}{\pi(\pi+\rho)(\pi+2\pi)} + \text{etc.}$$

The reader may want to pause while checking this calculation to admire how the $\rho q^{(\pi-o)/\rho}$ was chosen so that the number of factors in each denominator would decrease by one. This is part of Euler's larger plan to simplify the denominators. Now, he multiplies again, this time by ρ, and differentiates again. "Putting dq constant" eliminates another factor from the denominators, producing

$$\begin{aligned}\frac{\rho^2 dd(q^{(\pi-\rho)/\rho}S)}{dq^2} &= q^{(\pi-\rho)/\rho} + \frac{q^{(\pi+2\rho)/\rho}}{\pi(\pi+\rho)} + \text{etc.}\\ &= q^{(\pi-\rho)/\rho} + q^{(\pi-\rho)/\rho}\left(\frac{q^2}{\pi(\pi+\rho)} + \frac{q^4}{\pi(\pi+\rho)(\pi+2\rho)(\pi+3\rho)} + \text{etc.}\right).\end{aligned}$$

Something good has happened here. There is the series for S, in the last line between those big parentheses, so this gives us a second order differential equation satisfied by S:

$$\rho^2 dd(q^{(\pi-\rho)/\rho}S) = q^{(\pi-\rho)/\rho}\,dq^2 + q^{(\pi-\rho)/\rho}S\,dq^2.$$

Again, for the sake of brevity, put $q^{(\pi-\rho)/\rho}S = T$, and the differential equation becomes

$$\rho^2\,ddT = q^{(\pi-\rho)/\rho}\,dq^2 + T\,dq^2.$$

It is a little strange that Euler would make this substitution, because he immediately wants to substitute again, setting $T = rs$. Why did he not just use rs in the first place? And what are r and s? We will soon see.

This substitution makes

$$ddT = r\,dds + 2\,dr\,ds + s\,ddr$$

so the differential equation involving T becomes

$$\rho^2 r\,dds + 2\rho^2\,dr\,ds + \rho^2 s\,ddr = q^{(\pi-\rho)/\rho}\,dq^2 + rs\,dq^2.$$

Euler breaks this equation into two parts, one with r and one without it, so

$$\rho^2 r\,dds = rs\,dq^2$$

and

$$2\rho^2\,dr\,ds + \rho^2 s\,ddr = q^{(\pi-\rho)/\rho}dq^2.$$

Euler does not explain why he can split differential equations like this, but it works here. He attacks these two smaller differential equations one at a time. The first one he divides by r, multiplies by ds and integrates, ignoring constants of integration, giving

$$\rho^2\, ds^2 = s^2\, dq^2.$$

Take the square root of both sides to get $\rho\, ds = s\, dq$, which solves easily to give

$$s = c^{q/\rho}.$$

Here, as in previous papers, "c is the number whose logarithm is 1."

Now we turn to the other half of that split differential equation, the part that did not contain r:

$$2\rho^2\, dr\, ds + \rho^2 s\, ddr = q^{(\pi-\rho)/\rho}\, dq^2.$$

Substituting for s gives

$$2\rho c^{q/\rho}\, dq\, dr + \rho^2 c^{q/\rho}\, ddr = q^{(\pi-\rho)/\rho}\, dq^2.$$

Putting $dr = v\, dq$ so that $ddr = dv\, dq$ (remember dq is held constant so that the missing term containing ddq is zero) gives a simpler differential equation

$$2\rho c^{q/\rho} v\, dq + \rho^2 c^{q/\rho}\, dv = q^{(\pi-\rho)/\rho}\, dq.$$

Just a few years earlier in E-10, Euler had discovered how to solve such differential equations using integrating factors. Here, the appropriate integrating factor is $c^{q/\rho}$. He multiplies, integrates, and solves for v to get

$$v = \frac{1}{\rho^2} c^{-2q/\rho} \int c^{q/\rho} q^{(\pi-\rho)/\rho}\, dq.$$

Now some dominos fall. Since $dr = v\, dq$, integrating v gives r. Then $T = rs$, and $q^{(\pi-\rho)/\rho} S = T$ which implies that

$$S = \frac{1}{\rho^2} c^{q/\rho} q^{(\pi-\rho)/\rho} \int c^{-2q/\rho}\, dq \int c^{q/\rho} q^{(\pi-\rho)/\rho}\, dq$$

where, as often in Euler, we have to interpret this integral "correctly" regarding order and bounds of integration. And at last, since $R = 1 + S$, we get

$$R = 1 + \frac{1}{\rho^2} c^{q/\rho} q^{(\pi-\rho)/\rho} \int c^{-2q/\rho}\, dq \int c^{q/\rho} q^{(\pi-\rho)/\rho}\, dq.$$

We should pause again to see what progress we are making towards a solution. We know that $q = \sqrt{gQ}$, and we know P, Q and R in terms of π, ρ, μ, and ν, and we know that $dZ = PR\, dz$. Things are looking pretty good, if only Euler can show us how π, ρ, μ, and ν fit into the picture. In fact, they will turn out to depend on n in the original Ricatti equation $ax^n\, dx = dy + y^2\, dx$.

Euler takes us back to the series from much earlier,

$$1 + A\alpha g Q + AB\alpha\beta g^2 Q^2 + \text{etc.}$$

All this hard work has given Euler values for A, α, B, β, etc., so he substitutes these values and gets

$$1 + \frac{g}{b\mu\nu\pi(\pi+\rho)} + \frac{(\mu+1)g^2}{b^2\mu^2\nu(\nu+1)\pi(\pi+\rho)(\pi+2\rho)(\pi+3\rho)} + \text{etc.}$$

The pattern here may not be obvious, so Euler tells us that to go from term θ to term $\theta+1$, multiply by

$$\frac{g(1+(\theta-1)\mu)}{b\mu(\nu+\theta-1)(\pi+(2\theta-2)\rho)(\pi+(2\theta-1)\rho)}.$$

Euler compares this to the series for t several sections before, where the corresponding ratio was

$$\frac{f}{(\theta n+2\theta-1)(\theta n+2\theta)}.$$

Yet again, Euler matches components, matching $g = bf$ and

$$\frac{1}{(\theta n+2\theta-1)(\theta n+2\theta)} = \frac{\theta\mu-\mu+1}{(\mu\nu+\mu\theta-\mu)(\pi+2\theta\rho-2\rho)(\pi+2\theta\rho-\rho)}.$$

Now, this last equation does not look promising, but because it must be true for all θ, there really is more information here than it appears. Euler suggests writing it as a polynomial in θ and setting the coefficients equal to zero. Should the reader decide to check Euler's algebra, once the fractions are cleared and the third degree polynomial in θ is set equal to zero, the cubic term shows immediately that $\rho = \frac{n+2}{2}$, and the constant term leads to four cases corresponding to the factors of $\mu(\nu-1)(\pi-2\rho)(\pi-\rho)$. One of those four cases, $\mu = 0$, makes no sense, and the other three, $\pi = 2\rho$, $\pi = \rho$ and $\nu = 1$, lead to a total of four different solutions to the system of equations:

Taking $\nu = 1$ leads to a quadratic equation and gives rise to two solutions. The first is

$$\mu = \frac{2n+4}{3n+4},\ \nu = 1,\ \pi = n+1 \quad \text{and} \quad \rho = \frac{n+2}{2}.$$

The second solution arising from $\nu = 1$ is

$$\mu = \frac{2n+4}{n},\ \nu = 1,\ \pi = \frac{n}{2} \quad \text{and} \quad \rho = \frac{n+2}{2}.$$

If $\pi = \rho$, then

$$\mu = 2,\ \nu = \frac{n+1}{n+2},\ \pi = \frac{n+2}{2} \quad \text{and} \quad \rho = \frac{n+2}{2}.$$

Finally, if $\pi = 2\rho$, then

$$\mu = \frac{2}{3},\ \nu = \frac{n+1}{n+2},\ \pi = n+2$$

and, as always,

$$\rho = \frac{n+2}{2}.$$

We seem to be finished, for now we know how to find π, ρ, μ and ν in terms of n. With those, we can calculate P, Q and R, and then get PR. From that, since $PR\,dz = dZ$, we can find our answer, Z. But can this actually be done? And what are we to make of these four different sets of values for π, ρ, μ and ν in terms of n?

Euler answers the second question first. Recall that

$$P = \frac{1}{(1+bz^{\mu})^{\nu}} \quad \text{and} \quad Q = \frac{z^{\mu}}{1+bz^{\mu}}.$$

Euler advises us that these are easier to manipulate if, for example, $\nu > 0$, so for $-2 < n < -1$, we should avoid the third and fourth of these solutions because they would not give positive values for ν. His analysis of this can be summarized by saying all four solutions work, and we are free to choose the solution that is simplest for any particular value of n.

Does Euler's solution actually work out? He gives us several examples, beginning with the case $n = 2$. We are to solve $ax^2\,dx = dy + y^2\,dx$. Using the third set of solutions, we find that

$$\mu = 2,\ \nu = \frac{3}{4},\ \pi = \rho = 2.$$

This makes

$$\begin{aligned} S &= \frac{1}{\rho^2} c^{q/\rho} q^{(\pi-\rho)/\rho} \int c^{-2q/\rho}\,dq \int c^{q/\rho} q^{(\pi-\rho)/\rho}\,dq \\ &= \frac{1}{4} c^{q/2} \int c^{-q}\,dq \int c^{q/2}\,dq \\ &= \frac{k}{4} c^{q/2} - \frac{i}{4} c^{-q/2} - \frac{1}{4}, \end{aligned}$$

where i and k are constants of integration, $k = 1 + i$, and this particular antiderivative is chosen so that $S = 0$ when $q = 0$. Moreover, Euler wants $dS = 0$ if $q = 0$, and this gives particular values for i and k of $-\frac{1}{2}$ and $\frac{1}{2}$ respectively, so

$$S = \frac{c^{q/2} + c^{-q/2}}{8} - \frac{1}{4}.$$

Because $R = S + 1$, this gives

$$R = \frac{3}{4} + \frac{c^{q/2} + c^{-q/2}}{8}.$$

Now, because

$$q = \sqrt{\frac{bfz^2}{1+bz^2}}$$

we can substitute, yielding

$$R = \frac{3}{4} + \frac{1}{8} c^{\frac{1}{2}\sqrt{\frac{bfz^2}{1+bz^2}}} + \frac{1}{8} c^{\frac{-1}{2}\sqrt{\frac{bfz^2}{1+bz^2}}}.$$

So, finally, the answer we've been waiting for is

$$Z = \int PR\,dz = \frac{3}{4}\int \frac{dz}{(1+bz^2)^{3/4}} + \frac{1}{8}\int \frac{dz\left(c^{\frac{1}{2}\sqrt{\frac{bfz^2}{1+bz^2}}} + c^{\frac{-1}{2}\sqrt{\frac{bfz^2}{1+bz^2}}}\right)}{(1+bz^2)^{3/4}}.$$

Here, the integrals are to be taken from 0 to ∞.

The reader may not be entirely satisfied with this answer. Indeed, the integrals are intractable, but at least they are answers. The sharp-eyed reader may note that there are still variables b and f embedded in the answer as given, and that may seem unsatisfactory as well. It is easy to account for f. A long time ago, we set $ax^n + 2 = f$, "for the sake of brevity." We also had $g = bf$, and g was described only as a "parameter of the curve." For the elliptic arc length problem, g was the length of the major semi-axis of the ellipse, but for the Riccati equation, Euler does not explain what g means in a satisfying way.

Euler's next example is not quite so specific as the first one. He considers the case $-2 < n < 0$ and uses the third set of solutions for π, ρ, ν and ν in terms of n. His answer resembles the one above in complexity. Finally, he briefly does an example with $n = -4$.

Why was Euler (and his Editor as well) so happy about this solution to the Riccati equation? Indeed, the answer might politely be described as "messy," and the analysis seems to be a pointless and prolix sequence of substitutions, expansions and matching of terms.

People at the time were not so much concerned about what the answer was, but that the answer could be written down. A "messy" answer would do. The Riccati equation was a theoretical thorn. There was a general feeling that differential equations arising in a "normal" way should have solutions. This had been borne out time and again as the 17th century problems of the tractrix, the catenary, the brachistochrone, the isochrone and others succumbed to analysis. All these were based in geometry and in physics, and all had solutions. The Riccati equation also had a geometrical interpretation and physical origins. Mathematicians of the time sincerely believed that all equations' natural and geometric origins should be "good" and should have explicit solutions. If an equation as natural and geometric as the Riccati equation could not be solved, then a key aspect of their world view would be shaken.

Perhaps also people were optimistic that easier and more elegant solutions would eventually be found. Indeed, we now know 19th century solutions involving Bessel functions. Textbooks that mention the Riccati equation often consign it to an exercise in a chapter about special functions like those of Bessel.

Interlude: 1734

World events

The year 1734 seems to have been a quiet one in world history. Daniel Boone was born on November 2. English mathematician Edward Waring was born in 1734, or maybe 1736, depending on sources. Everyone seems sure he died in 1798. Also, the Austrian hypnotist Franz Mesmer (1734–1815) was born on May 23.

In Euler's life

Euler's big salary increase the previous year gave him the resources he needed to marry Katharine Gsell, the daughter of a Swiss painter living in St. Petersburg. They had 13 children altogether although only five survived to adulthood. They purchased a comfortable house near the river Neva close to the Academy.

Euler's other work

Euler wrote the first volume of his *Mechanica*, E-15, in 1734. It was published two years later.

The only other work he wrote that is not classified as "mathematical" was E-49, De oscillationibus fili flexilis quotcunque pondusculis onusti, "On the oscillations of flexible fibers laden by arbitrary weights."

Euler's mathematics

Five of Euler's mathematical papers were written in 1734:

E-42, De linea celerrimi descensus in medio quocunque resistente, "On curves of fastest descent in a resistant medium,"

E-43, De progressionibus harmonicis observationes, "Observations on harmonic progressions,"

E-44, De infinitis curvis eiusdem generis, "On an infinity of curves of a given kind,"

E-45, Additamentum ad dissertationem de infinitis curvis eiusdem generis, "Additions to the dissertation on infinitely many curves of a given kind," and

E-48, Investigatio binarum curvarum, quarum arcus eidem abscissae respondentes summam algebraicam constituant, "Investigation of two curves, the abscissas of which are corresponding arcs and the sum of which is algebraic."

16

E-42: De linea celerrimi descensus in medio quocunque resistente*

On curves of fastest descent in a resistant medium

Euler returns to calculus of variations to apply it to a specific problem in mechanics. In 1696 Johann Bernoulli had posed the famous Brachistochrone Problem about curves of quickest descent through a frictionless medium. The next year, *Acta Eruditorum* published solutions by Newton, Leibniz, Jakob Bernoulli and Johann Bernoulli.

A natural extension would be to solve the same problem taking into account friction. Euler's very first publication, E-1, Constructio linearum isochronarum in medio quocunque resistente, "Construction of isochronal curves in any kind of resistant medium," appeared in the *Acta Eruditorum* in 1726. This paper might be considered a sequel to E-1. Recall that Euler made some mistakes in E-1. He uses the new variational techniques he developed in E-27 and here completely avoids his errors of E-1.

Euler credits the posing of this problem to Jakob Hermann in 1727, and he credits the key technique to Huygens, probably from his work in optics. Hermann died in 1733, at least a year before Euler submitted this paper to be published in the *Commentarii*. Before he died, though, Hermann wrote to Euler discussing the results presented in this paper and also the results in E-31, which was published in the 1732/33 volume of the *Commentarii*. All this shows that sometimes a considerable interval elapsed between the time Euler was first exposed to a problem (in this case, 1727), the time he had enough results to talk about with his colleagues (early 1733), and when he eventually submitted the paper to the Academy for publication, 1734 or 1735.

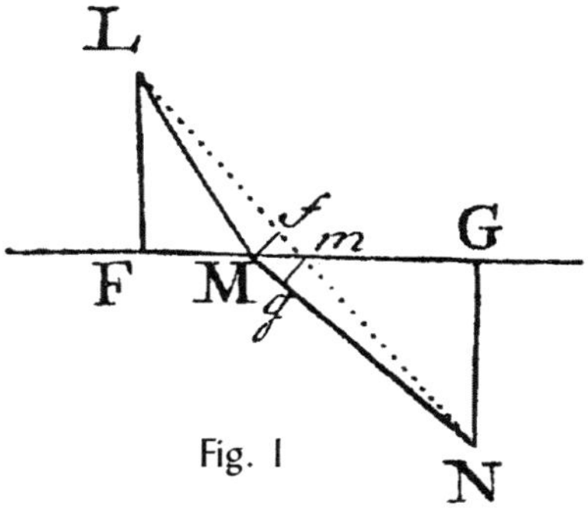

Fig. 1

Euler calls his first result in this paper Huygens's lemma. We are to consider a piece of the path of fastest descent, *LMN*, as it crosses a horizontal straight line *FG*, the path cutting the line at *M*, and a nearby path *LmN*, cutting the line at *m* (Figure 1.) The object is accelerating as it descends, so Euler lets its speed above the line *FG* be *m* (different *m* than the point on the line *FG*) and its speed below the line *FG* be *n*. Because the intervals involved are short, Euler follows Huygens

**Comm. Acad. Sci. Imp. Petropol.* 7 (1734/5) 1740, 135–149; *Opera Omnia* I.25, 41–53

and assumes that the speeds m and n are constant on the given segments. Then the time it takes to descend via path LMN is $\frac{LM}{m} + \frac{MN}{n}$. As we have seen before, by the principle of maxima and minima, if LMN is the optimal curve, and if LmN is a nearby curve, then the appropriate differential must be zero, so

$$\frac{LM}{m} + \frac{MN}{n} = \frac{Lm}{m} + \frac{mN}{n}$$

where, again, we must be aware that the symbol m has two different meanings.

Draw Mf and mg altitudes of triangles LmM and NmM respectively. Then Euler claims that $\frac{mf}{m} = \frac{Mg}{n}$ and that mf is to Mg as cosine LMF is to cosine GMN. Euler does not justify these ratios, but we will fill in his gaps to demonstrate how differential quantities were manipulated at the time. Because Mm is very small, it is almost true that $Lf = LM$, so we can take them to be equal. Likewise, $Nm = Ng$. Then

$$\frac{LM}{m} + \frac{MN}{n} = \frac{LM}{m} + \frac{Mg}{n} + \frac{gN}{n}$$

and

$$\frac{Lm}{m} + \frac{mN}{n} = \frac{Lf}{m} + \frac{fm}{m} + \frac{mN}{n}.$$

From here, it is only algebra. Since the outer terms cancel, the inner terms must be equal.

The claim about the cosines follows from the almost true assumption that $\angle LmF = \angle LMF$, and Huygens's lemma is complete.

Now, denote by q the speed of the object through the curve element LM and Lm, by $q + dt$ the speed through MN, and by $q + dt + dd\theta$ the speed through mN. Then the minimality condition becomes

$$\frac{LM}{q} + \frac{MN}{q+dt} = \frac{Lm}{q} + \frac{mN}{q+dt+dd\theta}.$$

Applying this and Huygens's lemma to the element mf gives us

$$\frac{mf}{q} = \frac{Mg}{q+dt} + \frac{mN \cdot dd\theta}{(q+dt)((q+dt+dd\theta)}$$

which is equivalent to

$$\left(q^2 + 2q\,dt + dt^2 + q\,dd\theta + dt\,dd\theta\right)mf = \left(q^2 + q\,dt + q\,dd\theta\right)Mg + q \cdot mN \cdot dd\theta.$$

Substituting

$$mf = \frac{FM \cdot Mm}{LM} \quad \text{and} \quad Mg = \frac{MG \cdot Mm}{LM}$$

and *neglectis negligendis* or "neglecting what ought to be neglected"

$$q\left(\frac{MG}{LM} - \frac{FM}{LM}\right) = \frac{FM}{LM}dt - \frac{LM\,dd\theta}{Mm}.$$

Euler tells us that $dd\theta$ depends only on Mm, that it is of the form $Z \cdot Mm$, and that it does not depend on any other quantities nor on the point M. He must have reasons for making this claim, but he does not tell us what they are.

Now, Euler supposes that $LF = GN$, and he calls this distance dx. Here, he is taking the x direction to be vertical rather than horizontal. Moreover, he lets $FM = dy$ and $LM = ds$. Then $MG = dy + ddy$ and $MN = ds + dds$, and the equations above become

$$\frac{q\,ds\,ddy - q\,dy\,dds}{ds^2} = \frac{dy\,dt}{ds} - \frac{ds\,dd\theta}{Mm}.$$

By another of Euler's differential equalities that are true *neglectis negligendis*, $ds\,dds = dy\,ddy$, and by holding dx constant, we get

$$\frac{q\,dx^2\,ddy}{ds^3} = \frac{dy\,dt}{ds} - \frac{ds\,dd\theta}{Mm}.$$

This is the differential equation that Euler will use to find the brachistochrone in a resistant medium.

Now, assume that the force of gravity is 1 and that it is perpendicular to the line FG. Call the curve described by the descending body p. Euler proposes that the resistance of the medium be a multiple of a power, $2n$, of the speed, n not necessarily an integer. Take the speed to be c (for *celeritas*, quickness) and the altitude through which the object falls to be v. (Be careful. We expect v to be velocity. To Euler, it is altitude.) Then the resistance slowing the motion between L and FG will be $= \frac{v^n}{c^n}$. The altitude will change by $p\,dx$, and Euler decides that, for the path through element LM, the resistance will be $= \frac{v^n}{c^n} LM$ and by Lm it will be $= \frac{v^n}{c^n} Lm$. So the corresponding altitude elements will be

$$v + p\,dx - \frac{v^n}{c^n} LM \quad \text{and} \quad v + p\,dx - \frac{v^n}{c^n} Lm.$$

Now, Euler compares this with his differential equation above. We have $q = \sqrt{v}$,

$$q + dt = \sqrt{v + p\,dx - \frac{v^n}{c^n} LM} = \sqrt{v} + \frac{p\,dx - \frac{v^n}{c^n} LM}{2\sqrt{v}} \quad \text{and} \quad dt = \frac{p\,dx - \frac{v^n}{c^n} ds}{2\sqrt{v}}.$$

This makes

$$q + dt + dd\theta = \sqrt{v + p\,dx - \frac{v^n}{c^n} Lm} = \sqrt{v} + \frac{p\,dx - \frac{v^n}{c^n} Lm}{2\sqrt{v}}.$$

This allows us to calculate that

$$dd\theta = \frac{v^n(LM - Lm)}{2c^n\sqrt{v}} = -\frac{v^m \cdot FM \cdot Mm}{2c^n \cdot LM \cdot \sqrt{v}}$$

and so

$$\frac{dd\theta}{Mm} = -\frac{v^n\,dy}{2c^n\,ds\sqrt{v}}.$$

Substituting this into the main differential equation and multiplying by $2\sqrt{v}$ shows that $2v\,dx\,ddy = p\,dy\,ds^2$, or, solving for v,

$$v = \frac{p\,dy\,ds^2}{2\,dx\,ddy},$$

which is the differential equation for the brachistochrone.

Now we want to take into account resistance. Divide $2v\,dx\,ddy = p\,dy\,ds^2$ by ds^3 to get

$$\frac{2v\,dx\,ddy}{ds^3} = \frac{p\,dy}{ds}$$

in which $\frac{p\,dy}{ds}$ represents the component of the force normal to the path p.

The other term,

$$\frac{2v\,dx\,ddy}{ds^3},$$

contains

$$-\frac{ds^3}{dx\,ddy}$$

which is the radius of curvature[1] of the curve *LMN* at the level F. In Euler's time, the radius of curvature was called *radius osculi*, or *radius of kissing*, and it was an important topic in calculus. His readers would have been considerably more familiar with this expression than are most today. Call that radius of curvature r, and call the normal force N. Note that we are using the symbol N twice, once as a point and once as a force. Then $\frac{2v}{r} = N$ is the centrifugal force.

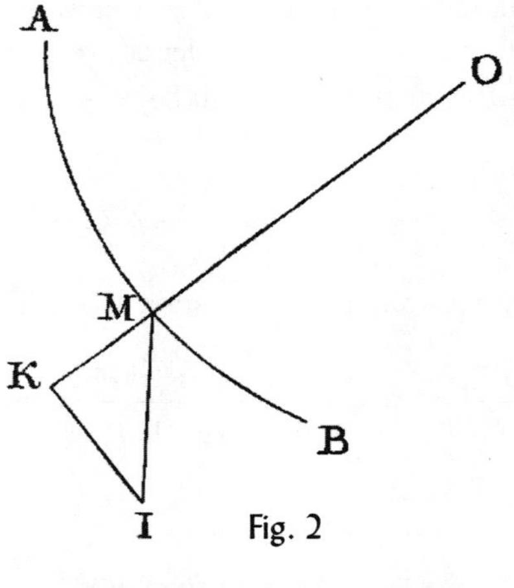

Fig. 2

Note that all solutions to the problem, whether in a resistant medium or in a vacuum, will be concave over any curved part *AMB*, as shown in Figure 2. Let *MI* be the force acting upon the body at the point M, and let it be resolved into two components, *MK* normal to the curve and *KI* parallel to its tangent.

Euler claims that if the two forces acting on the body, (the centrifugal and the force normal to that,) are equal, then the curve will have the brachistochrone property, whether there be a vacuum or a resistant medium.

In a vacuum, the second theorem of Huygens says that the speed should be proportional to the angle that the curve forms with the direction of the force, that is, $\frac{MK}{MI}$, so that

$$\frac{MK^2}{MI^2 \cdot MO}$$

will be to *MK* as $\frac{MK}{MI}$ is to $MI \cdot MO$.

[1] Most modern calculus books do not seem to dwell much on radius of curvature. See, for example, Courant, *Differential and Integral Calculus*, 2nd ed., 1937, p 283.

Thus, a brachistochrone in a vacuum has the property that the sine of the angle that the direction of the force makes with the curve is proportional to the radius of curvature and the disturbing force combined. From this, the brachistochrone in a vacuum is easily found.

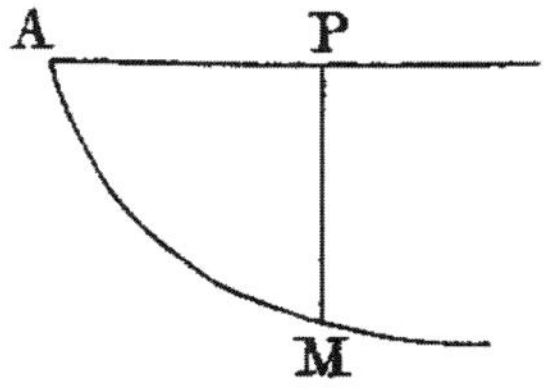

Fig. 3

This is challenging material. Euler helps us out a bit with some examples. His first is to show that his analysis is consistent with what is already known about the problem. Let a constant force g act in the vertical direction, parallel to PM (Figure 3). Call the brachistochrone curve we are seeking AM, and take AP as an axis. Assign variables $AP = y$, $PM = x$ and $AM = s$. Then the sine of the angle that PM makes with the curve will be $\frac{dy}{ds}$ and the radius of curvature will be

$$\frac{ds^3}{dx\,ddy}.$$

Hold dx constant. By the lemma,

$$\frac{ds^3}{dx\,ddy} = \frac{a\,dy}{ds}.$$

Since

$$ddy = \frac{ds\,dds}{dy},$$

this makes $ds^3 = a\,dx\,dds$. Dividing by ds^2 and integrating give

$$s = C - \frac{a\,dx}{ds}.$$

When $s = 0$, then it ought to be true that $dx = ds$. This tells us that $C = a$ and so $s\,ds = a\,ds - a\,dx$, which, integrated, gives $s^2 = 2as - 2ax$, which is the equation of the cycloid, as it should be.

Now that we have some confidence in the techniques, Euler does a less familiar example (Figure 4). Let C be a center of attraction pulling the object towards itself with a force "in proportion to some multiple of the distance, whose exponent is m." This is the wording used in Euler's time to say that the force obeys some power law. The most common of such laws is an inverse square law, but Euler does not rule out other laws of the same form.

Let curve AM be the brachistochrone. Let $CA = a$, $CM = y$, and draw MT tangent to the curve with the point T on the tangent so that MT is perpendicular to TC. Let $MT = z$. The force acting at M in the direction MC will be ym, and the sine of the angle that the curve makes with this direction

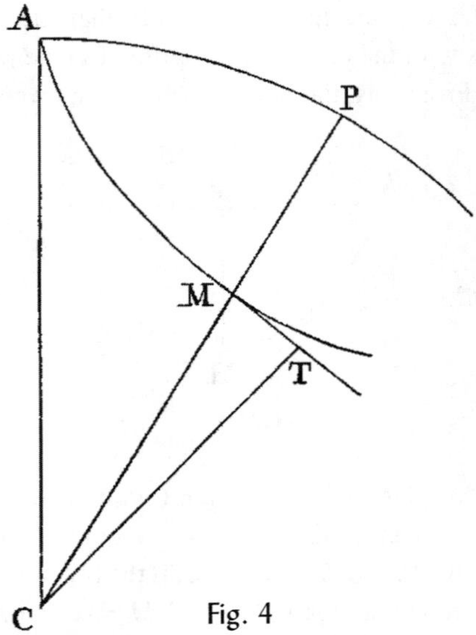

Fig. 4

will be $= \frac{z}{y}$. The radius of curvature will be $-\frac{y\,dy}{dz}$. The rule of forces says that $\frac{z}{y}$ is proportional to

$$\frac{y^{m+1}\,dy}{dz},$$

so, letting A be the proportionality constant, we get $Az\,dz = y^{m+2}\,dy$. Integrating gives $C + Az^2 = y^{m+3}$. To find what the constants should be, note that if $y = a$, then $z = 0$ and then $C = a^{m+3}$, and so $Az^2 = a^{m+3} - y^{m+3}$, and that constant A is assumed to be negative. "And so this is the equation for all brachistochrones for which there is a center of force."

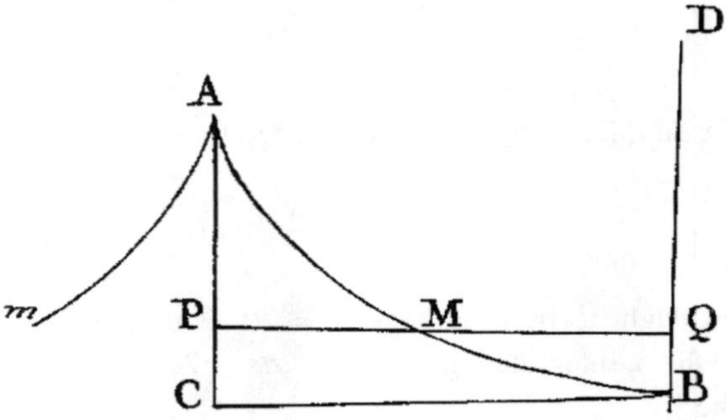

Fig. 5

Euler returns to media that provide a resistance in proportion to some power, $2n$, of the speed, and with a constant vertical force g, parallel to the axis AP (Figure 5). Call the curve we seek AMB, and let $AP = x$, $PM = y$ and $AM = s$. If the altitude of M is v, then the resistance at M will be $\frac{v^n}{c^n}$, where v is altitude and c is speed. The combined effect of resistance and force makes

$$dv = g\,dx - \frac{v^n\,ds}{c^n}.$$

From earlier, we know that $2v\,dx\,ddy = g\,dy\,ds^2$. Holding dx constant leads to

$$ddy = \frac{ds\,dds}{dy} \quad \text{and} \quad v = \frac{g\,ds\,dy^2}{2\,dx\,dds},$$

and so

$$dv = \frac{g\,dy^2\,dds^2 + 2g\,ds^2\,dds^2 - g\,ds\,dy^2 d^3 s}{2dx\,dds^2}.$$

Substituting this in

$$dv = g\,dx - \frac{v^n\,ds}{c^n}$$

and applying some algebra yield

$$ds\,d^3 s - 3\,dds^2 = \frac{g^{n-1}\,ds^{n+1}\,dy^{2n-2}}{2^{n-1} c^n\,dx^{n-1}\,dds^{n-2}}.$$

Euler checks his work again by noting that "if the resistance of the medium is infinitely thin, it is transformed into a vacuum, in which case it makes $c = \infty$, and it becomes $ds\,d^3 s = 3\,dds^2$, the integral of which is $a\,dx\,dds = 3\,ds^3$. This, as we have seen, is the cycloid again."

Euler does three pages of integral substitutions and calculations to get some general forms for the solutions to this differential equation, then applies it to the special case with the resistance proportional to the square of the speed. He gets a formula relating x to the arc length s,

$$e^{s/c}(c - ac) = s - ax + c - ac,$$

where, for a change, the symbol "e" appears in its modern usage.

Euler expands the exponential on the left into a power series, substitutes $\frac{1-a}{a} = k$ and solves for x to get

$$x = s - \frac{ks^2}{1 \cdot 2 \cdot c} - \frac{ks^3}{1 \cdot 2 \cdot 3 \cdot c^2} - \frac{ks^4}{1 \cdot 2 \cdot 3 \cdot 4 \cdot c^3} - \text{etc.}$$

where k ought to be positive, or it will lead to the absurd situation that $x > s$. Euler is happy with this, because this series converges "*vehementer*," exceedingly or forcefully.

Euler notes that the curve descending the other direction, to *Am* rather than to *AM*, would be the same shape as the one he has found.

Euler now considers extending the curve beyond the point B. He draws a vertical axis *BD* (still Figure 5) and locates M so that $BQ = PC$ and calls that length u. He lets the arc length $MB = t$. This, he says, makes $s = c\,l\frac{1}{1-a} - t$, where, as usual, l denotes the natural logarithm function, and

$$x = \frac{c}{a}\,l\frac{1}{1-a} - c - u.$$

Substitutions lead to the equation $ce^{-t/c} = au - t + c$, or, as a differential equation, $t\,dt - au\,dt = ac\,du$. Here, as before, c is speed, and because it is the next symbol that he has not used yet, e is "e." As a series, this gives

$$au = \frac{t^2}{1 \cdot 2 \cdot c} - \frac{t^3}{1 \cdot 2 \cdot 3 \cdot c^2} - \frac{t^4}{1 \cdot 2 \cdot 3 \cdot 4 \cdot c^3} - \text{etc.}$$

This, Euler points out, is the same curve he found in 1729 and described in E-6 as the tautochrone in a resistant medium.

Finally, Euler makes a remark on a peculiar kind of tautochronal curve in a resistant medium (Figure 6.) His curve goes from a cusp at A, down to B, then up to a cusp at C. From there, it can continue to D, on the same level as B, forming another brachistochrone from C to D. This strange curve is a hint that Euler will soon have to develop a more sophisticated idea of functions.

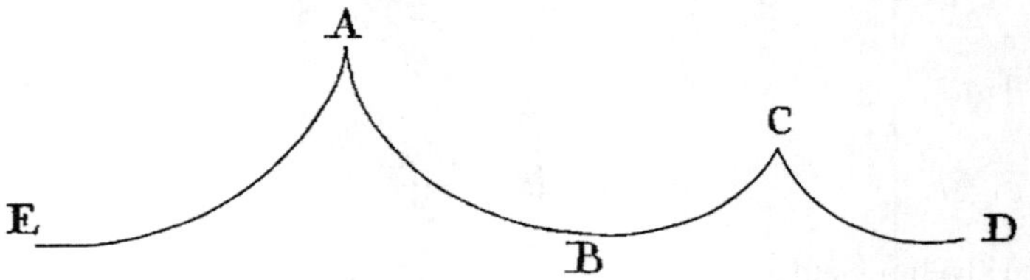

Fig. 6

In this paper, Euler has considerably extended two results of Huygens. We have seen him use difficult mathematics to attack physical problems, and we have watched as he checks his work whenever possible by making sure that the special cases of his results agree with known results. This paper is more important as a step in the sequence E-1, E-27, now E-42, later E-56 and E-99, and eventually leading to his landmark 1744 textbook on the calculus of variations, the *Methodus Inveniendi*.

17

E-43: De progressionibus harmonicis observationes*

Observations on harmonic progressions

In calculus we all learn about conditionally convergent series. One of the main points that is drilled into us is that we cannot rearrange the terms of a conditionally convergent series and expect the value of the series to remain the same. This paper may be one of the main reasons that fact is in the curriculum. As we will see, Euler rearranges series recklessly, for he was a man of his era, and at this point in his career, the dangers of this practice were as yet unknown. Only late in his career will Daniel Bernoulli begin to confront some of the paradoxes, as he saw them, of conditionally convergent series.

But, like the hero in an action movie, Euler never seems to falter in the face of danger. His fantastic manipulations almost never lead to false results. Impossible things seem to happen, but it almost always works out in the end.

So we shall suspend our modern sensibilities about convergence. Accept that if i is an infinite number, then ni is a bigger one. Enjoy Euler's fantastic manipulations, and marvel that, out of such chaos, truth can emerge.

This paper fits in a sequence of articles about harmonic series. We have seen Euler's earlier efforts on the subject in E-20 and E-25, and we will see more of it in E-47. Euler also includes a bit on the Basel Problem from E-41. Euler continued to work in the subject at least until E-583, written for the Petersburg Academy's journal, by then called the *Acta*, of 1781, two years before Euler's death in 1783, and published two years after his death. This paper is particularly noteworthy because it is Euler's first mention of what is now called the Euler-Mascheroni constant. That constant, approximately 0.577218, is now usually denoted by γ, but in this paper, Euler writes C.

To Euler, a progression or series, (they mean the same to him,) is harmonic if it is a "series of fractions the numerators of which are equal to each other and the denominators consist of an arithmetic progression." The general form of such a progression is

$$\frac{c}{a}, \frac{c}{a+b}, \frac{c}{a+2b}, \frac{c}{a+3b} \text{ etc.}$$

and the general term of index n is

$$\frac{c}{a+(n-1)b}.$$

Euler will use the observation that, three consecutive terms,

$$\frac{c}{a+b}, \frac{c}{a+2b}, \frac{c}{a+3b},$$

Comm. Acad. Sci. Imp. Petropol. 7 (1734/5) 1740, 150–161; *Opera Omnia* I.14, 87–100

have the property that "the differences between the extremes and the middle term are in proportion to the extremes." To some of his readers, this wording would have made more sense than symbolic formulas. To others, though, symbols would be more natural, so Euler interprets this for us as

$$\frac{c}{a+b} - \frac{c}{a+2b} : \frac{c}{a+2b} - \frac{c}{a+3b} = \frac{c}{a+b} : \frac{c}{a+3b}.$$

Euler finds it remarkable, as we still do today, that harmonic progressions always decrease, yet their sums "continued to infinity are always infinite." He proves this, and uses his proof to introduce some of his techniques for manipulating infinite series and infinite quantities.

Take i to be an infinite number, and continue the progression from the first term up to the term

$$\frac{c}{a+(i-1)b},$$

which is the term of index i. Now, consider the series that begins with the next term,

$$\frac{c}{a+ib}$$

and continues up to the term

$$\frac{c}{a+(ni-1)b}.$$

This is the term Euler says has exponent ni. Euler is using the words index and exponent interchangeably. It reminds us that the word "exponent" meant something "put out," encompassing subscripts, superscripts, indices and what we now call exponents.

Continuing his analysis, Euler notes that the progression has $(n-1)i$ terms and it decreases so the sum of the series that begins

$$\frac{c}{a+ib}$$

and ends

$$\frac{c}{a+(ni-1)b}$$

is less than $(n-1)i$ times the first term and greater than $(n-1)i$ times the last. That is to say, it is between

$$\frac{(n-1)ic}{a+ib} \quad \text{and} \quad \frac{(n-1)ic}{a+(ni-1)b}.$$

Now, Euler claims that "since i is an infinite number, the value a vanishes in each denominator," so the sum is between

$$\frac{(n-1)c}{nb} \quad \text{and} \quad \frac{(n-1)c}{b}.$$

"From this it is seen that this sum is finite, and consequently the sum of the proposed series $\frac{c}{a}$, $\frac{c}{a+b}$ etc. continued to infinity is infinitely large." Here, "finite" means not infinitesimal and bounded away from zero. Some readers will recognize this as an adaptation of the standard proof that the harmonic series diverges. That proof takes $c = b = 1$, $a = 0$ and $n = 2$.

Euler hopes to sum the progression between the terms of index i and index ni, and he plans to use that fact about ratios, so he needs three terms of a harmonic progression. He takes the first term to be

$$\frac{c}{a+ib}$$

and the third to be

$$\frac{c}{a+(ni-1)b}$$

and concludes that the middle term must be

$$\frac{c}{a+\frac{ni+i-1}{2}b}.$$

Euler does not state this explicitly, but the middle term of a harmonic progression is less than the mean of the terms, so the number of terms times the middle term is less than the sum of the terms. That is

$$\frac{(n-1)ic}{a+\frac{ni+i-1}{2}b}$$

is less than the sum of the terms, but because i is infinite, the value of this is much simpler,

$$\frac{2(n-1)c}{(n+1)b}.$$

Similarly, the sum will be less than the arithmetic mean of the extremes, which works out to be

$$\frac{(2a+(ni+i-1)b)c}{2(a+ib)(a+(ni-1)b)}$$

or, since i is infinite,

$$\frac{(n+1)c}{2nib}.$$

The sum is between

$$\frac{2(n-1)c}{(n+1)b} \quad \text{and} \quad \frac{(n+1)c}{2nib},$$

so Euler takes the "mean proportional" or geometric mean of these,

$$\frac{(n-1)c}{b\sqrt{n}}$$

as his estimate of the sum.

Euler turns to a slightly different progression,

$$\frac{c}{a}, \frac{c}{a+b}, \frac{c}{a+2^{\alpha}b} \text{ etc.},$$

where the general term is

$$\frac{c}{a+(n-1)^{\alpha}b}$$

and the infinite term is

$$\frac{c}{a+i^{\alpha}b}.$$

Using the same methods as before, the sum of the terms from index i to index ni will be less than

$$\frac{(n-1)c}{i^{\alpha-1}b}$$

and greater than

$$\frac{(n-1)c}{n^{\alpha}i^{\alpha-1}b}.$$

Euler uses a kind of convergence test here to conclude that if $\alpha > 1$, then the sum of these terms of the sequence will be equal to zero, and so the sum of the series is finite. Also, if $\alpha < 1$, then the sum is infinite, and if $\alpha = 1$, then the sequence is harmonic.

This is all we see of this progression, for Euler returns to his general harmonic progression. He sets the sum of the first i terms of the progression equal to s and notes that s "is determined by the quantities a, b, c and i. Let i grow by 1, and s will be augmented by the succeeding term $\frac{c}{a+ib}$. Thus

$$di : ds = 1 : \frac{c}{a+ib} \quad \text{or} \quad ds = \frac{c\,di}{a+ib}."$$

Euler is treating finite changes in infinite quantities exactly the same way he treats infinitesimal changes in finite quantities, so the rules of calculus apply. Integrating this last expression gives

$$s = C + \frac{c}{b}l(a+ib).$$

Something big is starting to happen here. In modern notation, this value C is

$$\lim_{k\to\infty}\left(\int_0^k \frac{c\,dx}{a+bx} - \sum_{n=0}^{k}\frac{c}{a+nx}\right).$$

This is the first inkling in this paper of the Euler-Mascheroni constant. We will get more details in a few paragraphs.

Returning to our infinite sums, if we take the difference between s evaluated at ni and s evaluated at i, then we get the sum from i to ni,

$$\frac{c}{b}l\frac{a+nib}{a+ib} = \frac{c}{b}l\,n$$

where l is the natural logarithm again, and where also, "a vanishes." But we have already estimated the value of the sum from i to ni, so we can deduce that

$$\frac{2(n-1)}{n+1} < l\,n < \frac{n^2-1}{2n}.$$

Euler concludes from this that if s is evaluated at i, then the result is $\frac{c}{b} l\,i$, which is smaller than i, but it is still infinite. This demonstrates that Euler had some idea of the rate of growth of functions that diverge. We now recognize that he did not have a modern understanding of limits and divergence, and he had not yet developed the concepts of functions necessary to deal with this any better than he did.

Euler returns to the basic harmonic series by taking $a = b = c = 1$. From what he has done before, he writes

$$l\,n = \left(1 + \frac{1}{2} + \frac{1}{3} + \cdots + \frac{1}{ni}\right) - \left(1 + \frac{1}{2} + \frac{1}{3} + \cdots + \frac{1}{i}\right).$$

Modern sensibilities are offended, for the right-hand side is an indeterminate form $\infty - \infty$. Euler doesn't know that, though, so he naively rearranges the terms to suit his needs, subtracting one term in the right-hand series from each nth term in the left-hand series. He writes

$$l\,n = 1 + \frac{1}{2} + \cdots + \left(\frac{1}{n} - 1\right) + \frac{1}{n+1} + \cdots + \left(\frac{1}{2n} - \frac{1}{2}\right) + \frac{1}{2n+1} + \cdots + \left(\frac{1}{3n} - \frac{1}{3}\right) + \text{etc.}$$

To make this clearer, Euler gives us series for logarithms from 2 to 6 that look like

$$l\,2 = 1 - \frac{1}{2} + \frac{1}{3} - \frac{1}{4} + \frac{1}{5} - \frac{1}{6} + \frac{1}{7} - \frac{1}{8} + \frac{1}{9} - \frac{1}{10} + \frac{1}{11} - \frac{1}{12} + \text{etc.}$$

This is reassuring, because this is a well-known result. For the logarithm of 4 he gives

$$l\,4 = 1 + \frac{1}{2} + \frac{1}{3} - \frac{3}{4} + \frac{1}{5} + \frac{1}{6} + \frac{1}{7} - \frac{3}{8} + \frac{1}{9} + \frac{1}{10} + \frac{1}{11} - \frac{3}{12} + \text{etc.}$$

Euler notes that the denominators simply increase one step at a time and that the numerators come in groups that follow the pattern $1, 1, \ldots, 1 - n$.

New series can be built from these. For example, $l\,4 = 2\,l\,2$, so Euler subtracts the series to get

$$\begin{aligned} 0 &= 2\,l\,2 - l\,4 \\ &= \left(2 - \frac{2}{2} + \frac{2}{3} - \frac{2}{4} + \text{etc.}\right) - \left(1 + \frac{1}{2} + \frac{1}{3} - \frac{2}{4} + \text{etc.}\right) \\ &= 1 - \frac{3}{2} + \frac{1}{3} + \frac{1}{4} + \frac{1}{5} - \frac{3}{6} + \text{etc.} \end{aligned}$$

He does a similar example starting with $l\,2 + l\,3 = l\,6$.

These series all converge, *sed admodum tarde*, "but very slowly." Euler shows how they can be made to converge faster by taking the groups, n terms at a time. His calculations parallel those at the end of E-25, and give him a list of series for $l\,\frac{n+1}{n}$. Turning the series around a little bit he gets

$$\begin{aligned} 1 &= l\,2 + \frac{1}{2} - \frac{1}{3} + \frac{1}{4} - \frac{1}{5} + \frac{1}{6} - \frac{1}{7} + \text{etc.} \\ \frac{1}{2} &= l\,\frac{3}{2} + \frac{1}{2 \cdot 4} - \frac{1}{3 \cdot 8} + \frac{1}{4 \cdot 16} - \frac{1}{5 \cdot 32} + \text{etc.} \\ \frac{1}{3} &= l\,\frac{4}{3} + \frac{1}{2 \cdot 9} - \frac{1}{3 \cdot 27} + \frac{1}{4 \cdot 81} - \frac{1}{5 \cdot 243} + \text{etc.} \\ \frac{1}{4} &= l\,\frac{5}{4} + \frac{1}{1 \cdot 16} - \frac{1}{3 \cdot 64} + \frac{1}{4 \cdot 256} - \frac{1}{5 \cdot 1024} + \text{etc.} \end{aligned}$$

and, finally, for the infinite number i,

$$\frac{1}{i} = l\frac{i+1}{i} + \frac{1}{2\cdot i^2} - \frac{1}{3\cdot i^3} + \frac{1}{4\cdot i^4} - \frac{1}{5\cdot i^5} + \text{etc.}$$

Euler adds all of these equations together. The left-hand side forms the harmonic series, summed to the term $\frac{1}{i}$. The logarithms on the right telescope, leaving just a term $l(i+1)$, and the remaining terms converge fairly quickly. Euler writes it as

$$\begin{aligned}1 + \frac{1}{2} + \frac{1}{3} + \cdots + \frac{1}{i} = l(i+1) &+ \frac{1}{2}\left(1 + \frac{1}{4} + \frac{1}{9} + \frac{1}{16} + \text{etc.}\right)\\ &- \frac{1}{3}\left(1 + \frac{1}{8} + \frac{1}{27} + \frac{1}{64} + \text{etc.}\right)\\ &+ \frac{1}{4}\left(1 + \frac{1}{16} + \frac{1}{81} + \frac{1}{256} + \text{etc.}\right)\\ &\text{etc.}\end{aligned}$$

Note that some of the series Euler evaluated in the Basel Problem paper, E-41, appear in the parentheses on the right. Euler can approximate the series on the right, though he does not say exactly how he does it, and he tells us

$$1 + \frac{1}{2} + \frac{1}{3} + \cdots + \frac{1}{i} = l(i+1) + 0,577218, \text{ approximately.}$$

The number on the right is the Euler-Mascheroni constant and gives a particular value to the constant of integration, C, that we saw earlier. Note that Euler gets it wrong in the sixth decimal place. To ten decimal places, it is 0.57721 56649. Euler will get it right two years later in E-47, where he gives the correct value to 15 decimal places.

We should pause to remark on Lorenzo Mascheroni (1750–1800). In 1790, Mascheroni published *Adnotationes ad calculum integrale Euleri*, where he explained and extended Euler's work on this constant. He gave the value to 32 decimal places, but 20 years later it was realized that only the first 19 of those decimal places were correct. The number is often called just "Euler's constant," but some people use that name to describe the number we call e.

Returning to the harmonic series, Euler notes that if its sum from i to $2i$ is $l\,2$ or $l\,\frac{2}{1}$, and if the sum from $2i$ to $3i$ is $l\,\frac{3}{2}$, etc., then by analogy, the sum from 1 to i can be said to be $l\frac{1}{0}$. "In this way we will obtain the following scheme which is not a little curious."

Series	$1 + \frac{1}{2} + \cdots + \frac{1}{i}$	$+ \cdots + \frac{1}{2i}$	$+ \cdots + \frac{1}{3i}$	$+ \cdots + \frac{1}{4i}$	$+ \cdots + \frac{1}{5i}$	etc.
Sums	$l\,\frac{1}{0}$	$l\,\frac{2}{1}$	$l\,\frac{3}{2}$	$l\,\frac{4}{3}$	$l\,\frac{5}{4}$	

Euler changes gears a bit and returns to some series that he had studied in E-25. He describes them as "more general" and "not simple harmonic progressions but combined with geometric ones." The general form is

$$s = \frac{cx}{a} + \frac{cx^2}{a+b} + \frac{cx^3}{a+2b} + \frac{cx^4}{a+3b} + \text{etc.}$$

Using calculations similar to what we saw in E-20 and E-25, Euler multiplies this by $bx^{(a-b)/b}$ and differentiates. There, he sees a geometric series, which he sums. Then he integrates again and solves for s to show that

$$s = \frac{c}{bx^{(a-b)/b}} \int \frac{x^{(a-b)/b}\,dx}{1-x}.$$

Similarly, Euler finds that if

$$t = \frac{fx^m}{g} + \frac{fx^{2m}}{g+h} + \frac{fx^{3m}}{g+2h} + \text{etc.},$$

then

$$t = \frac{fm}{hx^{m(g-h)/h}} \int \frac{x^{(mg-h)/h}\,dx}{1-x^m}.$$

Note that the expressions for t have a new variable m that makes t slightly different from s.

Euler uses these series and integral forms to rediscover and to extend some of his earlier results. He starts taking $a = b$ and $g = h$. Then he finds that

$$s = \frac{c}{b} \int \frac{dx}{1-x} = \frac{c}{b} l \frac{1}{1-x}$$

and

$$t = \frac{f}{h} \int \frac{mx^{m-1}\,dx}{1-x^m} = \frac{f}{h}\, l\, \frac{1}{1-x^m}$$

so that

$$s - t = l\,\frac{(1-x^m)^{f/h}}{(1-x)^{c/b}}.$$

"For this expression to be finite when $x = 1$, it ought to be true that $\frac{f}{h} = \frac{c}{b}$; so because of this, all these letters will be set $= 1$, and that makes"

$$s - t = l\frac{1-x^m}{1-x} = l\big(1 + x + x^2 + \cdots + x^{m-1}\big).$$

Recall that s and t are series. With our assumptions that $f = g = c = b = 1$, this expression $s - t$ is the difference between the two series

$$x + \frac{x^2}{2} + \frac{x^3}{3} + \frac{x^4}{4} + \frac{x^5}{5} + \text{etc.} \quad \text{and} \quad \frac{x^m}{1} + \frac{x^{2m}}{2} + \frac{x^{3m}}{3} + \text{etc.}$$

Taking $m = 2$, this makes

$$l(1+x) = x - \frac{x^2}{2} + \frac{x^3}{3} - \frac{x^4}{4} + \text{etc.}$$

Now setting $x = 1$, this gives the same series for the logarithm of 2 that we saw earlier. Likewise, taking $m = 3$ and then $x = 1$ produces the logarithm of 3, and so forth.

Next, Euler lets $h = 2g$, $a = b = g = 1$, $c = 1$, $f = 2$ and $m = 2n$ to get that the difference between the two series

$$x + \frac{x^2}{2} + \frac{x^3}{3} + \frac{x^4}{4} + \frac{x^5}{5} + \text{etc.} \quad \text{and} \quad \frac{2x^n}{1} + \frac{2x^{3n}}{3} + \frac{2x^{5n}}{5} + \text{etc.}$$

is

$$l\,\frac{1 - x^n}{(1-x)(1+x^n)}.$$

If $n = 2$ and $x = 0$ we find that

$$0 = 1 - \frac{3}{2} + \frac{1}{3} + \frac{1}{4} + \frac{1}{5} - \frac{3}{6} + \frac{1}{7} + \text{etc.},$$

another result we saw earlier.

Finally, Euler uses these techniques to sum what he calls "all irregular series of this type." His first example is

$$1 - \frac{2}{2} + \frac{1}{3} + \frac{1}{4} - \frac{2}{5} + \frac{1}{6} + \frac{1}{7} - \frac{2}{8} + \text{etc.}$$

He recognizes this as the difference between the two power series

$$x + \frac{x^2}{2} + \frac{x^3}{3} + \frac{x^4}{4} + \frac{x^5}{5} + \text{etc.}$$

and

$$\frac{3x^2}{2} + \frac{3x^5}{5} + \frac{3x^8}{8} + \text{etc.}$$

taken when $x = 1$. The first series sums to $l\frac{1}{1-x}$. Euler safely ignores the constant of integration. The second series sums to

$$\int \frac{3x\,dx}{1 - x^3},$$

which Euler expands as

$$l\frac{1}{1-x} + \frac{1}{2}l(x^2 + x + 1) + \frac{\sqrt{-3}}{2}l\frac{2x + 1 - \sqrt{-3}}{2x + 1 + \sqrt{-3}} - \frac{\sqrt{-3}}{2}l\frac{1 - \sqrt{-3}}{1 + \sqrt{-3}}.$$

He subtracts and sets $x = 1$ to evaluate the series to be

$$-\frac{1}{2}l\,3 + \frac{\sqrt{-3}}{2}l\,\frac{3 + \sqrt{-3}}{3 - \sqrt{-3}} - \frac{\sqrt{-3}}{2}l\,\frac{1 + \sqrt{-3}}{1 - \sqrt{-3}}.$$

Euler notes that the middle term here,

$$\frac{\sqrt{-3}}{2}l\,\frac{3 + \sqrt{-3}}{3 - \sqrt{-3}}$$

is "the periphery of the circle divided by $\sqrt{3}$, assuming its diameter = 1" and the last term is half of that. Knowing this, Euler evaluates the series to be approximately 0.3576.

Euler concludes the paper with one last example, which we omit, noting only that he gives an approximate answer again, and that example depends on his estimate of the Euler-Mascheroni constant.

In this paper, we have seen some of Euler's acrobatics with infinite series, and we have seen the first use of the Euler-Mascheroni constant, here denoted C, and misevaluated in the sixth decimal place. Moreover, we have seen how this fits into a sequence of articles about harmonic series and sequences. This topic, along with his work on mechanics in resistant media, forms one of Euler's main interests in his early years.

18

E-44: De infinitis curvis eiusdem generis seu methodus inveniendi aequationes pro infinitis curvis eiusdem generis*

On an infinity of curves of a given kind, or a method of finding equations for an infinity of curves of a given kind

174 DE INFINITIS CVRVIS

DE
INFINITIS CVRVIS
EIVSDEM GENERIS.
SEV
METHODVS INVENIENDI
AEQVATIONES PRO INFINITIS CVRVIS
EIVSDEM GENERIS.
AVCTORE
Leonh. Eulero.

From the title, the reader might not guess that this is an article about partial differential equations. As we have seen, some of Euler's work in differential equations has been motivated by specific problems in mechanics like E-1 on isochronal problems. Other times his motivation has been geometry, as in his work on reciprocal trajectories in E-3, E-5 and E-23. Still others have been based on applications in calculus of variations, as E-9, E-27 and E-42, or particular differential equations like the Riccati equation in E-31. This paper and its sequel, E-45, belong to a list of papers about differential equations in general.

Euler begins by addressing the ambiguity between parameters and variables in an equation. He notes that the equation $y^2 = ax$ produces infinitely many parabolic curves all having the same axis and vertex, one for each value of a. This symbol a is usually treated as a constant rather than as a variable. Most people in Euler's day call a a parameter, but Euler follows Jakob Hermann, the one who taught him the three-dimensional coordinate system he used in E-9, and he calls it a *modulus*. Euler plans to try to treat moduli as variables.

**Comm. Acad. Sci. Imp. Petropol.* 7 (1734/5) 1740, 174–189, 180–183; *Opera Omnia* I.22, 36–56

Euler says that if an algebraic equation is given relating variables x and z and involving a modulus a, then it determines a particular curve if a is held constant, but it gives "all curves" if a is made a variable. Euler really does seem to mean *algebraic* equations here, that is, equations built of sums, products, ratios and roots, and not involving any transcendental functions. It is not clear that Euler actually chose to restrict himself in this way, but rather it just did not occur to him that he was restricting himself. One's own times often make people blind to things that become completely obvious when viewed from a different time and place.

Suppose $z = \int P\,dx$, where P involves a, z and x. Then $dz = P\,dx$, in which expression a is to be considered as a constant. Euler works hard to explain that if a is considered a variable, then this last expression could be differentiated with respect to a, and also that if $dz = P\,dx$ is integrated, then the resulting expression might involve a function of a. All this leads to a conclusion that seems paradoxical when we first see it, that if a is a constant, then

$$dz = P\,dx$$

but if a is considered as a variable, then

$$dz = P\,dx + Q\,da.$$

Euler moves on to state a theorem: "If a quantity A composed of two variables t and u is differentiated first holding t constant, and then that differential is differentiated holding u constant and letting t be a variable, then the same result will occur if the order is reversed and A is first differentiated holding u constant and then that differential is differentiated holding t constant and letting u be a variable."

We recognize this as claiming that mixed partial derivatives are equal, regardless of the order of the differentiation. Today, this is the second thing we learn about partial derivatives when we encounter them in calculus class. There, we also learn about some continuity conditions that Euler does not yet know about.

The awkward wording and the lack of notation are dictated by Euler's times. He writes about differentials and not derivatives, so the very idea of partial derivatives and second order partial derivatives is more difficult to discuss. Our modern notations for partial derivatives have evolved over many years specifically so that it is easy to use them to write facts like this. Compare

$$\frac{\partial^2 f}{\partial x \partial y} = \frac{\partial^2 f}{\partial y \partial x}$$

or $f_{xy} = f_{yx}$ to the tools Euler had to make this same statement.

Before Euler offers a proof of his claim, he does an example using the formula $A = \sqrt{t^2 + nu^2}$, where the variables are t and u, while n is a modulus. Euler offers the following as a proof.

> Let A be a function of t and u. In place of t in a put $t + dt$ and call that B, and also, in place of u in A, put $u + du$, and call that C. Finally, put at the same time $t + dt$ in place of t and $u + du$ in place of u in A and call that D. This is the same as putting $u + du$ in place of u in B, and also the same as putting $t + dt$ in place of t in C. This done, the differential of A if t is held constant produces $C - A$. If $t + dt$ is put in place of t in $C - A$, it produces $D - B$, and so the differential will be
>
> $$D - B - C + A.$$

Reversing the order of things, put $t + dt$ in place of t in A to get B, so the differential in which only t varies will be $B - A$. In this differential, put $u + du$ in place of u to get $D - C$, and so the differential will again be

$$D - B - C + A,$$

which is the same as the other differential. Q. E. D.

Note that differentials are calculated by an algebraic substitution process and not by using any limit process. It is not how we do it today, and a modern reader may be suspicious of the technique. Henri Dulac, [D] editor of the volumes on differential equations of Euler's *Opera Omnia*, writes in the preface of Series I, Volume 22, "If certain reasonings of Euler appear incomplete, it will be because certain means of reasoning, little used today, were so common in the 18th Century that the authors were able to hope that their reasoning would be easily reconstructed by their readers. I have never come across a case where a claim of Euler was false, nor when in the course of a proof he indicated a condition was necessary without demonstrating it."

Remember that Euler considers $z = \int P\,dx$ as a function with modulus a. Then $dz = P\,dx + Q\,da$, and Euler wants to understand what Q is. Euler takes more differentials, $dP = A\,dx + B\,da$ and $dQ = C\,dx + D\,da$. Because of Euler's result about mixed partial derivatives, he knows that $C = B$, so $dQ = B\,dx + D\,da$, and finally, $Q = \int B\,dx$, and "in this integration, a is to be considered as a constant." So, if $B\,dx$ is integrable, Euler has the modular equation he is seeking.

On the other hand, if Euler cannot integrate $B\,dx$, he does not give up. Euler seems to have some specific examples in mind when he divides this hard-to-integrate situation into two cases, but he doesn't tell us what those examples are. In his first case, he supposes that the integral of $B\,dx$ depends on $\int P\,dx$, so that

$$\int B\,dx = a\int P\,dx + K$$

where K is an algebraic function of a and x. Now, because $\int P\,dx = z$, we get

$$\int B\,dx = az + K$$

and

$$dz = P\,dx + az\,da + K\,da.$$

Integrating this gives the modular equation Euler seeks.

On the other hand, perhaps the integral of $B\,dx$ cannot be reduced to depend on $\int P\,dx$. Then perhaps $\int B\,dx$ can be reduced to a form that does not involve a. Euler says this can lead to a case

$$\int P\,dx = h\int K\,dx$$

where h depends on a and K depends on x. Then

$$z = h\int K\,dx$$

and

$$dz = \frac{z\,dh}{h} + Kh\,dx.$$

If $\int B\,dx$ does not have one of these special forms, then the modular equation is not given by a first order differential equation, but one of higher degree. In the equation

$$dz = P\,dx + da \int B\,dx$$

Euler puts

$$dB = E\,dx + F\,da$$

and this makes the differential of $\int B\,dx$ equal to

$$B\,dx + da \int F\,dx,$$

by the same kinds of calculations that have led Euler to this point.

From $dz = P\,dx + da \int B\,dx$, Euler gets

$$\int B\,dx = \frac{dz}{da} - \frac{P\,dx}{da},$$

and, using this, he gets

$$ddz = P\,ddx + dP\,dx + \frac{dz\,dda}{da} - \frac{Pdx\,dda}{da} + Bda\,dx + da^2 \int F\,dx.$$

Now, if $\int F\,dx$ can be reduced to forms involving $\int B\,dx$ and $\int P\,dx$, then this leads to a representation of P. If the integral doesn't reduce, then Euler takes the differential of $\int F\,dx$ as $F\,dx + da \int H\,dx$ so that $dF = G\,dx + H\,da$, and he repeats the process. Eventually, he hopes that one integral can be re-written as forms involving previous integrals so that he will be able to substitute back and find his modular equation.

Euler offers examples, starting with the simplest. If P is a function of x alone and does not involve a, then call that function X so that $dz = X\,dx$, and this last equation does not involve a. Then, the equality of mixed partials gives $B = 0$, so $\int B\,dx = n$ and

$$z = \int X\,dx + na.$$

This means that the family of curves parameterized by the modulus a are just the curve given by $\int X\,dx$ translated vertically by the shift na.

Euler continues with a second example. He supposes that $P = AX$ where A is a function of a alone and X a function of x alone. Then $z = \int AX\,dx = A \int X\,dx$ and

$$A\,dz - z\,dA = A^2 X\,dx.$$

This is enough for Euler to find the modular equation he has been seeking, so he stops here.

Euler somewhat botches his third example, the case $P = A + X$, A and X as before (Figure 1). Then he says that $dz = A\,dx + X\,dx$ so that $z = Ax + \int X\,dx$. He should have said that

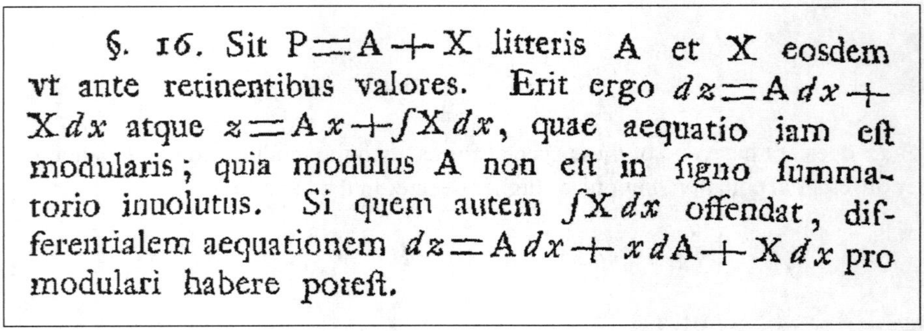
§. 16. Sit P $=$ A $+$ X litteris A et X eosdem vt ante retinentibus valores. Erit ergo $dz = Adx + Xdx$ atque $z = Ax + \int Xdx$, quae aequatio iam eſt modularis; quia modulus A non eſt in ſigno ſummatorio inuolutus. Si quem autem $\int Xdx$ offendat, differentialem aequationem $dz = Adx + xdA + Xdx$ pro modulari habere poteſt.

Fig. 1 (image from www.EulerArchive.org)

$z = \int (A + X)\,dx = Ax + \int X\,dx$ (because A does not depend on x). Nonetheless, he concludes correctly that

$$dz = A\,dx + x\,dA + X\,dx,$$

and this, as before, satisfies Euler as an answer.

Euler treats in similar ways the cases $P = AX + BY + CZ + \text{etc.}$, and $P = (A + X)^n$ where A, B and C are functions only of a, and X, Y, Z are functions only of x. This last example requires repeating the process and then substituting back, and, after n steps, results in

$$z = A^n x + \frac{n}{1}A^{n-1}\int X\,dx + \frac{n(n-1)}{1 \cdot 2}A^{n-2}\int X^2\,dx + \text{etc.}$$

Euler goes on to work examples with $X = bx^m$, where b may depend on a. First he takes $z = \int (A + bx^m)^n\,dx$ and then $z = \int x^m\,dx(A + bx^k)^n$.

Now, Euler wants to consider functions u that are what he calls "of no dimension in a and x." He gives examples

$$\frac{a}{x} \quad \text{and} \quad \frac{\sqrt{a^2 - x^2}}{a},$$

then says that he means the dimension of the numerator equals the dimension of the denominator. He seems to mean that the numerator and denominator have some kind of homogeneity property as well, but he does not describe that explicitly. Then, if $du = R\,dx + S\,dy$, Euler claims that

$$Rx + Sa = 0,$$

"for if in the function u we put $x = ay$, then every a will destroy itself [or, as we might say, "all the a's cancel"] and no other letters will remain except y and constants." So, in the differential of u, no differential elements will appear except dy. Now, because $x = ay$, we have $dx = a\,dy + y\,da$ and so $du = Ra\,dy + Ry\,da + S\,da$. Because du contains no differential elements except dy, it must be that $Ry + S = 0$, or, written another way using $x = ay$, $Rx + Sa = 0$, as claimed.

Euler applies this in the case u is a function of a and x of dimension m. He notes that $\frac{u}{x^m}$ is then a function of no dimension in a and x. He finds that $Rx + Sa = mu$ and that $a\,du = Ra\,dx - Rx\,da + mu\,da$.

Here, we have finished section 23 of the 41 sections in this article. Most of what remains is a litany of progressively more complicated cases that tax even the most faithful Euler enthusiast.

For example, in section 29, we consider $z = \int (P + Q + R)\,dx$, where P, Q, R are functions of dimensions $n - 1$, $m - 1$ and $k - 1$, respectively. There is a facet worthy of note in section 36, where Euler sets $R = e^K$, where e is the number whose logarithm equals 1. In one sense, this is Euler's first use of e to denote this number. Euler's "official" first use in a published work is in his two volume *Mechanica*, E-15 and E-16, published in 1736. This paper was published in 1740, but it had been written for the volume of the *Commentarii* for the years 1734 and 1735.

So, in this paper, we have seen that Euler derives an extremely important fact, that

$$\frac{\partial^2 z}{\partial x\,\partial y} = \frac{\partial^2 z}{\partial y\,\partial x},$$

but he does it for a purpose that we now might regard as just a bit silly, to find what $\frac{\partial z}{\partial y}$ might be if we know $\frac{\partial z}{\partial x}$. Maybe it is not silly, but the tool seems so much more important than the problem.

Another interesting facet of this paper is that Euler follows Hermann in his use of the term "modular." Usually, when Euler adopts a term, we still use the term. This is something of an exception. Though we still use the words "modular" and "modulus," the word "parameter," that Euler specifically declined to use, is the one we would usually use in this context.

References

[D] Dulac, Henri, Editor's preface to *Commentationes analyticae ad theoriam aequationum differentialium pertinentes, Opera Omnia* I.22, p. XII.

[E] Engelsman, Steven B., *Families of Curves and the Origins of Partial Differentiation*, North Holland Mathematics Studies **93**, Amsterdam, New York, Oxford, 1984.

[M] Miller, Jeff, *Earliest Uses of Various Mathematical Symbols*, `http://members.aol.com/jeff570/mathsym.html`, revision of December 29, 2001.

19

E-45: Additamentum ad dissertationem de infinitis curvis eiusdem generis*

Additions to the dissertation on infinitely many curves of a given kind

This "Addition" is mostly just a continuation of the previous paper, but it is quite noteworthy, indeed a landmark, in Euler's work. It marks the first use anywhere of the $f(x)$ notation for functions. As we saw in E-44 with Euler's result on mixed partial derivatives, important tools can be discovered and developed in pursuit of relatively unimportant goals. We will see that this is the case with $f(x)$ as well.

Rather than write one 35-page paper, Euler took the ideas that began in E-44 and broke them into two papers, a 20-page one and this, a fifteen-page paper. There is a certain loss of efficiency in doing this, for he spends the first two pages of E-45 recounting the main results of E-44. The two papers appeared one right after the other in the 1734/5 volume of the *Commentarii*. The sharp-eyed reader may have noticed that E-44 took pages 174–189 and 180–183 of that volume and wondered how that could happen. When the volume was being printed, it repeated the page numbers 180 to 189. Such errors by printers were not that uncommon at the time. It is ironic, though, that two papers, E-44 and E-45, whose content overlaps so much, also overlap in their page numbers.

As we mentioned, Euler begins with a review of his results from E-44. Recall that P is a function of x and a, and that $z = \int P\,dx$. Then $dz = P\,dx + Q\,da$, and Euler wants to know what Q could be. Euler further reminds us that he regards z as a family of curves, each one determined by a particular value of a.

Following the terminology in E-44, Euler considers the case when P is a function of degree -1 in the variables a and x. Then, he notes, that Px and Pa will be of "null dimension," and, from one of his results in E-44, $Px + Qa = 0$, so that $Q = -\frac{Px}{a}$. This makes

$$dz = P\,dx - \frac{Px\,da}{a}.$$

Euler looks for ways to integrate this, and he starts by factoring the function P and wondering if the part that remains, $dx - \frac{x\,da}{a}$, can be integrated. He notes that "if it is multiplied by $\frac{1}{a}$, then its integral will be $\frac{x}{a} + c$, denoting by c a constant quantity that does not depend on a." The reader might ask "what happened to the dx?" Euler is not at his clearest here, but saying that c does not depend on a does not rule out the possibility that it depends on x. The integral of dx becomes part of c.

Watch closely now as a pivotal event in the history of mathematics occurs.

> Because of this, if $f\left(\frac{x}{a} + c\right)$ denotes any kind of function of $\frac{x}{a} + c$, it makes $dx - \frac{x\,da}{a}$ integrable if it is multiplied by $\frac{1}{a} f\left(\frac{x}{a} + c\right)$. This value, given with its greatest generality, will be $P = \frac{1}{a} f\left(\frac{x}{a} + c\right)$ and $Q = -\frac{Px}{a}$.

Euler leaves it to his readers to check this, but, as Henri Dulac told us in a passage of the introduction to volume 22 of the *Opera Omnia*, it works out.

---

*Comm. Acad. Sci. Imp. Petropol. 7 (1734/5) 1740, 184–200; Opera Omnia I.22, 57–76

From this it is not clear *why* Euler needed a whole new kind of notation to describe these functions of $\frac{x}{a}+c$. Up to this point, he has done just fine writing things like "X is a function only of x and A is a function only of a." For what he does in this paragraph, he could just as well have taken F to be a function of $\frac{x}{a}+c$. Two sections later, though, he will want to apply this same function to the argument $\frac{x}{a^n}+c$. Later, he will take X to be a function of x, and A and E to be functions of a. Then he will apply this same function f to $x+A$, $X+A$, $\frac{X}{A}$ and even $x+\int E\,da$. With so many different arguments in front of him, Euler suddenly realizes that he does not need a new function name every time he has a new combination of variables. He can use just one function name and keep changing the arguments. Thus we see that the innovation was born to save symbols, to make the notation shorter. Only later will the notation help lead to a clearer understanding of the nature of functions themselves. Euler still does not distinguish clearly between curves and functions, and it will take a while before his new notation leads him to an understanding of the differences.

This is very much the climax of the paper, though Euler does not realize it at the time. He gets to this point only about one fifth of the way through, and little of note happens in the last four fifths. We will summarize the rest only briefly.

Euler notes that his function $f\left(\frac{x}{a}+c\right)$ will be a function of no dimensions in a and x, so he can use his theorems about such functions. He considers a number of special forms for P in addition to the ones we mentioned when we described his new function notation. For example, he finds that if

$$P = X + \frac{1}{a^n} f\left(\frac{a}{x^n}+c\right),$$

then

$$Q = A - \frac{nx}{a^{n+1}} f\left(\frac{a}{x^n}+c\right).$$

He studies forms that P might have if Q has the form PE, E a function only of a, or the form PY, where Y is a function only of x and does not involve a.

He moves on to a bit more generality and considers $Q = PR$, where R involves both x and a. This leads to a process that must be iterated until he arrives at a stage where R has one of his other forms. Many of his examples become complicated, and often they depend on earlier examples.

It is easy to see why this paper has been almost entirely forgotten today. Indeed, its content is rather obscure, and it addresses a problem that modern readers no longer find interesting. Still, it deserves its moment in the sun because it marks one of the turning points in modern mathematics, the origin of the $f(x)$ notation for functions.

20

E-48: Investigatio binarum curvarum, quarum arcus eidem abscissae respondents summam algebraicam constituant*

Investigation of two curves, the abscissas of which are corresponding arcs and the sum of which is algebraic

Surely, "Investigatio binarum curvarum" must be one of Euler's more forgotten papers. Even Henri Dulac, the editor of volume I.22 of the *Opera Omnia* on Differential Equations in which the article appears, says only that it deals with "questions of indeterminate analysis." Even in his own time, Euler can drum up little response to the paper. He mentions his results in a 1737 letter to Goldbach and again in one to Clairaut in 1741, but neither is moved to mention the paper in any of their replies.

Euler states his problem, numbering the conditions that must be satisfied. Required are

I. two algebraic curves
II. neither of which is rectifiable
III. the abscissas of which correspond to each other
IV. the sum is algebraic.

This is not clear, but Euler does not really have to be clear here. He is addressing an audience of his contemporaries, primarily Jakob Hermann and Johann Bernoulli, who have already worked on this problem, so they are already familiar with it. We are neither Bernoulli nor Hermann, so we need a bit of clarification.

Euler is looking for two curves. A curve can be the locus of solutions of an equation in two variables, but this time, Euler means a parametric curve, an abscissa and an application (ordinate) given as formulas involving some variable. Euler does not think of curves as being functions.

Euler does not want the curves to be algebraically rectifiable. Though the curves themselves are to be algebraic, their arc length integrals are not. Though we do not much dwell on it today, it is easy for algebraic integrands to give integrals that are not algebraic. Logarithms, arctangents and elliptic integrals are all non-algebraic integrals that arise from algebraic integrands.

Euler wants to add two curves together by adding their ordinates. If the curves were functions, this would be easy, but his curves are parameterized. So, to make sure that the ordinates he adds have the same abscissas, he has to make the abscissas match up. This is what he means when he says the abscissas must correspond. Today, we would say that the functions we add must have the same domain.

Finally, when Euler adds his ordinates, he wants to get a sum that is algebraic and rectifiable.

Though this does not seem at all like a modern problem, immediately after stating the problem, Euler makes some modern remarks. He shows that curves can be found satisfying any three of the four conditions, though eliminating the third condition presents problems that Hermann had been able to overcome "with difficulty." Euler's challenge is to satisfy all four conditions simultaneously.

---

*Comm. Acad. Sci. Imp. Petropol. 8 (1736) 1741, 23–29; *Opera Omnia* I.22, 76–82

We begin using formulas Euler gave in E-23 without describing how he had found them. Here, he mentions in passing that "they are obtained with a great deal of difficulty." There, he used these formulas to find algebraic curves with algebraic arc length integrals. Here, he is going to squeeze out curves that are algebraic but that have non-algebraic arc lengths. He defines two curves, A and B as follows:

	Curve A		Curve B
abscissa	$\dfrac{(dP^2 - dQ^2)^{3/2}}{dQ\,ddP - dP\,ddQ}$	abscissa	$\dfrac{(dp^2 - dq^2)^{3/2}}{dq\,ddp - dp\,ddq}$
ordinate	$P + \dfrac{dQ(dP^2 - dQ^2)}{dQ\,ddP - dP\,ddQ}$	ordinate	$p + \dfrac{dq(dp^2 - dq^2)}{dq\,ddp - dp\,ddq}$
arc length	$Q + \dfrac{dP(dP^2 - dQ^2)}{dQ\,ddP - dP\,ddQ}$	arc length	$q + \dfrac{dp(dp^2 - dq^2)}{dq\,ddp - dp\,ddq}$

Here, he wants the quantities p, P, dq and dQ to be algebraic, so that the abscissas and the ordinates will be algebraic, but the quantities Q and q are to be transcendental so that the arc length will not be algebraic. He further wants the quantity $Q + q$ to be algebraic.

Now he must make the abscissas equal to each other. To do this, he needs conditions relating P and Q to p and q. He sets

$$\frac{(dP^2 - dQ^2)^{3/2}}{dQ\,ddP - dP\,ddQ} = \frac{(dp^2 - dq^2)^{3/2}}{dq\,ddp - dp\,ddq}$$

and supposes that $dQ = R\,dP$ and $dq = r\,dp$. Note that this gives Q and q in terms of R, P, r and p. Substitution gives

$$\frac{(1 - R^2)^{3/2}\,dP}{-dR} = \frac{(1 - r^2)^{3/2}\,dp}{-dr}.$$

Solving this for dP gives

$$dP = \frac{(1 - r^2)^{3/2}\,dR\,dp}{(1 - R^2)^{3/2}\,dr}.$$

Now, for P to be algebraic, this quantity must be algebraic and "integrable." Note that to Euler, "integrable" means that the expression has a closed-form integral involving elementary functions and that he can find it. Riemann and other more modern mathematicians have not yet arrived on the scene.

Since R and r are also algebraic, it follows that

$$\frac{(1 - r^2)^{3/2}\,dR}{(1 - R^2)^{3/2}\,dr}$$

must also be algebraic. Set this last quantity equal to T, so that $dP = T\,dp$, so, with Euler's first explicit use of integration by parts, it follows that $P = Tp - \int p\,dT$. Now, set $N = \int p\,dT$, so that

$$p = \frac{dN}{dT} \quad \text{and} \quad P = \frac{T\,dN}{dT} - N.$$

So, the abscissas will correspond if p and P satisfy these conditions. Euler has taken care of three of his four conditions. He now turns to making the sum algebraic, and that happens if $Q+q$ is algebraic. He needs to find conditions on N that will guarantee that $Q+q$ will be algebraic. He starts with

$$\begin{aligned} Q+q &= \int R\,dP + \int r\,dp \\ &= RP + rp - \int P\,dR - \int p\,dr. \end{aligned}$$

Let $M = \int P\,dR + \int p\,dr$ so that

$$P = \frac{dM - p\,dr}{dR} \quad \text{and} \quad Q+q = RP + rp - M.$$

Note that M is algebraic. Substituting the values for P and p in terms of T and N into this gives

$$\frac{T\,dN\,dR}{dT} - N\,dR + \frac{dN\,dr}{dT} = dM.$$

Integrating yields

$$M = \frac{TN\,dR}{dT} + \frac{N\,dr}{dT} - \int N\left(dR + d\cdot\frac{T\,dR}{dT} + d\cdot\frac{dR}{dT}\right).$$

Let the integral in this last equation be equal to u, so that

$$N = \frac{du}{dR + d\cdot\frac{T\,dr}{dT} + d\cdot\frac{dr}{dT}}$$

where R, r and u can be any algebraic quantities.

Euler reviews for us. If R, r and u depend on some variable z, then he knows that

$$T = \frac{(1-r^2)^{3/2}\,dR}{(1-R^2)^{3/2}\,dr}$$

and that T also depends on z. Knowing T, R and r, Euler can find N. Then, knowing T, N, R, r and u he can get M as well. He knows P and p in terms of T and N, and finally, he remembers that $Q+q = RP + rp - M$, so $Q+q$ is algebraic.

Since he also knows Q and q in terms of P, p, R and r, the problem is solved. In particular

$$q = rp - \int p\,dr = rp - \int \frac{dr\,dN}{dT} \quad \text{and} \quad Q = RP - M + \int \frac{dr\,dN}{dT}.$$

Contrary to his usual practice, Euler does not do any examples here. He does do part of a couple of examples in his letter to Goldbach on 23 July 1737, but those examples amount to little more than suggesting some possible values for R, r and u. He does not finish them.

Euler hasn't quite reached the end of this paper. He notes that both Q and q depend on an integral $\int \frac{dr\,dN}{dT}$, and he suggests that one could ask that Q and q satisfy certain other properties. For example, one could try to make their arc length integrals be of certain forms by making $\int \frac{dr\,dN}{dT}$ be of different forms.

Finally, he says that if one wanted the difference of the two ordinates to be algebraically rectifiable instead of the sum, the calculations would be "*prolixissimos*," very, very long.

So, we come to the end of what seems to be one of Euler's less distinguished papers. He provides little motivation, and he works out no examples. At least it is short.

Interlude: 1735

World events

Horseman and silversmith Paul Revere, and the second US President John Adams were both born in 1735.

In Euler's life

Euler suffered a life-threatening fever in 1735. Calinger [C] suggests that this fever was the beginning of Euler's eye problems. Euler hid this illness from his family and friends and claimed that his eye problems began in 1738 due to eyestrain.

Starting about 1735, Euler was more often assigned particular scientific projects in magnetism, machines, ship construction and the writing of textbooks. He was appointed the Director of the Geography Section of the Academy, with the specific task of working with Delisle to map the Russian Empire.

Euler's younger brother Johann Heinrich, a painter, moved from Basel to live with Euler's family in St. Petersburg.

Euler's other work

Euler wrote two astronomical articles in 1735: E-37, De motu planetarum et orbitarum determinatione, "On the determination of the motion of planets and orbits," and E-50, Methodus computandi aequationem meridiei, "A method for computing the equation of the meridian." He also wrote a mechanics article, E-40, De minimis oscillationibus corporum tam rigidorum quam flexibilium, "On the smallest vibration of bodies both rigid and flexible."

Euler's mathematics

Euler wrote seven works of mathematics in 1735. The first to see publication was Volume 1 of his arithmetic textbook, E-17, known as the "Rechen-Kunst." He also discovered his famous solution to the Basel Problem, E-41, De summis serierum reciprocarum, "On the sum of series of reciprocals." Two other papers were about series, E-46, Methodus universalis serierum convergentium summas quam proxime inveniendi, "A universal method for finding sums which approximate convergent series," and E-47, Inventio summae cuiusque seriei ex dato termino generali, "Finding the sum of a series from a given general term." He wrote the perplexing article, E-51, De constructione aequationum ope motus tractorii, aliisque ad methodum tangentium inversam pertinentibus, "On the solution of equations from the motion of dragging, and also pertaining to the methods of inverse tangents." There was a paper on the boundary between elliptic integrals and differential equations, E-52, Solutio problematum rectificationem ellipsis requirentium, "Solution of a problem requiring the rectification of an ellipse." Finally, Eneström believes that in 1735 Euler wrote E-53, Solutio problematis ad geometriam situs pertinentis, "Solution of a problem relating to the geometry of position." This paper is about the Königsburg Bridge problem, and some sources say it was written in 1736. For consistency, we will follow Eneström.

References

[C] Calinger, Ronald, "Leonhard Euler: The First St. Petersburg Years (1727–1741)," *Historia Mathematica* 23 (1996) 121–166.

21

E-41: De summis serierum reciprocarum*

On sums of series of reciprocals

Progress in mathematics has always been motivated by problems. Progress in 20th Century mathematics was often measured against Hilbert's 23 problems. Through the 18th and 19th Centuries, the scientific academies posed prize problems to guide mathematical research. Now we have the seven Millennium Prize Problems, with the Clay Mathematics Institute offering a million dollars each for their solutions. In Euler's early years, there were only a few great problems waiting to be solved. Fermat's problems weren't taken very seriously yet. They were only a few decades old. Euler would eventually solve most of them anyway, and he would give a partial solution to the "Last" one as well, giving proofs in the cases $n = 3$ and $n = 4$. The other great problems of the 17th Century, the tractrix, DeBeaune's problem and other problems in inverse tangents had been solved with the invention of calculus.

The biggest prize that remained was the so-called Basel Problem, to find the sum of the reciprocals of the square numbers, that is, to evaluate the series $1 + \frac{1}{4} + \frac{1}{9} + \frac{1}{16} + \frac{1}{25} +$etc. The problem had been around at least since the time of Pietro Mengoli, (1625–1686). It got its name from a frustrated Jakob Bernoulli, in Basel, who wrote, "If anyone finds and communicates to us that which thus far has eluded our efforts, great will be our gratitude."

Jakob Bernoulli died in 1706, so he would not see Euler's efforts with the problem.

Euler began his assault on the Basel problem in 1730 in E-20, where he gave a six decimal place approximation to the value of the sum. For the next four years, he was silent on the subject, mentioning it neither in his papers nor in his letters. Then in 1734, he suddenly announces he has solved the problem that first makes him famous and launches him into the scientific elite of his times.

What could be better than solving the biggest mathematics problem in the world? Euler finds an even bigger result that contains the Basel problem as a special case. Euler sets out to sum the reciprocals of any powers, squares, cubes, whatever. He does not solve the whole extended problem, but he solves more than the Basel Problem.

He starts with the Basel problem. In the last four years, he has improved his six decimal place estimate for the value of the sum, using the same methods he used in E-20, to

$$1{,}64493\,40668\,48226\,4364$$

(still using a comma where Americans would use a decimal point), and, apparently by trial and error, he "found the sextuple of the sum of this series to be equal to the square of the periphery of a circle whose diameter is 1." In other words, the sum is $\frac{\pi^2}{6}$. He continues that he has found, by means he does not describe, that the sum of the reciprocals of the fourth powers, $1 + \frac{1}{16} + \frac{1}{81} + \frac{1}{256} + \frac{1}{625} +$etc. "also depends on the quadrature of the circle. The sum of this, multiplied by 90, gives the biquadrate

**Comm. Acad. Sci. Imp. Petropol.* 7 (1734/5) 1740, 123–134; *Opera Omnia* I.14, 73–87

of the periphery of the circle whose diameter is 1," that is $\frac{\pi^4}{90}$, "and by similar reasoning, for the subsequent series in which the exponents are even powers, I have been able to determine the sums."

As always, we will try to be faithful to Euler's notation in our exposition. Readers who appreciate the efficiency and economy of modern notation should compare this presentation with Dunham's presentation [D] of the same material beginning on page 45 of *Euler: The Master of Us All.*

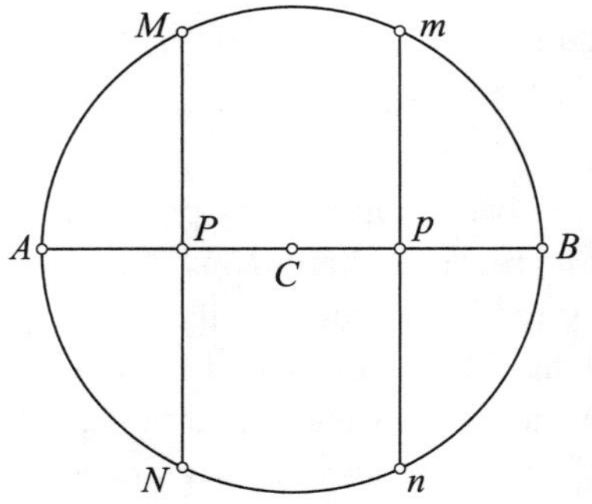

Fig. 1

For the first time that we have seen in Euler's writings, he introduces some trigonometric functions. In Figure 1, *AMBNA* is a circle of radius 1, with center C. If $AM = s$, then $PM = y = \sin s$ and $CP = x = \cos s$.

Euler knows series (but he doesn't use "!" for factorials), so

$$y = s - \frac{s^3}{1\cdot 2\cdot 3} + \frac{s^5}{1\cdot 2\cdot 3\cdot 4\cdot 5} - \frac{s^7}{1\cdot 2\cdot 3\cdot 4\cdot 5\cdot 6\cdot 7} + \text{etc.}$$

and

$$x = 1 - \frac{s^2}{1\cdot 2} + \frac{s^4}{1\cdot 2\cdot 3\cdot 4} - \frac{s^6}{1\cdot 2\cdot 3\cdot 4\cdot 5\cdot 6} + \text{etc.}$$

From the sine series, Euler writes

$$0 = 1 - \frac{s}{y} + \frac{s^3}{1\cdot 2\cdot 3y} - \frac{s^5}{1\cdot 2\cdot 3\cdot 4\cdot 5y} + \text{etc.}$$

and factors the right-hand side as if it were a polynomial. He gets

$$1 - \frac{s}{y} + \frac{s^3}{1\cdot 2\cdot 3y} - \frac{s^5}{1\cdot 2\cdot 3\cdot 4\cdot 5y} + \text{etc.} = \left(1 - \frac{s}{A}\right)\left(1 - \frac{s}{B}\right)\left(1 - \frac{s}{C}\right)\left(1 - \frac{s}{D}\right)\text{etc.}$$

For now, he doesn't worry about what these values A, B, C, D, etc., might be.

It will surprise nobody that Euler's next step is to match terms on the right with terms on the left. Matching terms that contain s gives

$$\frac{1}{y} = \frac{1}{A} + \frac{1}{B} + \frac{1}{C} + \frac{1}{D} + \text{etc.}$$

Matching terms that contain s^2, which equals 0 on the left, would be, as he said, "equal to the sum of factors taken two at a time from the series $\frac{1}{A}$, $\frac{1}{B}$, $\frac{1}{C}$, $\frac{1}{D}$ etc. Also, $-\frac{1}{1\cdot 2\cdot 3y}$ is the sum of the terms of the series taken three at a time. And in the same way, the sum of terms taken four at a time is 0, and the sum of factors taken five at a time is $+\frac{1}{1\cdot 2\cdot 3\cdot 4\cdot 5y}$, and so forth."

Now, referring to Figure 1, Euler takes $AM = A$ (using the same symbol, A, to denote two different things, a point and an arc length) the smallest arc that has its sine equal to $PM = y$. Also, he takes p to be the semiperiphery of a circle of radius 1. He still hasn't decided to call this value π. Euler notes that, by the periodicity of the sine function, the other arcs with the same sine are

$$A,\ p-A,\ 2p+A,\ 3p-A,\ 4p+A,\ 5p-A,\ 6p+A,\ \text{etc.}$$

and

$$-p-A,\ -2p+A,\ -3p-A,\ -4p+A,\ -5p-A,\ \text{etc.}$$

This gives Euler his values for A, B, C, D, etc., so he says that the terms of the sequence

$$\frac{1}{A},\ \frac{1}{p-A},\ \frac{1}{-p-A},\ \frac{1}{2p+A},\ \frac{1}{-2p+A},\ \frac{1}{3p-A},\ \frac{1}{-3p-A},\ \frac{1}{4p+A},\ \frac{1}{-4p+A}\ \text{etc.}$$

taken one at a time will sum to $\frac{1}{y}$, taken two at a time will sum to zero, taken three at a time will sum to $\frac{-1}{1\cdot 2\cdot 3y}$, and so forth.

Now we need some facts about coefficients. If $a+b+c+d+e+f+\text{etc.}$ is a series with sum $=\alpha$, and if the sum of the terms taken two at a time $=\beta$, taken three at a time $=\gamma$, four at a time $=\delta$, etc., then the sum of squares of the individual terms is

$$a^2+b^2+c^2+d^2+\text{etc.}=\alpha^2-2\beta$$

and

$$a^3+b^3+c^3+d^3+\text{etc.}=\alpha^3-3\alpha\beta+3\gamma$$

and the sum of the fourth powers is

$$=\alpha^4-4\alpha^2\beta+4\alpha\gamma+2\beta^2-4\delta.$$

For convenience, Euler denotes the sum of the terms by P, the sum of the squares by Q, of the cubes by R, fourth powers (which he calls here "quadroquatorum", or squared squares) by S, fifth by T and sixth by V. Then Euler knows some formulas of Newton:

$$\begin{aligned}
P &= \alpha \\
Q &= P\alpha-2\beta \\
R &= Q\alpha-P\beta+3\gamma \\
S &= R\alpha-Q\beta+P\gamma-4\delta \\
T &= S\alpha-R\beta+Q\gamma-P\delta+5\varepsilon \\
&\text{etc.}
\end{aligned}$$

We now apply these facts to the sequence that interests Euler,

$$\frac{1}{A},\ \frac{1}{p-A},\ \frac{1}{-p-A},\ \frac{1}{2p+A},\ \frac{1}{-2p+A},\ \frac{1}{3p-A},\ \frac{1}{-3p-A},\ \frac{1}{4p+A},\ \frac{1}{-4p+A}\ \text{etc.}$$

Ignoring the fact that the first series might not converge, Euler tells us that

$$\begin{aligned} \alpha &= \frac{1}{y} \\ \beta &= 0 \\ \gamma &= \frac{-1}{1\cdot 2\cdot 3y} \\ \delta &= 0 \\ \varepsilon &= \frac{+1}{1\cdot 2\cdot 3\cdot 4\cdot 5y} \\ \zeta &= 0 \\ &\text{etc.} \end{aligned}$$

and so

$$\begin{aligned} P &= \frac{1}{y} \\ Q &= \frac{P}{y} = \frac{1}{y^2} \\ R &= \frac{Q}{y} - \frac{1}{1\cdot 2y} \\ S &= \frac{R}{y} - \frac{P}{1\cdot 2\cdot 3y} \\ T &= \frac{S}{y} - \frac{Q}{1\cdot 2\cdot 3y} + \frac{1}{1\cdot 2\cdot 3\cdot 4y} \\ V &= \frac{T}{y} - \frac{R}{1\cdot 2\cdot 3y} + \frac{P}{1\cdot 2\cdot 3\cdot 4\cdot 5y} \\ W &= \frac{V}{y} - \frac{S}{1\cdot 2\cdot 3y} + \frac{Q}{1\cdot 2\cdot 3\cdot 4\cdot 5y} - \frac{1}{1\cdot 2\cdot 3\cdot 4\cdot 5\cdot 6y} \end{aligned}$$

Euler makes his next step a bit awkwardly. "We put the sine, $PM = y$ equal to the radius, that is $y = 1$. That makes the minimum arc, A, the sine of which is 1, equal to the quarter part of a periphery, that is $= \frac{1}{2}p$, or, denoting by q the quarter part of the periphery, that makes $A = q$ and $p = 2q$. The sequence above therefore becomes"

$$\frac{1}{q},\ \frac{1}{q},\ \frac{-1}{3q},\ \frac{-1}{3q},\ \frac{1}{5q},\ \frac{1}{5q},\ \frac{-1}{7q},\ \frac{-1}{7q},\ \frac{1}{9q},\ \frac{1}{9q}\ \text{etc.}$$

This sums to

$$\frac{2}{q}\left(1 - \frac{1}{3} + \frac{1}{5} - \frac{1}{7} + \frac{1}{9} - \frac{1}{11} + \text{etc.}\right).$$

But this sum is the definition of P. Because $P = \frac{1}{y}$ and because $y = 1$, this gives

$$1 - \frac{1}{3} + \frac{1}{5} - \frac{1}{7} + \frac{1}{9} - \frac{1}{11} + \text{etc.} = \frac{p}{4}.$$

In modern notation, this is $\frac{\pi}{4}$. This fact was already well known, and Euler attributes it to Leibniz. Euler probably included this observation just to check that his calculations had not led him astray, or it may have been to add to the suspense, for the golden ring is in sight.

Now, Euler turns to the sum of the squares of his sequence, which he writes with a wayward "+" sign at the beginning:

$$+\frac{1}{q^2}+\frac{1}{q^2}+\frac{1}{9q^2}+\frac{1}{9q^2}+\frac{1}{25q^2}+\frac{1}{25q^2}+\text{etc.}=\frac{2}{q^2}\left(\frac{1}{1}+\frac{1}{9}+\frac{1}{25}+\frac{1}{49}-\text{etc.}\right).$$

This sum equals Q. However, we already know that $Q=\frac{1}{y^2}$ and that $y=1$, so $Q=1$. A bit of algebra leads to

$$1+\frac{1}{9}+\frac{1}{25}+\frac{1}{49}+\text{etc.}=\frac{q^2}{2}=\frac{p^2}{8}.$$

For a moment, this looks great. A second glance reveals that this is the wrong series! Euler has summed the reciprocals of the *odd* squares, not all of the squares! Besides that, he got the answer $\frac{\pi^2}{8}$ and we were expecting an answer $\frac{\pi^2}{6}$. What is going on here?

Euler only tells us that the series he wants, the Basel problem, is one third larger than this sum, and that would be $\frac{p^2}{6}$. Indeed, this is the solution to the Basel problem but Euler does little to tell us why it is correct. We can fill the gap.

Euler has summed the reciprocals of the odd squares. All of the even ones are missing. Because every even number is a power of 2 times an odd number, every even square will be a power of four times an odd square. Knowing this and using modern notation, we can rearrange series (and this time it's legal, because every series involved is absolutely convergent) as follows:

$$\begin{aligned}\sum_{n=1}^{\infty}\frac{1}{n^2}&=\sum_{n\text{ odd}}\frac{1}{n^2}+\frac{1}{4}\sum_{n\text{ odd}}\frac{1}{n^2}+\frac{1}{4^2}\sum_{n\text{ odd}}\frac{1}{n^2}+\text{etc.}\\&=\left(1+\frac{1}{4}+\frac{1}{4^2}+\text{etc.}\right)\sum_{n\text{ odd}}\frac{1}{n^2}\\&=\frac{4}{3}\frac{\pi^2}{8}.\end{aligned}$$

Euler was right, and the problem is solved.

Of course, Euler doesn't stop there. He has a lot of information in those values P, Q, R, S, etc., that he has not yet used. First, he evaluates the formulas above in the case $y=1$. They are

$$\begin{aligned}P&=1\\Q&=1\\R&=\frac{1}{2}\\S&=\frac{1}{3}\\T&=\frac{5}{24}\\V&=\frac{2}{15}\end{aligned}$$

$$W = \frac{61}{720}$$
$$X = \frac{17}{315}$$
etc.

The fact that $R = \frac{1}{2}$ tells us that

$$\frac{2}{q^3}\left(1 - \frac{1}{3^3} + \frac{1}{5^3} - \frac{1}{7^3} + \frac{1}{9^3} - \text{etc.}\right) = \frac{1}{2}.$$

This is equivalent to a fact about the alternating sum of reciprocals of odd cubes:

$$1 - \frac{1}{3^3} + \frac{1}{5^3} - \frac{1}{7^3} + \frac{1}{9^3} - \text{etc.} = \frac{p^3}{32}.$$

The same kind of calculation, starting with $S = \frac{1}{3}$, leads to

$$1 + \frac{1}{3^4} + \frac{1}{5^4} + \frac{1}{7^4} + \frac{1}{9^4} + \text{etc.} = \frac{p^4}{96}.$$

This is a fact about an even power of odd numbers, and, as before, it can be extended to a fact about even powers of all integers. In this case, the interesting sum is $1/15$ more than this sum, giving

$$1 + \frac{1}{2^4} + \frac{1}{3^4} + \frac{1}{4^4} + \frac{1}{5^4} + \frac{1}{6^4} + \text{etc.} = \frac{p^4}{90}.$$

Euler continues like this, up to eighth powers, giving values for alternating sums for odd powers of odd numbers and for non-alternating sums for odd numbers and for all numbers for even powers, and he tells us that the same methods find values for all higher exponents as well.

Euler has solved much more than the Basel Problem, but he isn't finished yet. These calculations began "put the sine, $PM = y$ equal to the radius, that is $y = 1$." Euler now tries other values of y, beginning with $y = \frac{1}{\sqrt{2}}$, which corresponds to an arc length $A = \frac{1}{4}p$, a 45° angle. This makes $P = \frac{1}{y} = \sqrt{2}$. P is the sum of the reciprocals of all the angles with sine equal to y, so we get the sum

$$\frac{4}{p} + \frac{4}{3p} - \frac{4}{5p} - \frac{4}{7p} + \frac{4}{9p} + \frac{4}{11p} - \text{etc.} = P = \sqrt{2}.$$

Here, we note that Euler would be scolded today for arbitrarily choosing an order in which to sum a conditionally convergent series, but as we have mentioned before, in Euler's time, the pitfalls of such manipulations were not yet well understood.

This series transforms into

$$\frac{p}{2\sqrt{2}} = 1 + \frac{1}{3} - \frac{1}{5} - \frac{1}{7} + \frac{1}{9} + \frac{1}{11} - \frac{1}{13} - \frac{1}{15} + \text{etc.}$$

Euler notes that this result was known both to Leibniz and to Newton.

Euler continues to reprise known results, as he shows that when $y = \frac{1}{\sqrt{2}}$, $Q = 2$, and that leads to his earlier discovery that

$$1 + \frac{1}{9} + \frac{1}{25} + \frac{1}{49} + \text{etc.} = \frac{p^2}{8}.$$

In the same manner, taking $y = \frac{\sqrt{3}}{2}$, corresponding to a 60° angle leads to a second calculation that

$$1 + \frac{1}{4} + \frac{1}{9} + \frac{1}{16} + \frac{1}{25} + \text{etc.} = \frac{p^2}{6}.$$

Finally, Euler takes $y = 0$. He recognizes the problems with this argument, because the series from which this all began

$$0 = 1 - \frac{s}{y} + \frac{s^3}{1 \cdot 2 \cdot 3y} - \frac{s^5}{1 \cdot 2 \cdot 3 \cdot 4 \cdot 5y} + \text{etc.}$$

has division by zero in almost every term. Euler does not despair. First, he goes back to his sine series,

$$y = s - \frac{s^3}{1 \cdot 2 \cdot 3} + \frac{s^5}{1 \cdot 2 \cdot 3 \cdot 4 \cdot 5} - \frac{s^7}{1 \cdot 2 \cdot 3 \cdot 4 \cdot 5 \cdot 6 \cdot 7} + \text{etc.}$$

Here, he takes $y = 0$ and gets

$$0 = s - \frac{s^3}{1 \cdot 2 \cdot 3} + \frac{s^5}{1 \cdot 2 \cdot 3 \cdot 4 \cdot 5} - \frac{s^7}{1 \cdot 2 \cdot 3 \cdot 4 \cdot 5 \cdot 6 \cdot 7} + \text{etc.}$$

Now, he divides by s to get

$$0 = 1 - \frac{s^2}{1 \cdot 2 \cdot 3} + \frac{s^4}{1 \cdot 2 \cdot 3 \cdot 4 \cdot 5} - \frac{s^6}{1 \cdot 2 \cdot 3 \cdot 4 \cdot 5 \cdot 6 \cdot 7} + \text{etc.}$$

and says that the roots of the right-hand side must be

$$p,\ -p,\ +2p,\ -2p,\ 3p,\ -3p,\ \text{etc.}$$

(with no explanation for that one lonesome plus sign there). This tells him that the factors of the right-hand side are

$$1 - \frac{s}{p},\ 1 + \frac{s}{p},\ 1 - \frac{s}{2p},\ 1 + \frac{s}{2p}\ \text{ etc.}$$

Grouping, and using difference of squares, Euler realizes that

$$\begin{aligned} 1 &- \frac{s^2}{1 \cdot 2 \cdot 3} + \frac{s^4}{1 \cdot 2 \cdot 3 \cdot 4 \cdot 5} - \frac{s^6}{1 \cdot 2 \cdot 3 \cdot 4 \cdot 5 \cdot 6 \cdot 7} + \text{etc.} \\ &= \left(1 - \frac{s^2}{p^2}\right)\left(1 - \frac{s^2}{4p^2}\right)\left(1 - \frac{s^2}{9p^2}\right)\left(1 - \frac{s^2}{16p^2}\right) \text{ etc.} \end{aligned}$$

It is time to match coefficients again. Matching the coefficients of s^2 gives

$$\frac{1}{1 \cdot 2 \cdot 3} = \frac{1}{p^2} + \frac{1}{4p^2} + \frac{1}{9p^2} + \frac{1}{16p^2} + \text{etc.}$$

which is, of course, another way of saying that

$$1 + \frac{1}{4} + \frac{1}{9} + \frac{1}{16} + \frac{1}{25} + \text{etc.} = \frac{p^2}{6},$$

giving another solution to the Basel Problem. This calculation is often cited as Euler's "first" solution to the problem, but, as we have seen, it is actually his third one in this paper.

The formulas earlier in the paper, in the case $y = 0$, give

$$\begin{aligned} P &= \frac{1}{6} \\ Q &= \frac{1}{90} \\ R &= \frac{1}{945} \\ S &= \frac{1}{9450} \\ T &= \frac{1}{93555} \\ V &= \frac{691}{6825 \cdot 93555} \end{aligned}$$

and Euler reminds us that these will be the values of the coefficients for sums of reciprocals of higher even powers. For example,

$$1 + \frac{1}{2^4} + \frac{1}{3^4} + \frac{1}{4^4} + \frac{1}{5^4} + \text{etc.} = Qp^4 = \frac{p^4}{90}$$

and

$$1 + \frac{1}{2^{12}} + \frac{1}{3^{12}} + \frac{1}{4^{12}} + \frac{1}{5^{12}} + \text{etc.} = Vp^{12} = \frac{691p^{12}}{6825 \cdot 93555}.$$

Euler calls these sums P' for the sums of the reciprocals of the squares, Q' for the fourth powers, R' for the sixth powers, etc., and notes that

$$p^2 = 6p' = \frac{15Q'}{P'} = \frac{21R'}{2Q'} = \frac{10S'}{R'} = \frac{99T'}{10S'} = \frac{6825V'}{691T'}.$$

It is not clear why Euler makes this observation, whether to give relations among sums of reciprocals of higher powers, or to give ways to approximate π^2. If it is for the sake of the approximations, then they do converge rapidly, but they involve large numbers.

Finally, Euler returns to some of his earlier results to give seven approximations for what he calls p, what we call π. He starts with the one known to Leibniz and Newton

$$p = 4\left(1 - \frac{1}{3} + \frac{1}{5} - \frac{1}{7} + \frac{1}{9} - \frac{1}{11} + \text{etc.}\right).$$

His last approximation is

$$p = \frac{192}{61} \cdot \frac{1 - \frac{1}{3^7} + \frac{1}{5^7} - \frac{1}{7^7} + \frac{1}{9^7} - \frac{1}{11^7} + \text{etc.}}{1 + \frac{1}{3^6} + \frac{1}{5^6} + \frac{1}{7^6} - \frac{1}{9^6} + \frac{1}{11^6} + \text{etc.}}.$$

Again, these series converge rapidly, but they feature large denominators.

And so ends one of Euler's most famous papers. We see that Euler actually solves the Basel problem three times and that he does much more. He also shows his first interest in approximations to π, a topic to which he returns many times.

Euler will refine his solution to the Basel problem many times, in E-61, E-63 in 1743 and in E-130 in 1740. Moreover, he will use this solution to solve a number of other problems, most spectacularly in E-72. In that paper, as Dunham puts it, Euler practically invents analytic number theory.

References

[D] Dunham, William, *Euler: The Master of Us All*, MAA, Washington, DC, 1999.

22

E-46: Methodus universalis serierum convergentium summas quam proxime inveniendi*

A universal method for finding sums which approximate convergent series

Today, we are used to the idea of using series to approximate, or even to define integrals. A glance at the illustrations in this article leads the modern reader to expect that is what Euler is doing here. In this case, though, first impressions are wrong, and Euler is working in the other direction. He is using integrals to approximate sums of series.

In the course of this approximation, Euler makes a step that very closely resembles Simpson's rule. Because Euler wrote this paper in 1736 and Simpson's work was published in 1743 [K, p. 562], people might think that Simpson's rule should be called Euler's rule. Neither Euler nor Simpson was the first to discover the rule named after Simpson. It was known even to Cavalieri as early as 1639 [C]. Euler wasn't the first one to discover *everything*.

Euler begins by reminding us of his earlier work in E-43. There he worked with the harmonic series and found that

$$1+\frac{1}{2}+\frac{1}{3}+\cdots+\frac{1}{n-1}=\int_1^n \frac{dx}{x}+C=l\,n+C.$$

As usual, Euler uses l to denote the natural logarithm, so we would write $\ln n$ where he writes $l\,n$. Euler uses C to denote this particular constant of integration, the value we now usually write as γ. Its value is approximately 0.577218, and it becomes more and more important. Julian Havil has written a whole book [H] devoted to the constant γ. In this paper, Euler will estimate similar constants for other series.

Euler considers an arbitrary series given by

$$a+b+c+d+e+\text{etc.}$$

Let x be the term of index n, and suppose that x can be given in terms of n.

To sum the series geometrically, Euler draws Figure 1. There, the intervals AB, BC, CD, etc. are all taken to be 1, and the heights of the rectangles are taken to be the values of the terms of the series, that is $Aa=a$, $Bb=b$, $Cc=c$, and so forth up to $Pp=x$. Euler calls these heights *applicata*, or *applications*. Euler takes y to be a curve that "naturally expresses" the series. Then Euler gets a first approximation of the series as

$$a+b+c+\cdots+x>\int y\,dn.$$

Note that in Figure 1, Euler completed his rectangles with points $\beta, \gamma, \delta, \ldots \rho$. In Figure 2, Euler draws a new curve he calls x that naturally expresses these points labeled with Greek letters. To the

Comm. Acad. Sci. Imp. Petropol. 8 (1736) 1741, 3–9; *Opera Omnia* I.14, 101–107

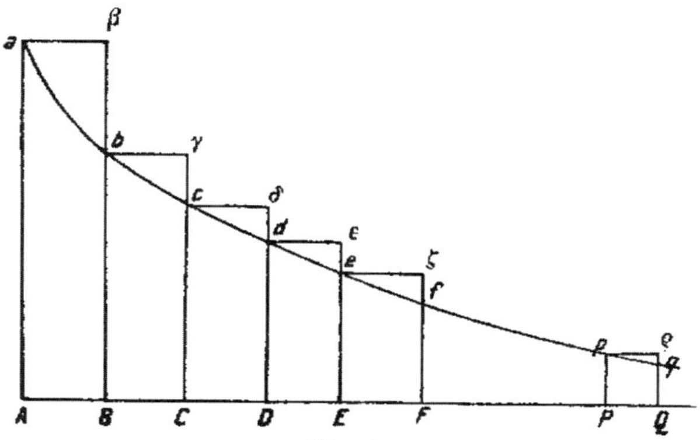

Fig. 1.

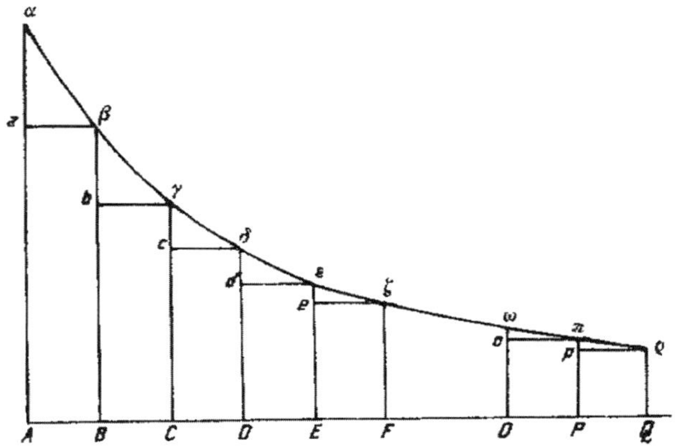

Fig. 2.

modern reader, these figures look like Riemann's upper sums and lower sums, though Bernhard Riemann wasn't even born until 1826. Riemann, though, takes one curve and draws two sets of rectangles, one inside the curve and one outside the curve, and he does this because he thinks that finding the total areas of the rectangles is an easy problem compared to the problem of defining the integral of the curve.

Looking closely here, though, we see that Euler only has one set of rectangles, the ones corresponding to the series he wants to sum. Then Euler constructs two different curves, one outside the rectangles and one inside the rectangles, and he uses the integrals of the two curves to bound the sum of the series. Euler thinks that evaluating the integrals is easy compared to summing the series.

Clearly Riemann's ideas are closely related to, and probably strongly influenced by Euler's, but the two ideas are different. We would be confused if we fail to distinguish between them.

In Figure 2, Euler let x be the curve that naturally expresses the points marked in Greek letters, and then notes that

$$a + b + c + d + \cdots + x < \int x\, dn$$

and so he has bounded the sum of his series between two integrals. Note that here, and also in formulas to come, Euler uses x in two different senses at the same time, once as a term in a series and again as the variable of integration.

Now he works to improve those bounds. First, he does something like the trapezoid rule. In Figure 1, he notes that if the areas of all the triangles $ab\beta$, $bc\gamma$, $cd\delta, \ldots, pq\rho$ are added to the integral, then the sum is still less than the sum of the series. Recalling that in Figure 1 $Aa = a$, $Qq = y$ and $AB = BC = \cdots = PQ = 1$, the sum of the areas of the triangles turns out to be $\frac{a-y}{2}$, and so he gets an improved lower bound for the sum of his series

$$a + b + c + \cdots + x > \int y\,dn + \frac{a-y}{2}.$$

In exactly the same way, the upper bound improves to be

$$a + b + c + \cdots + x < \int x\,dn - \frac{a-y}{2}.$$

Euler does not mention that it is best if the curves x and y are convex in order to be sure that these inequalities actually hold.

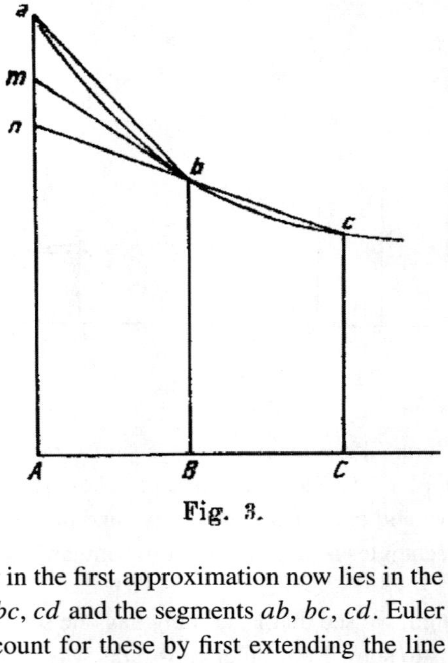

Fig. 3.

The only remaining error in the first approximation now lies in the lens shaped sectors bounded by the arcs of the curve ab, bc, cd and the segments ab, bc, cd. Euler calls these sectors aba, bcb, cdc, etc. Euler begins to account for these by first extending the line segment bc to n on the line Aa. Calculation shows that $an = (a-b) - (b-c) = a - 2b + c$. Because $AB = 1$, the area of triangle abn is $\frac{an}{2}$ or $\frac{a-2b+c}{2}$. Taking m the midpoint of the segment an, then the segment mb is approximately the tangent to the curve abc at the point b. If the curve abc were a parabola, then the area of the parabolic sector aba would be one third of the area of the triangle abm, which, in turn, is one half the area of triangle abn. So, taking a Simpson-like approximation, the area of the sector aba is approximately $\frac{a-2b+c}{12}$. Likewise, the area of the next sector is $\frac{b-2c+d}{12}$, and so on down to the last sector, which will have area $\frac{x-2y+z}{12}$, where y and z denote the terms of index $n+1$ and $n+2$, respectively. Note that these terms come beyond x, the last term in Euler's series. Summing the areas of all these sectors gives $\frac{a-b}{12} - \frac{y-z}{12}$, and so Euler gets his best and last approximation for the sum of his series

$$a + b + c + \cdots + x = \int y\,dn + \frac{a}{2} - \frac{y}{2} + \frac{a-b}{12} - \frac{y-z}{12}.$$

In the special case of an infinite series, the terms involving y and z vanish and this reduces to $\int y\,dn + \frac{7a}{12} - \frac{b}{12}$.

It is time for examples. Euler tries to sum the first million terms of the harmonic series. To reduce his error, he sums the first ten terms by hand to get 2.928968. To use his approximation to sum another n terms, he takes

$$a = \frac{1}{11}, \quad b = \frac{1}{12}, \quad x = \frac{1}{n+10}, \quad y = \frac{1}{n+11} \quad \text{and} \quad z = \frac{1}{n+12}.$$

Then $\int y\,dn = l\frac{n+11}{11}$, using l, as Euler always does, to denote the natural logarithm function. When Euler takes $n = 999990$; then his approximation gives

$$l\frac{1000001}{11} + \frac{1}{22} + \frac{1}{132} - \frac{1}{144} - \frac{1}{2000002} - \frac{1}{12000012} + \frac{1}{12000024} + 2.928968$$

or approximately 14.392669.

For his second and final example, Euler returns to the Basel problem to sum the series $1 + \frac{1}{4} + \frac{1}{9} + \frac{1}{16} + \frac{1}{25} +$ etc. Again, he sums the first ten terms by hand to get 1.549768. To sum the rest of the series he takes

$$a = \frac{1}{121}, \quad b = \frac{1}{144}, \quad x = \frac{1}{(n+10)^2} \quad \text{and} \quad y = \frac{1}{(n+11)^2}.$$

This makes $\int y\,dn = \frac{1}{11} - \frac{1}{n+11}$. Now, he takes $n = \infty$ so that $\int y\,dn = \frac{1}{11}$. He applies his formula to get the sum of the infinite series to be

$$\frac{1}{11} + \frac{7}{12 \cdot 121} - \frac{1}{12 \cdot 144} + 1.549768$$

or approximately 1.644920.

This little paper, only seven pages long, is by no means one of his great articles, but it does show some of his style and his continuing interest in numerical, as well as symbolic answers. Its sequel, E-47, follows this article directly in the 1736 volume of the *Commentarii*. It is twice as long, and Euler uses higher order polynomials to get much more accurate approximations.

References

[C] Chabert, Jean-Luc, et al. *A History of Algorithms: from the Pebble to the Microchip*, Springer, Berlin, 1999.

[H] Havil, Julian, *Gamma: Exploring Euler's Constant*, Princeton University Press, Princeton, 2003.

[K] Katz, Victor, *A History of Mathematics: An Introduction*, 2nd ed., Addison-Wesley, Reading, MA, 1998.

23

E-47: Inventio summae cuiusque seriei ex dato termino generali*

Finding the sum of a series from a given general term

Euler seems to like the idea of using integration and other calculus tools to sum series, both finite and infinite. After all, his greatest success to date has been the solution to the Basel problem and he used calculus tools to do that. Here, he continues the approximation techniques he introduced three years earlier in E-25 and refines the approach he calls "geometric" that he used in the preceding article E-46.

First, Euler recalls his key result from E-25. If x is a formula that gives the term of index n of a series, then the sum of that series is

$$\int x\,dn + \frac{x}{2} + \frac{dx}{12\,dn} - \frac{d^3x}{720\,dn^3} + \text{etc.}$$

where the constant of integration is to be taken "so that the total expression vanishes if $n = 0$." Note that the first two denominators, 2 and 12, arise for the same reasons those same denominators were in the approximation in E-46.

Because Euler wants a higher degree approximation for the correction terms than he used in E-46, he uses Taylor series. Euler specifically cites Taylor's 1717 book *Methodus incrementorum directa et inversa*, by now 20 years old. In fact, when he writes of *seriem Taylorus*, it may be the first time those series are called by the name they bear today. Taylor, though, derived his results in the language of fluxions, so Euler presents the same results in the language of differentials. It is worth noting that Euler does not use the notation of functions he introduced the previous year in E-45. He writes:

> First, if y is given in any way in terms of x and of constants, and if in place of x is put $x + dx$, then $y + dy$ will go in place of y. If in place of x we put $x + 2\,dx$, then y becomes $y + 2\,dy + ddy$, and if x grows by another differential element, then y is changed into $y + 3\,dy + 3\,ddy + d^3y$. All the coefficients are binomial coefficients.

In this passage, Euler has not taken measures to simplify his language. He uses different expressions, "will go," "y becomes," "y is changed into," to describe the results of the substitution. Euler's Latin is very good. After all, he had majored in languages at the University of Basel. As he matures, though, he will tend to simplify his grammar and his vocabulary in routine passages such as this one, because he knows that Latin is a second or third, or sometimes even fourth language for many of his readers.

If x is replaced by $x + m\,dx$, then y becomes

$$y + \frac{m}{1}dy + \frac{m(m-1)}{1\cdot 2}ddy + \frac{m(m-1)(m-2)}{1\cdot 2\cdot 3}d^3y + \text{etc.}$$

**Comm. Acad. Sci. Imp. Petrol.* 8 (1736) 1741, 9–22; *Opera Omnia* I.14, 108–123

Euler could have saved some words here if he had simply used his function notation to write $f(x + m\,dx)$.

Now, in a step that would not be tolerated today, Euler takes m to be "an infinitely large number that makes $m\,dx$ signify a finite quantity." In Euler's time, finite quantities are neither infinite nor infinitesimal nor zero. Then when $x + m\,dx$ replaces x, y becomes

$$y + \frac{m\,dy}{1} + \frac{m^2\,d^2y}{1\cdot 2} + \frac{m^3\,d^3y}{1\cdot 2\cdot 3} + \frac{m^4\,d^4y}{1\cdot 2\cdot 3\cdot 4} + \text{etc.}$$

Now, let $m\,dx = a$, so that $m = \frac{a}{dx}$, and, in place of x put $x + a$. Then y becomes

$$y + \frac{a\,dy}{1\,dx} + \frac{a^2\,ddy}{1\cdot 2\,dx^2} + \frac{a^3\,d^3y}{1\cdot 2\cdot 3\,dx^3} + \text{etc.}$$

This is Taylor's theorem.

Euler does an example, expanding $y = x^m$, where m is not an infinite number as it was in the previous paragraph. Then he suggests that Taylor's theorem may be useful in finding roots of equations, and he does an example. He wants to find a value of a that makes

$$0 = y + \frac{a\,dy}{1\,dx} + \frac{a^2\,ddy}{1\cdot 2\,dx^2} + \text{etc.}$$

A good place to start would be to take $a = -\frac{y\,dx}{dy}$, particularly if the remaining terms of the Taylor series are small.

For example, to solve $z^3 - 3z - 20 = 0$, take y to be the function $y = x^3 - 3x - 20$ so that $\frac{dy}{dx} = 3x^2 - 3$. Note that Euler distinguishes between solving for z, on the one hand, and finding a root of an equation on the other hand. This is a distinction we seldom bother to make today.

Then it is approximately true that

$$z = x - \frac{x^3 - 3x - 20}{3xx - 3} = \frac{2x^3 + 20}{3xx - 3}.$$

Taking $x = 3$ gives $z = 3\frac{1}{12}$, which Euler says is a good approximation.

This is clearly the first step in the Newton-Raphson method for finding roots, but Euler does not suggest taking $x = 3\frac{1}{2}$ and repeating the process. Euler mentions neither Newton nor Raphson, who did similar things in about 1671 and in 1697, respectively, but neither had Taylor series available to provide a theoretical foundation to their work. Thomas Simpson did have Taylor series when he developed the method into the form we see today, but he did not do that until 1740 [C, p. 178], four years after Euler wrote this paper.

Euler takes a few lines to show that any function y that vanishes when $x = 0$ can be written as

$$y = \frac{x\,dy}{1\,dx} - \frac{x^2\,ddy}{1\cdot 2\,dx^2} + \frac{x^3\,d^3y}{1\cdot 2\cdot 3\,dx^3} - \text{etc.}$$

Euler does not regard discontinuous functions as really being functions, and he does not know that some continuous functions are not analytic. Moreover, in general, Euler seems unconcerned with "exceptions" to his theorems, though sometimes he makes a point of saying that a theorem is "always true." This may leave open the possibility that he proves some theorems that are not always true.

Euler plans to use this result because integrals are among those functions that vanish when $x = 0$, and he wants to evaluate integrals. He takes $y = \int z\,dx$ so that $dy = z\,dx$, $ddy = dz\,dx$, $d^3y = d^2z\,dx$, etc. Euler has not mentioned that we should take dx to be constant, as he has in other papers, but clearly he means to, for otherwise $ddy = dz\,dx + z\,d^2x$.

Substituting these values for dy, ddy, etc. into his formula for y from the previous paragraph gives

$$\int z\,dx = \frac{xz}{1} - \frac{x^2\,dz}{1\cdot 2\,dx} + \frac{x^3\,ddz}{1\cdot 2\cdot 3\,dx^2} - \text{etc.}$$

Euler says that this general series for quadrature is from a 1694 paper by Johann Bernoulli.

Now Euler denotes his series and its sum S as

$$A + B + C + D + \cdots + X = S$$

where X is the term of index x. Usually for Euler, capital letters denote functions, but that is not the case here, though he does consider S as a function of x. Such series S are also functions that vanish when $x = 0$, so Euler can use the result again to get

$$X = \frac{dS}{1\,dx} - \frac{ddS}{1\cdot 2\,dx^2} + \frac{d^3S}{1\cdot 2\cdot 3\,dx^3} - \frac{d^4S}{1\cdot 2\cdot 3\cdot 4\,dx^4} + \text{etc.}$$

Euler repeats some rather intricate calculations from E-25. Rather than repeat them here, we skip right to the results and say

$$S = \alpha\int X\,dx + \beta X + \frac{\gamma\,dX}{dx} + \frac{\delta\,ddX}{dx^2} + \text{etc.}$$

where the coefficients, slightly different from what they were in E-25, are

$$\begin{aligned}
\alpha &= 1\\
\beta &= \frac{\alpha}{2}\\
\gamma &= \frac{\beta}{2} - \frac{\alpha}{6}\\
\delta &= \frac{\gamma}{2} - \frac{\beta}{6} + \frac{\alpha}{24}\\
\varepsilon &= \frac{\delta}{2} - \frac{\gamma}{6} + \frac{\beta}{24} - \frac{\alpha}{120}\\
\zeta &= \frac{\varepsilon}{2} - \frac{\delta}{6} + \frac{\gamma}{24} - \frac{\beta}{120} + \frac{\alpha}{720}\\
&\text{etc.}
\end{aligned}$$

Euler does the substitution for us and gets the rather formidable sum

$$\begin{aligned}
S = \int X\,dx &+ \frac{X}{1\cdot 2} + \frac{dX}{1\cdot 2\cdot 3\cdot 2\,dx} - \frac{d^3X}{1\cdot 2\cdot 3\cdot 4\cdot 5\cdot 6\,dx^3} + \frac{d^5X}{1\cdot\cdots\cdot 7\cdot 6\,dx^5}\\
&- \frac{3\,d^7X}{1\cdot\cdots\cdot 9\cdot 10\,dx^7} + \frac{5\,d^9X}{1\cdot\cdots\cdot 11\cdot 6\,dx^9} - \frac{691\,d^{11}X}{1\cdot\cdots\cdot 13\cdot 210\,dx^{11}}\\
&+ \frac{35\,d^{13}X}{1\cdot\cdots\cdot 15\cdot 2\cdot dx^{13}} - \frac{3617\,d^{15}X}{1\cdot\cdots\cdot 17\cdot 30\,dx^{15}} + \text{etc.}
\end{aligned}$$

We rarely see formulas involving 13th and 15th derivatives! Because this one alternates, involves only odd numbered derivatives, and has rapidly increasing denominators, it will tend to converge very rapidly, and it will give a very accurate description of how the sum of the series S differs from the integral approximating that sum, $\int X\,dx$.

Euler's first example is to take $X = x$, making the sum $S = 1 + 2 + 3 + \cdots + x$. Then

$$\int X\,dx = \frac{x^2}{2},$$

and all but the first term after the integral in the series for S vanishes. That leaves the sum

$$\frac{x^2 + x}{2},$$

as it should be.

For a second example, Euler takes $X = x^2$, so the series is $S = 1 + 4 + 9 + \cdots + x^2$. Then

$$\int X\,dx = \frac{x^3}{3}, \frac{X}{1 \cdot 2} = \frac{x^2}{2} \quad \text{and} \quad \frac{dX}{dx} = 2x$$

so

$$\frac{dX}{1 \cdot 2 \cdot 3 \cdot 2\,dx} = \frac{x}{6}.$$

All this makes $S = \frac{x^3}{3} + \frac{x^2}{2} + \frac{x}{6}$, which is also correct.

Euler turns to the more general series

$$S = 1^n + 2^n + 3^n + \cdots + x^n$$

and gets explicit formulas for all the values of n up to $n = 16$. This is the last example he can do with his series for S, where he stopped at the 15th derivative.

Euler's table shows that he is still not quite modern in his notation. He gives, for example, the formula

$$\int x^3 = \frac{x^4}{4} + \frac{x^3}{2} + \frac{x^2}{4},$$

using an integral sign where we would use a $\sum$. In Euler's day, the people who used the differential notation instead of Newton's notation of fluents and fluxions regarded an integral as a sum of differentials, so they did not need a separate notation for sums of finite objects, like x^3, and infinitesimal objects, as for integrals. We will not get confused if we note that integrals always contain a differential, like dx, whereas sums of finite objects will not.

Euler turns to issues that break some new ground. He considers the harmonic series, $1 + \frac{1}{2} + \frac{1}{3} + \frac{1}{4} + \text{etc.}$ In his notation, $X = \frac{1}{x}$ and his results tell him that $\int X\,dx = \text{Const.} + lx$, where, for a change, he writes "Const." to denote the constant of integration. He wants to estimate the constant. He calculates the necessary derivatives and gets

$$S = \text{Const.} + lx + \frac{1}{2x} - \frac{1}{12x^2} + \frac{1}{120x^4} - \frac{1}{252x^6} + \frac{1}{240x^8} - \frac{1}{132x^{10}} + \frac{691}{32760x^{12}} - \frac{1}{12x^{14}} + \text{etc.}$$

In most circumstances, Euler would find the constant by setting $x = 0$, but he notes that would not work here because "all of the terms are infinite magnitudes and the constant cannot be determined."

Euler can work around this, though. He sums the first ten terms by hand, to 16 decimal places. Then, his formula says that sum equals

$$\text{Const.} + l\,10 + \frac{1}{20} - \frac{1}{1200} + \frac{1}{1\,200000} - \frac{1}{252\,000000} + \frac{1}{24000\,000000} - \frac{1}{1\,320000\,000000} + \text{etc.}$$

With this, and knowing the logarithm of 10 to at least 18 decimal places, he is able to estimate the value of his constant to 16 decimal places as

$$= 0.57721\,56649\,01532\,9.$$

This is considerably more accurate than any of his previous estimates. This accurate value of the Euler-Mascheroni constant is the high point of the paper.

Euler next uses this improved value of his constant to give sums of the first 10, 100, 1000, 100000 and 1000000 terms of the harmonic series. Then he makes a remark that seems like nonsense to a modern reader. If the harmonic series is summed until $x = \infty$, then the sum will be "$l\,\infty + 0{,}57721\,56649\,01532\,9$." We are not allowed to write things like that any more.

For Euler, each series has its own constant. The Euler-Mascheroni constant pertains to just one particular series, the harmonic series. Euler finds constants for several other series as well.

He uses a shortcut to find the constant for the sum of the reciprocals of the odd numbers, the case $X = \frac{1}{2x-1}$, so the series is $1 + \frac{1}{3} + \frac{1}{5} + \frac{1}{7} + \frac{1}{9} + \text{etc.}$ Then $\int X\,dx = \text{Const.} + \frac{1}{2}l(2x-1)$. Euler sets up the series that, if he chose to evaluate it, would give him the value of the constant, but he would rather use his shortcut. First, he states that $1 + \frac{1}{3} + \frac{1}{5} + \frac{1}{7} + \frac{1}{9} + \text{etc.}$ is half the sum of the harmonic series, and so he can write it as $\frac{1}{2}\,l\,\infty$.

Euler knows that the alternating harmonic series sums to the logarithm of 2, so $l\,2 = 1 - \frac{1}{2} + \frac{1}{3} - \frac{1}{4} + \text{etc.}$ Putting these facts together, he writes

$$l\,2 = 2\,\text{const.} + l, \infty - l\,\infty - 0,577215\,\text{etc.}$$

so the constant for the odd harmonic series will be

$$0.63518\,14227\,30739\,2.$$

Euler uses these same techniques to calculate the sums of the reciprocals of the squares, of the cubes and of the fourth powers, values we now call $\zeta(2)$, $\zeta(3)$ and $\zeta(4)$ to 16 decimal places.

As his final example, he sets up but does not evaluate the sum $1 - \frac{1}{3} + \frac{1}{5} - \frac{1}{7} + \frac{1}{9} - \text{etc.}$ He knows this particular series sums to $\frac{\pi}{4}$, so he could give a good approximation to π, but he does not do that here.

This brings us to the end of this paper, but Euler will revisit these ideas later this year in the same volume of the *Commentarii* in E-55.

In this paper, we have seen how Euler makes the step from the geometric methods and the explanations he used in E-46 to the calculus-based methods and insights of E-47. Here he seems to view calculus as an extension of or a step beyond geometry, though in other papers, calculus seems to be a more algebraic tool.

On the way, Euler has given a new proof of Taylor's theorem and stopped one step short of rediscovering the Newton-Raphson method for finding roots, and he has more than doubled the accuracy of his approximation of the Euler-Mascheroni constant. Perhaps the most significant aspect of this paper, though, is that it is the first demonstration of Euler's excellent skills at calculations. For the first time, we see him doing calculations to 16 decimal places of accuracy. He will continue to do such accurate calculations throughout his life, even after he becomes blind and has to perform the calculations in his head.

References

[C] Chabert, Jean-Luc, et al. *A History of Algorithms: from the Pebble to the Microchip*, Springer, Berlin, 1999.

24

E-51: De constructione aequationum ope motus tractorii aliisque ad methodum tangentium inversam pertinentibus*

On the solution of equations from the motion of pulling and other equations pertaining to the method of inverse tangents

There is an area of mathematics, rather like the center part of a Venn diagram, where differential equations, geometry and mechanics all overlap. This paper, E-51, lies in that area, though the editors of the *Opera Omnia* have chosen to call it a paper in differential equations. The paper comes in two distinct parts. In the first part, Euler solves the problem of the generalized tractrix. The second part is devoted to trajectories described in terms of various unusual coordinate systems.

A tractrix is the path of an object dragged by an inelastic string as the string is pulled along a given curve. In the original tractrix, the given curve was taken to be a straight line. Euler solves the problem in the case where the given curve is more general.

The problem has a delightful history, so we will pause before reviewing Euler's work to tell a bit of the story of the tractrix as told by Fred Rickey [R] and Steven Jay Gould [G]. In 1676, more than 30 years before Euler was born, Gottfried Wilhelm Leibniz (1646–1716) was in Paris trying, with little success, to be a diplomat. His notebooks show that he did most of his work to invent calculus while he was there, work that he would publish in *Acta Eruditorum* in 1684. One wonders whether Leibniz failed as a diplomat because he was spending too much time on mathematics, or if he had time to spend on mathematics because he was so unsuccessful as a diplomat. When he probably should have been trying to meet with government officials, he instead met a physician and architect named Claude Perrault (1613–1688), and Perrault posed the problem to him.

In his time, Perrault was important enough to have been a founding member of the French Académie des Sciences in 1666, but today it is his younger brother Charles (1628–1703) whose fame endures. Charles wrote children's stories like *Cinderella*, *Puss 'n Boots*, *Little Red Riding Hood*, and others. Claude, though, died of an infection he incurred dissecting a camel.

At the time, Leibniz was in the process of inventing calculus, and he welcomed problems on which to use his new methods. Many years later, he claimed that he had solved the problem at once, but he did not select it as one of the examples for his first paper on calculus, and there seems to be no surviving trace of his work on the problem. Newton also claimed to have solved the problem, but his work does not survive, either. The earliest surviving solution seems to be that of Christian Huygens (1623–1695) published in 1693. Huygens is said to have called the tractrix the "dog curve" because it is the path traced by a reluctant dog being dragged by a leash.

The problem of the simple tractrix, where the given curve is a straight line, is still a classic in differential equations. It reflects the geometric, as opposed to the mechanical, origins of calculus. If dogs were really dragged by their leashes, the editors of the *Opera Omnia* might have put this particular paper in the series on Mechanics. As it is, they could have put it in the volumes on

**Comm. Acad. Sci. Imp. Petropol.* 8 (1736) 1741, 66–85; *Opera Omnia* I.22, 83–107

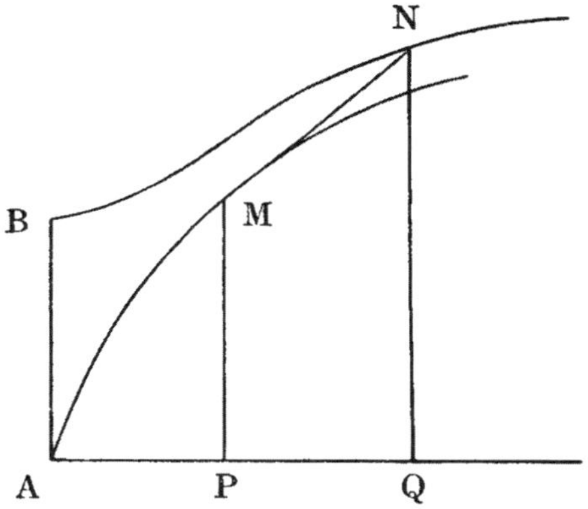

Fig. 1

Geometry along with the papers on reciprocal trajectories, but instead they decided to classify it as a paper on Differential Equations. The lines that divide the subjects are not always clear.

Euler considers the more general, and considerably more difficult case that the given curve, the path of the person walking the dog, is not a straight line. In Figure 1, a string of length b extends from A to B. One end of the string moves along the curve BN, and we are to find the curve AM traced by the other end of the string.

Even before he begins to assign variables, Euler tells us that the solution of this equation will depend on the differential equation

$$ds + ss\,dz = Z\,dz$$

where Z is any function of z. He sees that this is a more general form of the Riccati equation

$$ds + ss\,dz = z^m\,dz.$$

Euler has not yet told us what s and z are, and his first task in this paper is to derive the equation. The derivation is intricate, but it has some interesting and illuminating twists.

For the given curve BN, he lets the abscissa $AQ = t$, and the application, what we call the ordinate, be $QN = u$. He takes u to be given in terms of t and of constants. For the curve AM that is to be found, he takes $AP = x$, $PM = y$ and supposes that $dy = p\,dx$.

When the tractrix has reached the point M and the leading end of the string is at the point N, then $MN = AB = b$ and the segment MN is tangent to the curve AM. Euler is becoming more adventurous in his invention of notation, and he writes some of the ratios that follow from this geometry as

$$\sqrt{1+pp} : 1 = MN(b) : PQ(t-x).$$

The notation $MN(b)$ means that, in the equation, $MN = b$. Likewise $PQ(t-x)$ means that $PQ = t - x$. It is a convenient notation, for it saves writing the equation two ways, once geometrically and once algebraically, but the innovation never seems to have caught on. Euler adds, using the same notation, that

$$\sqrt{1+pp} : p = MN(b) : QN - PM(u-y).$$

Note the use of p in this equation. Because we know that $dy = p\,dx$, we see that p is the derivative of the tractrix at the point M, but Euler will never refer to this value as the slope of a tangent line. Rather, it is the length of the vertical side of a differential triangle with base 1 and hypotenuse $\sqrt{(1+pp)}$. This use of the derivative as a differential lets us glimpse the way people thought about differentials in Euler's day.

Euler continues, using two observations that follow immediately from these ratios, that

$$\frac{b}{\sqrt{1+pp}} = t - x \quad \text{and that} \quad \frac{bp}{\sqrt{1+pp}} = u - y,$$

to get that $pt - px = u - y$. Differentiating this and substituting $dy = p\,dx$ gives

$$p\,dt + t\,dp - x\,dp = du$$

so

$$x = t + \frac{p\,dt}{dp} - \frac{du}{dp}.$$

Because we already know that

$$\frac{b}{\sqrt{(1+pp)}} = t - x,$$

substituting into $du = pp\,dt + (t - x)\,dp$ gives

$$du = p\,dt + \frac{b\,dp}{\sqrt{1+pp}}.$$

Before continuing with his calculations, Euler remarks that the value of "p is the cotangent of the angle MNQ, if the total sine is set = 1" and that $\sqrt{1+pp}$ is the cosecant of that angle. This remark reveals two facets of how mathematics in Euler's day differs from today. First, to Euler, the derivative p is not slope, but it is the cotangent of an angle. What we learn today about derivatives being rates of change or slopes are interpretations that came after Euler's time. Second, Euler's remark about the total sine is rooted in ancient mathematics. Sine tables in Euler's day assumed a triangle with a fixed hypotenuse and, for any angle, gave the length of the side opposite that angle. To avoid using decimals, the length of the hypotenuse was often taken to be a large number, say 60,000 or 100,000. Using this system, they called the sine of a right angle the "total sine." This value is the same as the length of the hypotenuse, but they never described the total sine in terms of the hypotenuse. So, when Euler says he is taking the cotangent, if the total sine is 1, he means the same cotangent we mean today, but he means it as a geometric length based on a triangle with hypotenuse 1 and not as a ratio of sides of that triangle.

Getting back to the calculations, we see that there is no evident way to solve this differential equation for p, so Euler begins a chain of unlikely substitutions. He introduces the variable q as

$$q = \sqrt{1+pp} - p.$$

This quantity q is a curious one. As a length, it is a hypotenuse minus the length of a side. Nobody but Euler would suspect that such a quantity would ultimately simplify his expression.

A little algebra shows that

$$p = \frac{1-qq}{2q} \quad \text{and} \quad dp = \frac{-dq(1+qq)}{2qq},$$

so

$$\frac{dp}{\sqrt{1+pp}} = \frac{-dq}{q}.$$

This leads easily to the differential equation

$$2q\,du = dt - qq\,dt - 2b\,dq.$$

Euler makes another surprise substitution, introducing r defined in $du = \frac{b\,dr}{r}$. This amounts to saying that $b\,l\,r = QN = u$, where, as always, $l\,r$ means the natural logarithm of r. This yields

$$2bq\,dr + 2br\,dq = r\,dt - rqq\,dt.$$

Euler is still working towards that differential equation involving s and Z. He finally introduces these variables as $s = qr$ and $Z = rr$. The calculations continue

$$2b\,ds = r\,dt - \frac{ss\,dt}{r}.$$

Now let

$$\frac{dt}{r} = 2b\,dz$$

and

$$r\,dt = 2bZ\,dz.$$

Then

$$dt^2 = 4b^2 Z\,dz^2$$

and

$$t = 2b\int dz\sqrt{Z}$$

so the curve BN is given by

$$AQ = 2b\int dz\sqrt{Z} \quad \text{and} \quad QN = \frac{b}{2}\,l\,Z$$

and finally, the curve BN is given in terms of Z and z by

$$ds + ss\,dz = Z\,dz.$$

This is the main result of this paper and gets us through seven of the 46 sections of the paper.

Euler recognizes that some of the relationships among the variables may have been obscured in his blizzard of substitutions, so he does some algebra to illuminate them. He gives t and u, the functions that describe the path along which the string is pulled in terms of Z and the parameter z. Then he gives x and y, the functions that define the tractrix, in terms of t and u, and he verifies that these all solve his fundamental differential equation. Finally, he gives t and u in terms of x and y. This amounts to a solution for the inverse problem: If the dog is to follow a given path, described by t and u, where must its owner walk, given by x and y?

It is time for examples. Euler's first example begins with the differential equation

$$ds + s^2\,dz = a^2 z^{2n}\,dz,$$

which is a form of the Riccati equation. For this problem, it means that $Z = a^2 z^{2n}$, so that

$$\int dz\sqrt{Z} = \frac{az^{n+1}}{n+1} \quad \text{and} \quad t = AQ = \frac{2abz^{n+1}}{n+1}.$$

Also, $l\,Z = 2\,l\,a + 2n\,l\,z$ so that $u = b\,l\,a + nb\,l\,z$. These describe t and u, which give the path along which the string is pulled.

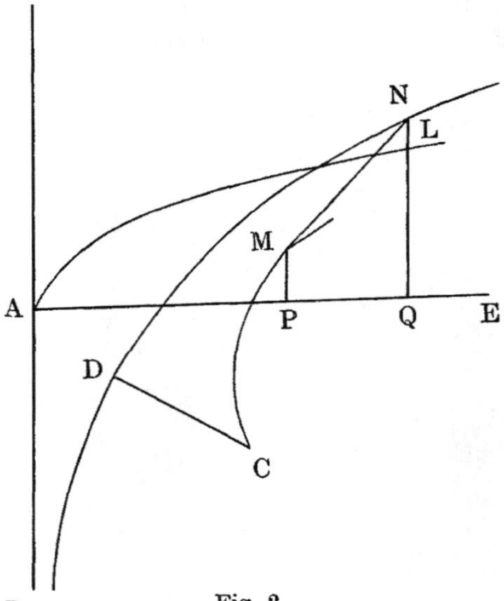

Fig. 2

Solving for u in terms of t, Euler deduces that the curve BN (he means DN) is logarithmic with a constant subtangent $\frac{nb}{n+1}$. The subtangent is one of several geometric ideas now lost in our modern approaches to calculus. A subtangent to a curve may be found with respect to either axis. If it is found with respect to the horizontal, or x-axis, then it is given by $-\frac{f(x)}{f'(x)}$, that is the step size in the Newton-Raphson method. Geometrically, it measures the distance between a point x on the axis and the point where the tangent to the curve at that value intersects the axis.

Everyone in Euler's audience knows that logarithmic curves are characterized by the property that their subtangents with respect to the vertical axis are constant, and that exponential curves have constant subtangents with respect to the horizontal axis. Rather than distinguish among logarithmic and exponential curves by their multiplicative factors or by the bases of the logarithm, Euler distinguishes them by their subtangents.

The curve DN is the logarithmic curve along which the string is to be pulled, and the axis AB is the asymptote of that curve. Then the curve CN is the tractrix. Euler also shows in Figure 2 the curve AL, the graph of $Z = a^2 z^{2n}$. He does not solve the Riccati equation to get an explicit solution, but contents himself with showing the relationship between the Riccati equation and this particular tractrix. Euler does not yet think of functions the way we do, so he is not prepared to note that the tractrix here is a curve, but not a function.

Euler moves on to the second part of the paper. He describes a trajectory in terms of an unusual coordinate system, though he would not recognize that as what he is doing. In Figure 3, the curve AM is to be given as an equation relating MN, the normal to the curve, to N, the point where that normal intersects the axis AN. He takes $u = MN$ and $t = AN$, and considers the case where the curve AM is given by the equation $u^2 = at$. Note that, though AM has the equation of a parabola, it is not a parabola because of the unusual coordinate system involved.

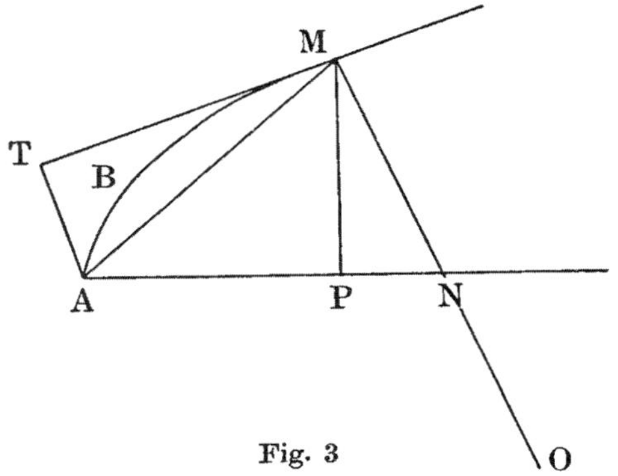

Fig. 3

This length, MN, is another of those quantities from the early days of calculus that is mostly forgotten today. It was known as the *normal*. We have seen that the problem of the tractrix was framed in terms of one of those forgotten quantities, what they called the *tangent*. Problems of logarithmic terms arose from the subtangent. Similarly, some problems were described in terms of a normal. Descartes [D] makes a brilliant use of the normal to calculate tangent lines in Book Two of his *Géométrie*. While we are on the subject, we mention the fourth geometric object of this type from early calculus, the *subnormal*, shown in Figure 3 as the segment PN.

Returning to Euler, he sets $x = AP$ and $y = PM$, and measures arc length with $ds = \sqrt{dx^2 + dy^2}$. Then he finds that the x-y coordinates are related to the t-u coordinates by

$$MN = u = \frac{y\,ds}{dx} \quad \text{and} \quad AN = t = x + \frac{y\,dy}{dx}.$$

Substituting these values into $u^2 = at$ gives

$$y^2\,ds^2 = ax\,dx^2 + ay\,dx\,dy.$$

This differential equation describes the curve $u^2 = at$ in ordinary x-y coordinates. Euler substitutes $ds^2 = dx^2 + dy^2$ and applies the quadratic formula to get

$$2y\,dy = a\,dx \pm dx\sqrt{a^2 + 4ax - 4y^2}.$$

Euler returns to his calculations relating t, u, x, y and s. Setting $dy = p\,dx$, he takes a few steps to find that $t = x + py$ and $y = \frac{u}{\sqrt{1+pp}}$ so

$$dy = p\,dx = \frac{du}{\sqrt{1+pp}} - \frac{up\,dp}{(1+pp)^{3/2}}.$$

Then differentiating $t = x + py$ and substituting gives $dt = dx + pp\,dx + y\,dp$, and a few calculations yield

$$\frac{p\,dt}{\sqrt{1+pp}} = du.$$

Euler writes, "from this equation, it is immediately obtained that"

$$p = \frac{du}{\sqrt{dt^2 - du^2}} \quad \text{and} \quad \sqrt{1+pp} = \frac{dt}{\sqrt{dt^2 - du^2}}.$$

This expresses x and y in terms of t and u as

$$y = \frac{u\sqrt{dt^2 - du^2}}{dt} \quad \text{and} \quad x = t - py = t - \frac{u\,du}{dt}.$$

This finally tells us something interesting. If the equation that relates t and u is algebraic, and if t and u are given using that peculiar length where u is measured normal to the curve rather than perpendicular to an axis, then the trajectory given in x and y is also algebraic.

Returning to the special case, and for the first time, Euler does call it a *casu speciali*, $u^2 = at$. Euler wants to find y in terms of x, that is, the equation of the tractrix when the string is pulled along this curve $u^2 = at$. First,

$$t = \frac{u^2}{a} \quad \text{and} \quad dt = \frac{2u\,du}{a}.$$

Then, $\sqrt{dt^2 - du^2} = \frac{du}{a}\sqrt{4u^2 - a^2}$. Substitution gives

$$y = \frac{1}{2}\sqrt{4u^2 - a^2} \quad \text{and} \quad x = \frac{u^2}{a} - \frac{a}{2}.$$

Eliminating u from these equations results in

$$y^2 = ax + \frac{a^2}{4}.$$

This is the first trajectory Euler has explicitly calculated. It is appropriate that a quadratic equation in Euler's peculiar "normal" coordinate system gives a trajectory that is a quadratic equation in the ordinary coordinate system. It makes one wonder what other secrets his peculiar coordinate system might hide.

Euler turns to another custom-designed coordinate, described in Figure 4. This one is based on tangents to the given curve, in the same way that the previous coordinate system was based on normals. The given curve is AM described in the usual way with coordinates $x = AP$ and $y = PM$. The line MT is tangent to the curve AM. The point T is where the tangent cuts the axis AP, and the point V is where the tangent cuts the axis perpendicular to AP at the point A. Euler sets $t = AT$ and $u = AV$, and supposes that the curve AM is described by a relation between t and u.

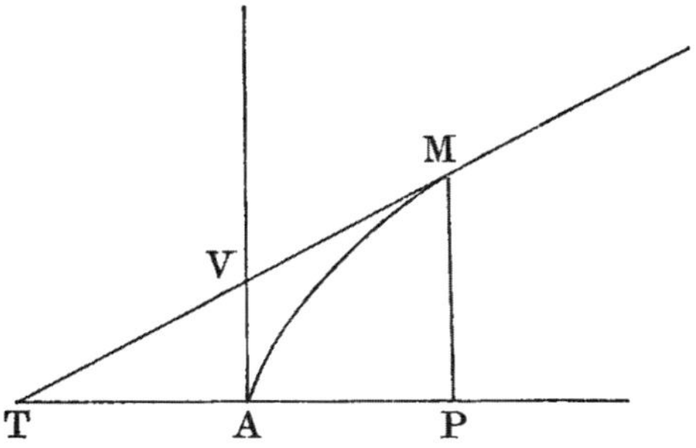

Fig. 4

The geometry of the situation immediately yields

$$t = AT = \frac{y\,dx}{dy} - x \quad \text{and} \quad u = AV = y - \frac{x\,dy}{dx}.$$

Taking, as before, $dy = p\,dx$ shows

$$t = \frac{y}{p} - x \quad \text{and} \quad u = y - px.$$

After a few substitutions and differentiations, he finds that

$$x = \frac{tt\,du}{u\,dt - t\,du} \quad \text{and} \quad y = u + \frac{ut\,du}{u\,dt - t\,du}.$$

As with the previous coordinate system, if the relation between u and t is algebraic, then the relation between x and y will also be algebraic.

Euler turns to yet another unusual coordinate system, this time without drawing a new figure. He modifies the system in Figure 4 by supposing that *TVM* instead is a parabola tangent to AM at M and with axis AP. Again he takes $x = AP$, $y = PM$, $dy = p\,dx$, $t = AT$ and $u = AV$. Then ordinary properties of the parabola yield

$$t : u^2 = t + x : y^2 \quad \text{and} \quad y^2 = u^2 + \frac{u^2 x}{t}.$$

Another property of parabolas, less well known today than in Euler's time, is that the subtangent to the parabola *TVM* at M is $2PT$, which equals $2t + 2x$. Because $dy = p\,dx$ so that $\frac{y\,dx}{dy} = \frac{y}{p}$, the property of subtangents tells us that

$$y = 2pt + 2px. \tag{1}$$

Euler divides

$$y^2 = u^2 + \frac{u^2 x}{t}$$

by $y = 2pt + 2px$ to get

$$y = \frac{u^2}{2pt}.$$

Then he substitutes this value for y into (1) to get

$$x = \frac{u^2}{4p^2t} - t.$$

Now, differentiating

$$y = \frac{u^2}{2pt} \quad \text{and} \quad x = \frac{u^2}{4p^2t} - t$$

gives

$$dy = p\,dx = \frac{u\,du}{pt} - \frac{u^2\,dt}{2ptt} - \frac{u^2 dp}{2p^2t}$$

and

$$dx = \frac{u\,du}{2p^2t} - \frac{u^2\,dt}{4p^2tt} - \frac{u^2\,dp}{2p^2t} - dt.$$

Combining these two equations so as to eliminate dx produces

$$\frac{u\,du}{2pt} + p\,dt = \frac{u^2\,dt}{4pt^2} \quad \text{or} \quad pp = \frac{u^2}{4tt} - \frac{u\,du}{2t\,dt}.$$

Now he substitutes this into his values for x and y to get

$$x = \frac{2tt\,du}{u\,dt - 2t\,du} \quad \text{and} \quad y = \frac{u^2\sqrt{dt}}{\sqrt{u^2\,dt - 2tu\,du}}.$$

As before, this shows that, if the relation between t and u is algebraic, then x and y will be algebraic as well.

Euler claims, but does not prove, that if *TVM* is any algebraic curve determined in this way by a relation between t and u, then the standard coordinates x and y will be algebraic and can be determined by using differential equations in the same way.

He provides yet another way of describing a curve (Figure 5). He says this a bit awkwardly. Lines RN are to be infinitely many straight lines between angle A. They are described by an equation involving t and u, where $AN = t$ and $AR = u$. The problem is to find a curve BM that always intersects these straight lines RN at right angles.

Euler is not concerned with whether or not such a curve exists for a given equation involving t and u, but rather with finding the curve assuming that it does exist. As usual, he takes

$$AP = x, \quad PM = y \quad \text{and} \quad dy = p\,dx.$$

Because RN is normal to BM at M, the geometry tells us that

$$PN = \frac{y\,dy}{dx} = py$$
$$t = x + py$$
$$dy : dx = p : 1 = t : u$$

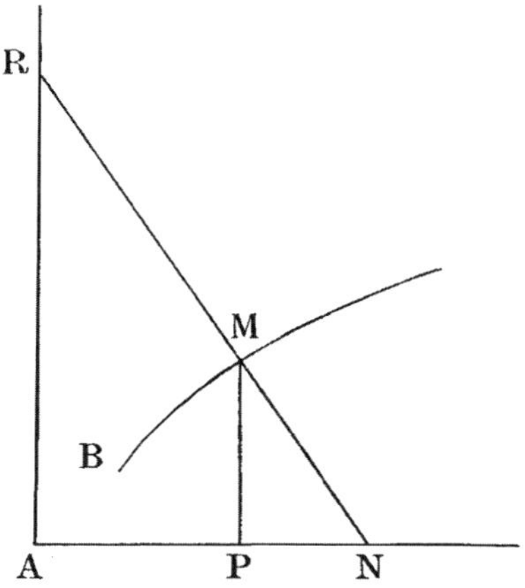

Fig. 5

so that

$$t = pu, \quad p = \frac{t}{u}, \quad dy = \frac{t\,dx}{u} \quad \text{and} \quad y = u - \frac{ux}{t}.$$

As usual, he differentiates, substitutes, multiplies and integrates to get

$$x = \frac{t}{\sqrt{tt + uu}} \int \frac{u\,du}{\sqrt{tt + uu}}$$

and he tells us that similarly

$$y = u - \frac{u}{\sqrt{tt + uu}} \int \frac{u\,du}{\sqrt{tt + uu}}.$$

This demonstrates that, if the integral is algebraic, then the curve BM will also be algebraic.

This is worth an example. Euler considers what sometimes appears in modern calculus textbooks as the "sliding ladder" problem, in which the segment RN is of constant length, say a. Then the algebraic equations relating t and u are

$$\sqrt{tt + uu} = a \quad \text{or} \quad u = \sqrt{a^2 - t^2}.$$

Then

$$\int \frac{u\,du}{\sqrt{tt + uu}} = \frac{-tt}{2a},$$

which is algebraic. Substituting gives

$$x = \frac{-t^3}{2a^2} \quad \text{and} \quad y = \frac{-(x + \sqrt[3]{2a^2x})}{\sqrt[3]{2a^2x}} \sqrt{a^2 - \sqrt[3]{4a^4x^2}}.$$

These, Euler assures us, lead to a sixth degree relation between x and y given by

$$(a^2 - x^2 - y^2)^3 = \frac{27}{4} a^2 x^4.$$

Euler generalizes this to a case when the curve connecting the sides of angle A is not just a straight line (Figure 6). Suppose that instead some other kind of curve is chosen to join points R and S, and it is required to find a curve that always cuts RS at right angles. Euler chooses the curve RS to be a quarter of an ellipse, and denotes $t = AS$ and $u = AR$. Suppose that BM is the curve required and that it cuts RS at M. Then, as usual, take $AP = x$, $PM = y$ and $dy = p\,dx$.

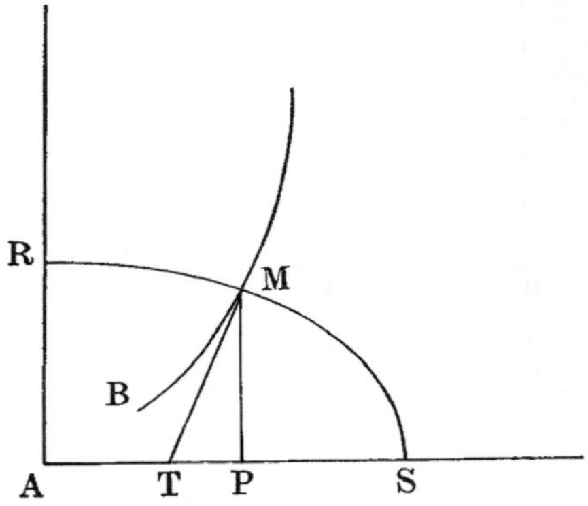

Fig. 6

The familiar formulas for ellipses give that

$$y = \frac{u}{t}\sqrt{tt - xx} \quad \text{or} \quad y^2 = u^2 - \frac{u^2x^2}{t^2}.$$

Euler's analysis proceeds as before, but the differentials are more complicated and so the whole derivation is about twice as long as the previous one. It concludes with a substitution $p = \frac{qtt}{uu}$ and a differential equation describing the curve being sought being

$$\frac{tu\,dq}{q} = \frac{(tt - uu)(q^2t^3\,du + u^3\,dt)}{q^2t^4 + u^4}.$$

He finishes this aspect of his analysis with an interesting example, though. He considers the case where the ellipses are all similar to each other, that is $u = nt$, for some fixed value of n. Then the defining differential equation becomes

$$\frac{dq}{q} = \frac{(1 - nn)(q^2\,dt + n^2\,dt)}{q^2t + n^4t}.$$

This, he finds, leads to the solution

$$x = b^{1-n^2}y^{n^2}$$

which Euler describes as a family of parabolas. To Euler, a variable to any positive even power gives a parabola.

This article is starting to wind down. We have reached section 37 of its 46 sections, and Euler is obligated to provide some motivation for this work on trajectories. He suggests that such curves determined by perpendiculars to other curves might be useful in describing centrifugal forces in mechanics. That is all the motivation we get for this class of problems.

Euler refers back to Figure 3 for his next example. He takes A to be a center of force and BM to be a trajectory. He sets $AM = t$ and lets TM be a tangent to BM at M, with T chosen so that AT is perpendicular to TM. Then he takes $TM = u$ and supposes that there is an equation relating t and u.

As always, he lets $AP = x$, $PM = y$ and $dy = p\,dx$ and gets relations giving x and y in terms of t and u. These relations involve an arctangent, so they are not necessarily algebraic, even if the relation between t and u is algebraic.

Still using Figure 3, Euler next draws MN perpendicular to the curve BM at B with N on the axis AP, and with O on the line MN being the center of curvature of the curve at M. He sets $MN = t$ and $MO = u$ and yet again supposes there is an equation relating t and u. The equations giving x and y involve $\int \frac{dt}{t-u}$ and so are not likely to be algebraic.

Euler expands on this example, supposing that MO is a constant multiple of MN, that is that the equation relating u and t is of the form $u = nt$.

Finally, Euler considers one more trajectory based on Figure 3. He lets $AM = s$ and $MO = r$ and continues as before, supposing there to be a relation between s and r, and seeking equations relating x and y.

This has been a rambling paper. It started with a discourse on tractrix curves, but, curiously for Euler, he did no examples in this part of the paper. Then he did a long series of examples relating trajectories described in terms of dynamic coordinate systems to ordinary x-y coordinates. This paper could as well have been classified as a geometry paper as a differential equations paper. It may be worth noting that it is the longest of Euler's papers we have yet seen, and the one with the most illustrations.

References

[D] Descartes, René, *The Geometry of René Descartes with a facsimile of the first edition*, tr. D E. Smith and M. L. Latham, Dover, NY, 1954.

[G] Gould, Steven Jay, *The Hedgehog, the Fox, and the Magister's Pox: Mending the Gap Between Science and the Humanities*, Harmony Books, NY, 2003.

[R] Rickey, V. Frederick, *Historical Notes for the Calculus Classroom*, unpublished manuscript, 1995.

25

E-52: Solutio problematum rectificationem ellipsis requirentium*

Solution of a problem requiring the rectification of an ellipse

Mathematicians of the 18th Century called themselves "Geometers". Geometers of the previous century had been, as Euler says, "agitated" by problems of arc lengths. Today, we consider areas and slopes and the relation between them to be the foundations of calculus. In Euler's day, though, there were not many problems involving slope or area. Instead of slope, as we saw in E-51, problems dealt with objects like subtangents and normals, objects that seem today like strange permutations on the concept of slope. Integration, also, was less often seen as leading to area. Instead, it was used as an algebraic operation as the inverse to differentiation, a process performed to rid an equation of its differentials. As a geometric action, it seemed to arise more often in problems of arc length than in problems of quadrature.

By Euler's time, geometers were pretty well convinced that the arc length of ellipses cannot be expressed in terms of well-known functions. Newton and Leibniz had written a bit on the subject. People still speculate why they did not write more. Perhaps they had interesting results but were too busy to write them down, or perhaps they just didn't get around to learning more. Wallis and Fagnano also worked on the topic.

Euler used series and differential equations to study ellipses in E-28, written in 1733. The Swedish mathematician Samuel Klingenstierna had published some related results in 1728, but it seems that Euler was not aware of Klingenstierna's work.

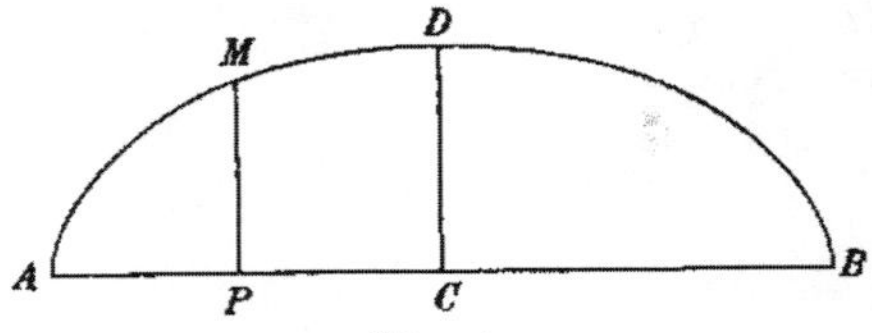

Fig. 1.

Euler first considers the family of ellipses described in Figure 1. The curve *AMDB* is a semi-ellipse over the axis *APCB*, and he takes the point *A* to be his origin. He assigns variables $AP = t$, $PM = u$, $AM = z$. He means this to be a family of infinitely many ellipses so he takes the length $AC = a$ to be a parameter, or, as he calls it, a *modulus*, which will be allowed to vary, and he takes $CD = c$ to be a constant. All the ellipses he considers will have the same vertical semiaxis. From the nature of the ellipse we get $u = \frac{c}{a}\sqrt{(2at - tt)}$. Euler has a particular problem in mind that becomes simpler if he sets $t = ax$, so that $u = c\sqrt{2x - xx}$. His differentials are $dt = a\,dx$ and

$$du = \frac{c\,dx - cx\,dx}{\sqrt{2x - xx}},$$

Comm. Acad. Sci. Imp. Petropol. 8 (1736) 1741, 86–98; *Opera Omnia* I.20, 8–20

so the arc length differential becomes

$$dz = \frac{dx\sqrt{2a^2x - a^2x^2 + c^2 - 2c^2x + c^2x^2}}{\sqrt{2x - xx}}.$$

If $a^2 - c^2 = b^2$, b would be the distance between the point C and a focus of the ellipse, not shown in the diagram. Then substituting into the arc length differential and integrating, Euler finds that the arc length is given by

$$z = \int \frac{dx\sqrt{c^2 + b^2(2x - xx)}}{\sqrt{2x - xx}}.$$

In this expression, he wants to treat both x and b as variables, so that

$$dz = \frac{dx\sqrt{cc + bb(2x - xx)}}{\sqrt{2x - xx}} + db \int \frac{b\,dx\sqrt{2x - xx}}{\sqrt{cc + bb(2x - xx)}}.$$

He notes that the integral in this expression ought to vanish when $x = 0$. Recall that this is how Euler treats constants of integration. He sets this integral equal to a new function, R, so that

$$\begin{aligned} R &= \frac{dz}{db} - \frac{dx\sqrt{cc + bb(2x - xx)}}{db\sqrt{2x - xx}} \\ &= \int \frac{b\,dx\sqrt{2x - xx}}{\sqrt{cc - bb(2x - xx)}}. \end{aligned}$$

Then

$$dR = \frac{b\,dx\sqrt{2x - xx}}{\sqrt{cc + bb(2x - xx)}} + db \int \frac{cc\,dx\sqrt{2x - xx}}{[cc + bb(2x - xx)]^{3/2}},$$

and here Euler again reminds us that the integral ought to be chosen so that it vanishes if $x = 0$. Euler introduces another new variable, S, so that

$$\begin{aligned} S &= \frac{dR}{db} - \frac{b\,dx\sqrt{2x - xx}}{db\sqrt{cc + bb(2x - xx)}} \\ &= \int \frac{cc\,dx\sqrt{2x - xx}}{[cc + bb(2x - xx)]^{3/2}}. \end{aligned}$$

This last formula itself cannot be integrated, but, as Euler says, it remains to be seen if it can be shown to depend on something that we have already seen, or will see, in this analysis. To look for such a relationship, Euler sets $S + \alpha R + \beta z = Q$, where α and β do not involve x and z, and where Q depends only on x and on constants, and where, again, Q ought to vanish when $x = 0$. We will learn more about α and β a little later. Then, holding b constant,

$$dQ = dS + \alpha\,dR + \beta\,dz.$$

This makes

$$dS = \frac{cc\,dx\sqrt{2x - xx}}{[cc + bb(2x - xx)]^{3/2}} \quad \text{and} \quad dR = \frac{b\,dx\sqrt{2x - xx}}{\sqrt{cc + bb(2x - xx)}}$$

$$\text{and} \quad dz = \frac{dx\sqrt{cc + bb(2x - xx)}}{\sqrt{2x - xx}}.$$

From this, he finds

$$\frac{dQ}{dx} = \begin{bmatrix} cc(2x - xx) + \alpha bcc(2x - xx) + \alpha b^3(2x - xx)^2 \\ + \beta c^4 + 2\beta b^2c^2(2x - xx) + \beta b^4(2x - xx)^2 \end{bmatrix} : [cc + bb(2x - xx)]^{3/2}\sqrt{2x - xx}.$$

Note the use of the ratio notation, because Euler regards a derivative as the ratio of differentials. Euler now makes a bit of a leap and supposes that Q has the form

$$Q = \frac{(\gamma x + \delta)\sqrt{2x - xx}}{\sqrt{cc + bb(2x - xx)}}.$$

He differentiates this form and matches terms with his expression for $\frac{dQ}{dx}$ to get equations relating the Greek letters:

I. $\gamma bb = \alpha b^3 + \beta b^4$
II. $\gamma b^2 = \alpha b^3 + \beta b^4$
III. $4\gamma bb = 2\gamma cc = 4\alpha b^3 + 4\beta b^4 - cc - \alpha bcc - 2\beta b^2c^2$
IV. $3\gamma cc - \delta cc = 2cc + 2\alpha bcc + 4\beta b^2c^2$
V. $\delta cc = \beta c^4$

The first two relations are repeated because they come from two different coefficients in the equation, though there seems to be no rationale for the first to contain the form bb while the second contains the form b^2. From these, we get values for the Greek variables

$$\alpha = \frac{1}{b}, \quad \beta = \frac{-1}{b^2 + c^2}, \quad \gamma = \frac{cc}{bb + cc}, \quad \text{and} \quad \delta = \frac{-cc}{bb + cc}.$$

Substituting these values into $S + \alpha R + \beta z = Q$ gives

$$\frac{cc(x - 1)\sqrt{2x - xx}}{(bb + cc)\sqrt{cc + bb(2x - xx)}} = S + \frac{R}{b} - \frac{z}{b^2 + c^2}.$$

Then a raft of substitutions for x, bb, dx and db lead, after *nimis prolixos calculos*, "not too many long calculations", to

$$\frac{z}{bb + cc} = \frac{cc(1 - x)\sqrt{2x - xx}}{(bb + cc)\sqrt{cc + bb(2x - xx)}} - \frac{dx}{b\,db}\sqrt{\frac{cc + bb(2x - xx)}{2x - xx}}$$
$$- \frac{2b\,dx}{db}\sqrt{\frac{2x - xx}{cc + bb(2x - xx)}} + \frac{cc\,dx^2(1 - x)}{db^2(2x - xx)^{3/2}\sqrt{cc + bb(2x - xx)}}$$
$$+ \frac{dz}{b\,db} + \frac{1}{db}d.\frac{dz}{db} - \frac{1}{db}d.\frac{dx}{db}\sqrt{\frac{cc + bb(2x - xx)}{2x - xx}}.$$

Note the d-dot notation to denote second differentials.

Now this formidable expression is a differential equation of the second degree in which z, x and b are the variables. Euler will make use of it in the following problem.

Problem 1. If the curve *EMN* (Figure 2) is constructed on the axis *APQ* and its application (ordinate) PM at a point P is equal to the arc length of a quarter of an ellipse given by the curve AF, whose semiaxis is equal to the abscissa AP, holding the other axis constant as AE or PF, the problem is to find an equation between the abscissa AP and the application PM that expresses the nature of the curve *EMN*.

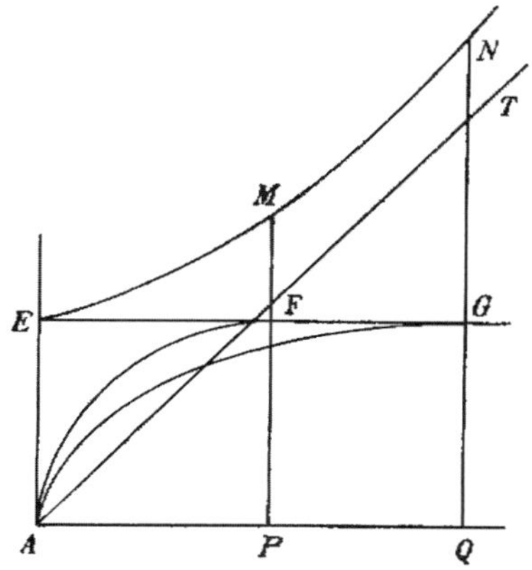

Fig. 2.

Clearly, the curve *EMN* does, in fact, pass through the point E, for if the semiaxis is zero, then the quarter ellipse is only the line segment AE. Euler draws the straight line AT, at an angle to the axis that Euler describes as *semirectum*, or "half-right." Euler notes that the curve *EMN* will be asymptotic to this line AT.

Euler assigns variables as $AE = c$, $AP = t$ and $PM = AF = z$. In the special case when the two semiaxes are equal, that is when $t = a$, it follows that $x = 1$. Setting $x = 1$ greatly simplifies the formidable differential equation above which reduces to

$$\frac{z}{bb + cc} = \frac{dz}{b\,db} + \frac{1}{db} d.\frac{dz}{db}.$$

A couple of substitutions change this to the form Euler wants. First he substitutes $bb = a^2 - c^2 = t^2 - c^2$ to obtain $b\,db = t\,dt$ and

$$db = \frac{t\,dt}{\sqrt{tt - c^2}},$$

then, holds dt constant so that

$$ddb = -\frac{cc\,dt^2}{(tt - cc)^{3/2}}.$$

All this makes

$$d.\frac{dz}{db} = \frac{ddz\sqrt{tt - cc}}{t\,dt} + \frac{cc\,dz}{tt\sqrt{tt - cc}}$$

which leads to

$$\frac{z}{tt} = \frac{dz}{t\,dt} + \frac{ddz(tt - cc)}{tt\,dt^2} + \frac{cc\,dz}{t^3\,dt}$$

and then to

$$tz\,dt^2 = (tt + cc)dtdz + t\,ddz(tt - cc).$$

This is the differential equation Euler has been looking for, so he closes his solution with "Q. E. I."

Most readers will be familiar with the notation "Q. E. D." to end a proof. It stands for the Latin *Quod Erat Demonstrandum*, "Which was to be shown." Some readers will be familiar with Euclid's Elements, where he ends propositions that are constructions with "Q. E. F.", *Quod Erat Faciendum*, "Which was to be done." This conclusion, "Q. E. I." is less common. Euler uses it to end "Problems" whose statement usually asks us to "find" something. "Q. E. I." stands for *Quod Erat Inveniendum*, "Which was to be found."

Before moving on to the next Problem, Euler wants to discuss this equation a bit. He remembers from E-10 how to reduce this second degree equation to a first degree equation. He takes $z = e^{\int s\,dt}$, where he reminds us again that $l\,e = 1$. He has adopted the symbol e permanently now. This substitution makes $dz = e^{\int s\,dt} s\,dt$ and $ddz = e^{\int s\,dt}(dsdt + ss\,dt^2)$, so that

$$t\,dt = (t^2 + c^2)s\,dt + t(tt - cc)\,ds + ts^2(t^2 - c^2)\,dt.$$

He adds that this equation can be reduced to the Ricatti equation using the methods from E-31, but does not explain how.

Now he is ready to move on to:

> **Problem 2.** Given infinitely many ellipses *AOF*, *ANG*, *AMH* (Figure 3), which all have the same semiaxis AE, but with the other semiaxes variables, AI, AK and AL, to find an equation for the curve *BONMC*, which all have arc lengths on their respective ellipses equal to AO, AN, AM, etc.

From the statement of this problem, we know to expect a "Q. E. I." at the end of the solution.

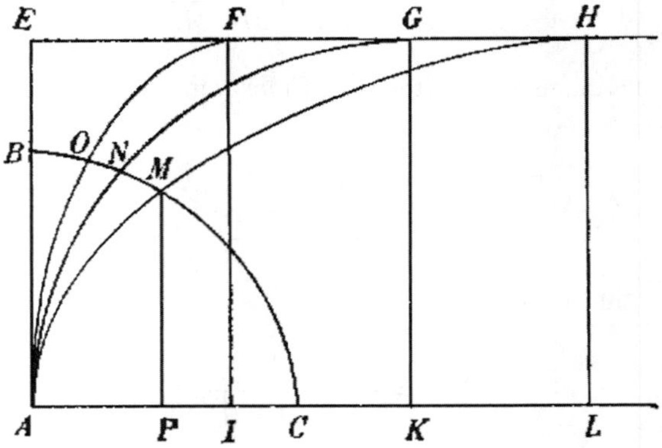

Fig. 3.

Euler chooses from the axis AC any "application" MP. What Euler called an "application," we now call an "ordinate," the vertical line segment connecting a point on a curve to the x-axis. Euler assigns variables $AP = t$, $PM = u$ and $AE = c$. Take the semiaxis of the ellipse *AMH* to be equal to a and take the arc length AM of that segment of the ellipse to be equal to the constant f. Now, set $x = \frac{t}{a}$ and $b = \sqrt{a^2 - c^2}$, as before. Then $z = f$ and $u = c\sqrt{2x - xx}$.

Euler substitutes all these into the differential equation he found above, just before he began Problem 1, a messy but straightforward calculation that we will omit. Then he multiplies by

$$\sqrt{cc + bb(2x - xx)} = \frac{\sqrt{c^4 + bbuu}}{c}$$

and reduces to get

$$\frac{f\sqrt{c^4 + bbuu}}{a^2 c} = \frac{cu(1-x)}{a^2} - \frac{c^3\,dx}{bu\,db} - \frac{3ubu\,dx}{c\,db} + \frac{c^5\,dx^2(1-x)}{u^3\,db^2} - \frac{(c^4 + bbuu)}{cu\,db} d.\frac{dx}{db}.$$

Euler asks the reader to do one last round of substitutions. He tells us that if we substitute

$$\frac{\sqrt{tt - ccxx}}{x}$$

for b and also

$$x = \frac{c - \sqrt{cc - uu}}{c},$$

then it will produce a second degree differential equation in the variables t and u that will describe the curve $BONMC$. Q. E. I.

Note that, in these problems, Euler regards a differential equation as a sufficient answer. He does not require any further information about that differential equation, for example the form it may have. This represents another step in the evolution of Euler's thought towards fully exploiting the ideas and notations of functions. He isn't there yet, though.

In Figure 4, Euler turns the previous problem on its side and supposes that infinitely many ellipses AMF, ANG and AOH have the horizontal axis AC and the center C in common, but that their vertical axes CF, CG, CH differ. Again, he would like to find the curve MNO such that the arc lengths AM, AN, AO are all equal.

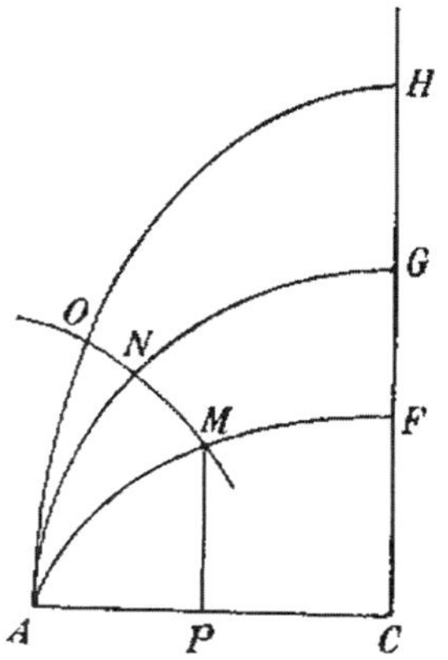

Fig. 4.

He assigns variables $AC = c$, $AP = t$, $PM = u$ and the arc $AM = z$. With this, he repeats, almost word for word, the descriptions of the functions Q, R, S, the matching of the coefficients α, β, γ and the derivation of the large differential equation. Using these results, he can state and solve the problem illustrated in Figure 4.

Because the problem and its solution are virtually identical to Problem 2, one wonders why Euler chose to do it again.

And this gets us to the end of E-52. Euler will set aside his studies of ellipses until 1750, though once he returns to them, he will publish seven papers on the subject in the 1750s, four more in the 1760s, two in the 1770s, three more in the 1780s before his death in 1783, and fifteen more will be published posthumously, the last in 1862, 79 years after his death.

It is not clear why Euler did not write about ellipses for fourteen years. It was apparently not because he became aware of Klingenstierna's results. They exchanged letters in 1745, but those seem to have dealt with questions of electricity and not of ellipses. Perhaps Euler thought in 1736 that he had exhausted the subject, but when he returns to it in 1750, he makes a series of beautiful discoveries that eventually lead to results about elliptic functions and elliptic integrals that are still important today.

26

E-53: Solutio problematis ad geometriam situs pertinentis*

Solution of a problem relating to the geometry of position

The Königsberg Bridge Problem is Euler's most famous work. It has been completely translated into English at least twice [BLW, N], and a translation of the first 13 of its 21 sections appears in [C] and [St]. It is a little surprising that David Eugene Smith [Sm] does not include it in his *Source Book*.

This paper marks the birth of graph theory. Its publication date became part of the title of the book by Biggs, Lloyd and Wilson, *Graph Theory 1736–1936*.

The paper itself is somewhat different from its reputation, both in what it contains and in what it does not. For example, there is no mention of graphs, vertices, edges or degrees. Instead, it keeps the vocabulary of geography and uses bridges and land areas. Euler explains his result in two different ways, but his explanations are a little too sketchy to call them proofs. Only the second explanation is mentioned in most of today's descriptions of the paper. The conscientious reader is encouraged to study the complete paper rather than be content with the summary and discussion here. The work by Newman [N] is easier to find, but the work by Biggs, Lloyd and Wilson [BLW] includes a nice 17th century map of Königsberg, as well as some good commentary. Both of these translations capture the delightful spirit of the original Latin.

Euler opens the paper discussing not Königsberg, but Leibniz. He credits Leibniz with the idea of *Geometriam situs*, or "geometry of position," a branch of geometry that Leibniz envisioned would be concerned only with the properties of position and not properties of distances. The idea of *Geometriam situs* plays a minor role in the works of Leibniz. He wrote nothing to develop the concept nor to explain what kinds of problems he had in mind. It is likely that Euler encountered the idea not by reading Leibniz directly but through his conversations with Johann Bernoulli. Euler writes that when he encountered this problem (he has not yet stated the problem), he recognized that it must be what Leibniz meant, because it seemed to be geometrical, but did not require the measurement of distances.

Euler now describes the problem (Figure 1).

Figure 1 is a map of the Seven Bridges of Königsberg. Euler tells us that the island A is called the *Kneiphof* but does not name the other landmasses, nor does he mention the name of the River *Pregel*. He denotes the bridges a, b, c, d, e, f and g, and states the question as whether anyone could arrange a route in such a way that he would cross each bridge exactly once. Euler does not mention whether or not he requires the person to be back at the starting point at the end of the route. He says that one could list all possible routes and then examine them for a solution, but that would be a long task for Königsberg, and virtually impossible for problems with more bridges.

Euler sees a notation that will facilitate his analysis. He will use AB to indicate that the traveler crosses from area A to area B, whether the traveler uses bridge a or bridge b. Later, he will distinguish by writing AaB or AbB, but for now he does not need to specify particular bridges. Then, a

---

**Comm. Acad. Sci. Imp. Petropol.* 8 (1736) 1741, 128–140; *Opera Omnia* I.7, 1–10

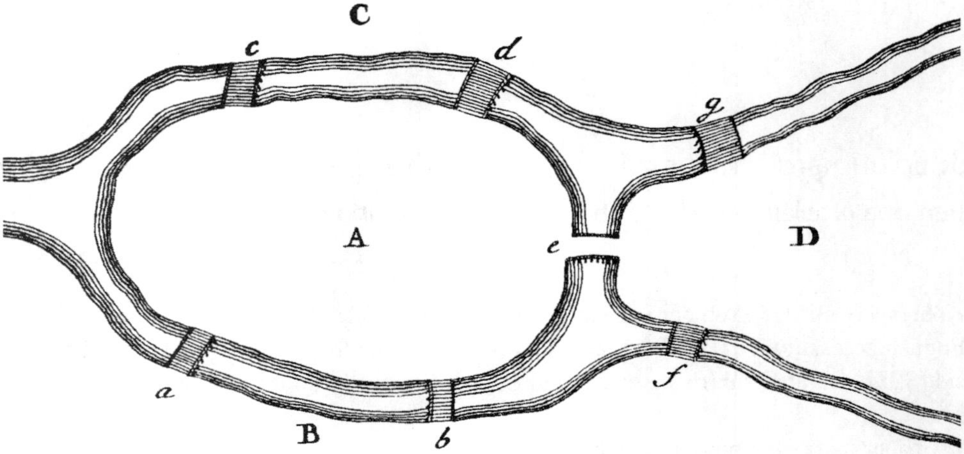

Fig. 1

sequence of three crossings can be represented using four letters. Euler gives the example *ABDC*, which is to mean that the traveler starts in *A*, crosses to *B* by whichever bridge, then goes to *D* and stops in *C*. He notes that a solution to the Königsberg Bridge Problem would require eight letters because there are seven bridges. Thus, the problem is reduced to finding such a sequence of eight letters.

There are constraints on the sequences given by the bridges. Because there are two bridges between *A* and *C*, the two letters *A* and *C* must be adjacent twice in the sequence, once for each crossing. Euler asks whether such a sequence is possible, and observes that if it is impossible, then there can be no solution. He turns to seeking a rule to tell him whether or not such a sequence exists. To find such a rule, he considers a simpler problem, illustrated in Figure 2.

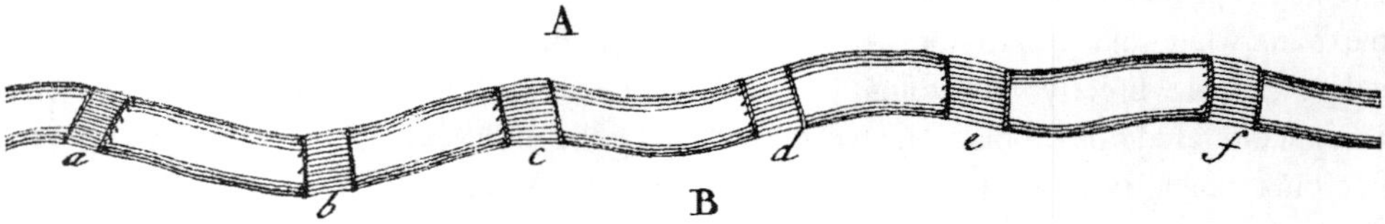

Fig. 2

In Figure 2, Euler considers area *A* and any number of bridges *a*, *b*, *c*, etc. Euler notes that if there is only one bridge, then the letter *A* must occur exactly once in the sequence, whether the traveler starts in *A* or finishes in *A*. If there are three bridges, then *A* must occur twice, and if there are five bridges, *A* must occur three times. In general, if there are any odd number of bridges leading to an area *A*, then the number of times *A* must occur in the sequence is half of one more than the number of bridges.

Applying this to the case of Königsberg, the letter *A* must occur three times, because there are five bridges to the area *A*. Likewise, *B*, *C* and *D* must occur twice each, because they each have three bridges. This means the length of the sequence giving a solution, if it exists, must be $3 + 2 + 2 + 2 = 9$. But there are only seven bridges, so a sequence crossing each bridge only once must have length 8. Thus no solution exists.

This analysis is perfectly valid, and it is quite different from the analysis usually described in modern sources.

Euler now seeks to generalize his results. He notes that his argument applies to any arrangement of areas and bridges in which the number of bridges to each area is odd, and he seeks to extend it to arrangements involving even numbers of bridges. He does not consider issues of connectivity.

Euler notes that if the number of bridges to A is even, then the number of times A occurs in the sequence of letters depends on whether the journey begins at A. He explains that if A has two bridges and the journey begins at A, then A must occur twice, but if the journey begins elsewhere, then A will occur only once. Euler patiently explains for four and for six bridges as well, before observing that, if the journey begins elsewhere, then the number of times A must occur is half the number of bridges, and if the journey begins at A, then A must occur one more time than that.

Euler takes a paragraph to state these two results together, the one about areas with an odd number of bridges, and the one about areas with an even number of bridges.

Euler describes his algorithm for determining whether or not a solution exists. He makes a table of areas and the corresponding numbers of bridges. He adds a third column, the minimum number of times the area must occur in a solution. At the top of the third column, he says to write the number that is one more than the number of bridges, but his worksheet for the Königsberg Problem omits this. For the Königsberg Problem, he gets the following table:

	Bridges	
A,	5	3
B,	3	2
C,	3	2
D,	3	2

He puts an asterisk next to any area with an even number of bridges, but in this example, there are none. He adds the numbers in the last column and sees that the sum is more than 8, the number of bridges, so a solution is impossible.

Euler proposes a second example (Figure 3), involving two islands, A and B, and four rivers. He does not claim that this example represents any real place, as the first represents Königsberg. His

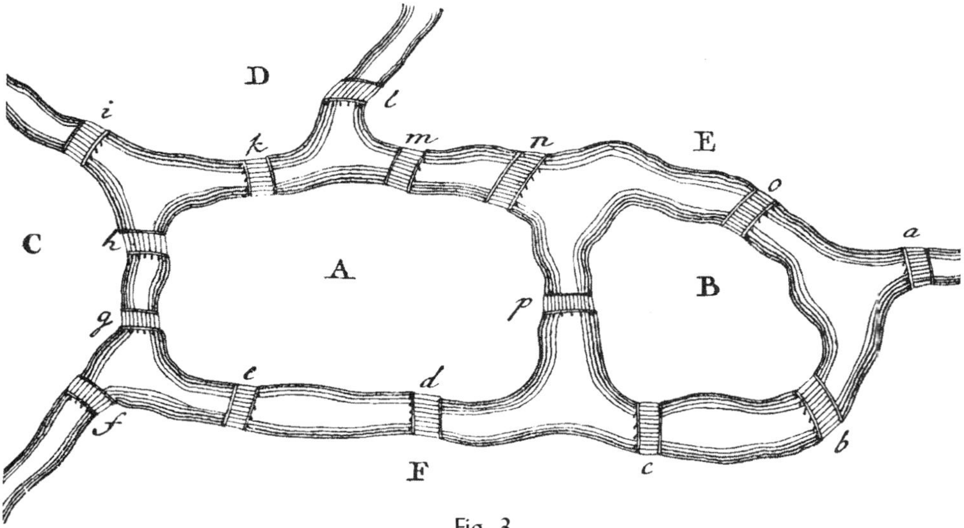

Fig. 3

worksheet is as follows:

		16
A^*,	8	4
B^*,	4	2
C^*,	4	2
D,	3	2
E,	5	3
F^*,	6	3
		16

Because the number of symbols available, the 16 at the top of the last column, is not less than the minimum number of symbols necessary, the 16 at the bottom of the last column, Euler concludes that a solution must exist. He is confusing necessary conditions with sufficient conditions here. He is also making the unstated assumption that the underlying graph is connected.

He notes that the journey cannot begin with any region marked with an asterisk, since then that symbol would occur one more time, and there would no longer be enough symbols available. Therefore, the journey must begin at D or at E. Euler proposes one solution as

$$EaFbBcFdAeFfCgAhCiDkAmEnApBoElD.$$

In this notation, he describes both the bridges and the land areas.

Euler has what he calls a much simpler method. Before telling it to us, he makes a few preliminary observations. First, the numbers of bridges to each area given in the second column sum to twice the total number of bridges, because each bridge is counted twice. Today we would say that the sum of the degrees of the vertices is twice the number of edges.

From this, the total number of bridges must be an even number, and this is impossible if exactly one of these numbers is odd, or three are odd, etc. Thus, the number of areas having an odd number of bridges must be an even number. In the Königsberg Problem, that even number is four, areas A, B, C and D. In the example in Figure 3, that even number is two, areas D and E.

In the case that all the numbers in the second column are even, Euler deduces that one of the areas must occur both at the beginning and at the end of the sequence in order to use all the available symbols. In passing, Euler notes that if the problem requires each bridge to be crossed twice, then the problem always has a solution, since each bridge can be treated as two bridges, and each area will then have an even number of bridges.

If two of the numbers in the second column are odd, he tells us that the journey must start at one of the areas with an odd number of bridges and end at the other in order to have the correct number of characters in the sequence. Finally, if more than two of those numbers are odd, then there will not be enough symbols for all the crossings necessary and the problem will have no solution.

Having said this in narrative, Euler says it again as a list of three rules, each depending on the number of regions with an odd number of bridges.

Finally, Euler turns to the question of actually finding the required journey, once one knows that it exists. He offers only one shortcut. If two regions are connected by two or more bridges, he suggests mentally removing them, in pairs, and solving the problem using the remaining bridges. Then that solution is easily extended to a complete solution.

So, we see that Euler solved the problem twice and that neither solution is the one seen so often today. Euler never used graphs, though most of what he says is easily translated into the language of graphs. Also, the general image of the content of this paper has been distorted somewhat by what is omitted in the abbreviated translations.

It is sometimes said that this is Euler's only work in graph theory. This is not true. In 1759, Euler will publish E-309, "Solution d'une question curieuse que ne paroit soumise à aucune analyse" in which he solves the problem of the Knight's Tour [S]. This is a problem we would now describe as a "Hamiltonian tour" problem, named after William Rowan Hamilton, who worked on such problems in the 1850s. It is much more difficult and still the subject of active research.

References

[BLW] Biggs, Norman L, E. Keith Lloyd and Robin J. Wilson, *Graph Theory 1736–1936*, Clarendon Press, Oxford, 1976.

[C] Calinger, Ronald, ed., *Classics of Mathematics*, Prentice Hall, Englewood Cliffs, 1995.

[N] Newman, James R., *The World of Mathematics*, v. 1, Simon and Schuster, New York, 1956.

[S] Sandifer, Ed, Knight's Tour, in *How Euler Did It*, MAAOnline (www.maa.org), April 2006.

[Sm] Smith, David Eugene, *A Source Book in Mathematics*, Dover, New York, 1959.

[St] Struik, D. J., ed., *A Source Book in Mathematics 1200–1800*, Princeton U. P., 1986.

Interlude: 1736

World events

In 1736, England amended its Witchcraft Act to eliminate capital punishment for witches and sorcerers. After 1736, they were only imprisoned and fined.

Joseph Louis Lagrange (1736–1813) was born. He would succeed Euler as Chair of the Mathematics Department when he left the Berlin Academy in 1766. Also born were James Watt (1736–1819), who perfected the steam engine and maybe English mathematician Edward Waring (1736–1798). Some sources say he was born in 1734.

The fourth Emperor of the Qing Dynasty in China, Qianlong (1711–1799) ascended to the throne in 1736. He ruled until 1796.

In Euler's life

Euler's great two-volume physics book, *Mechanica*, E-15 and E-16, was published in 1736.

Euler's other work

Euler finished writing the second volume of the *Mechanica*, E-16, in 1736. He also wrote a paper in dynamics, E-69, De communicatione motus in collisione corporum sese non directe percutientium, "On the communication of motion in the collision of bodies when the collision is not direct."

Euler's mathematics

Euler produced only three mathematics papers in 1736. The first, E-54, Theorematum quorundam ad numeros primos spectantium demonstratio, "Proof of a theorem about looking at prime numbers," contains his first proof of Fermat's Little Theorem. He wrote another paper on series, E-55, Methodus universalis series summandi ulterius promota, "Further universal methods for summing series," a kind of sequel to E-46 and E-47. He began to rekindle his interest in the calculus of variations with E-56, Curvarum maximi minimive proprietate gaudentium inventio nova et facilis, "A new and easy way of finding curves enjoying properties of maximum or minimum."

27

E-54: Theorematum quorundam ad numeros primos spectantium demonstratio*

Proof of some theorems about looking at prime numbers

Perhaps Euler's most important result in number theory is the theorem we now call the Euler-Fermat Theorem. Its modern statement requires the use of what is now called the Euler phi function. If n is a positive integer, then $\varphi(n)$ is defined as the number of positive integers less than n and relatively prime to n. Now that we have the phi function, we can state the Euler-Fermat Theorem:

Theorem. *Let n be a positive integer greater than 1 and let a be any positive integer relatively prime to n. Then n divides $a^{\varphi(n)} - 1$.*

Sometimes, the conclusion of this theorem is stated as

$$a^{\varphi(n)} \equiv 1 \pmod{n}.$$

This notation, however, is due to Gauss in 1801. He introduced "mod n" notation, the congruence symbol, and even the φ to denote Euler's phi function. Euler will define the phi function and derive most of its important properties in 1760 in E-271, in the same paper in which he proves the Euler-Fermat Theorem in its full generality. However, he will not give the function a name.

If n is a prime number, then it is usually denoted p. Then $\varphi(p) = p - 1$. In the case of prime numbers, the theorem is usually known as Fermat's Little Theorem, and can be stated as

Theorem. *If p is a prime number, then the expression $a^{p-1} - 1$ is divisible by p unless p divides a.*

Euler will prove Fermat's Little Theorem and the Euler-Fermat Theorem several times in his career. He first mentions the theorem in E-26, written in 1732. The main result in E-26 is to show that the fifth Fermat number, $2^{2^5} + 1$, is not a prime number, thus disproving a conjecture of Fermat. He lists several special cases of the Euler-Fermat Theorem among the six "theorems" at the end of the paper, saying they "can be proved by testing" and that "for which I do not have a proof but their truth is most certain."

Euler gives proofs of this theorem at approximately ten-year intervals for several decades. Here, in E-54, Euler gives a proof by mathematical induction. It is his first use in print of this important technique.

Eleven years later in 1747, Euler will give another proof of Fermat's Little Theorem that is only slightly different from the induction proof of E-54. In that paper, E-134, Euler tries to explain how he came to find the factor of the fifth Fermat number, and the proof of Fermat's Little Theorem comes up in that explanation.

**Comm. Acad. Sci. Imp. Petropol.* 8 (1736) 1742, 141–146; *Opera Omnia* I.2, 33–58

In 1758 Euler produces a completely different proof of Fermat's Little Theorem. In that paper, E-262, he uses the remainders mod p of the geometric progression $1, a, a^2, a^3, \ldots$. He shows that there can be at most $p-1$ distinct remainders, and that each remainder must occur the same number of times and in an order that repeats among the first $p-1$ terms of the progression. As a consequence, the $(p-1)$ term of the sequence must be a 1 (calling the 1 at the beginning of the sequence the 0th term), and the theorem follows immediately.

Just two years later in E-271, Euler states and proves the full Euler-Fermat Theorem by first developing the properties of the Euler phi function, and then applying techniques like those he used in E-262.

The whole exciting story spans almost 30 years of Euler's career. It follows the classic plot of mathematics, from evidence and conjecture, by Fermat and Euler and perhaps even Diophantus, to the search for proofs, followed by an ever-improving sequence of proofs in special cases, and finally generalization.

Now we must return to 1736 and Euler's first proof of the Euler-Fermat Theorem. It is a short paper, only six pages long, and begins with an acknowledgement of Fermat.

> Formerly, most of the arithmetic theorems of Fermat were given without proofs, and perhaps they were true, and not just exceptional properties of certain numbers. Still, the science of numbers is itself valid, and seeing what the limits of Analysis of numbers are ought to be vigorously sought.

Until about 1900, what we now call "number theory" was called "arithmetic." It spent a few years under the name "higher arithmetic" before the name "number theory" evolved. We see that Euler is not sure how much in number theory can actually be proved and how much must be left to observation, and he reminds us of the shortcomings of the latter: "I speak of course of that theorem whose falsity has only this year been announced, in which Fermat asserted that all numbers of the form $2^{2^n}+1$ are prime numbers."

Euler's chronology here is confusing. He had published this result in E-26 in the 1732/3 volume of the *Commentarii*, which was published in 1738. Now, he is writing E-54 for the 1736 volume of the *Commentarii*, which will be published in 1741. This suggests that either the result in E-26 was announced sometime between when it was presented to the *Commentarii* in 1732/3 or that he was writing E-54 before 1736. His correspondence with Goldbach and others gives us no clues.

So, Euler proclaims his plan to prove results in number theory, not only for the results themselves, but also to explore what can be established in the subject and to lessen the need for observation and speculation. He goes on to state Fermat's Little Theorem in the form given above. He skips over the case $p = 2$, which really is trivial, and takes p to be an odd prime. Then he considers the case $a = 2$, and writes:

> If p is an odd prime number, then the formula $2^{p-1} - 1$ can always be divided by p.

His proof goes as follows:

> In the place of 2 is put $1 + 1$ and we get
>
> $$(1+1)^{p-1} = 1 + \frac{p-1}{1} + \frac{(p-1)(p-2)}{1\cdot 2} + \frac{(p-1)(p-2)(p-3)}{1\cdot 2\cdot 3} + \frac{(p-1)(p-2)(p-3)(p-4)}{1\cdot 2\cdot 3\cdot 4} + \text{etc.}$$

Subtract the first term 1 from both sides, and it will be

$$(1+1)^{p-1} - 1 = 2^{p-1} - 1$$
$$= \frac{p-1}{1} + \frac{(p-1)(p-2)}{1 \cdot 2} + \frac{(p-1)(p-2)(p-3)}{1 \cdot 2 \cdot 3} + \text{etc.}$$

Note that there are an even number, $p-1$, of terms in this series, and that they are binomial coefficients, $\binom{p-1}{k}$, $k = 1, 2, \ldots, p-1$. Pair them up, and apply the familiar binomial identity,

$$\binom{p-1}{k-1} + \binom{p-1}{k} = \binom{p}{k}$$

to get, in Euler's notation,

$$2^{p-1} - 1 = \frac{p(p-1)}{1 \cdot 2} + \frac{p(p-1)(p-2)(p-3)}{1 \cdot 2 \cdot 3 \cdot 4} + \text{etc.}$$

Note that the last term of this series will be p, so that each term on the right contains the factor p. Hence the formula on the left, $2^{p-1} - 1$, is divisible by p. Q. E. D.

Euler offers us a second proof of this result. Write

$$2^p = (1+1)^p = 1 + \frac{p}{1} + \frac{p(p-1)}{1 \cdot 2} + \cdots + \frac{p}{1} + 1.$$

Subtract $1 + 1$ from both sides to get

$$2^p - 2 = \frac{p}{1} + \frac{p(p-1)}{1 \cdot 2} + \cdots + \frac{p}{1}.$$

The right-hand side is obviously divisible by p (and all terms are integers), so the left must also be divisible by p. But p and 2 are relatively prime, so p divides $2^{p-1} - 1$. Q. E. D.

Instead of going straight to the induction step, Euler, as usual, makes it easy for us and does the case $a = 3$ as another theorem.

Theorem. *Denoting by p a prime number other than three, then the formula* $3^{p-1} - 1$ *can always be divided by that prime number.*

Euler's proof of this parallels his second proof of the previous result, rather than the first. He begins by writing 3^p as $(1+2)^p$ and then expands the second formula using the Binomial Theorem to get:

$$3^p = 1 + \frac{p}{1} \cdot 2 + \frac{p(p-1)}{1 \cdot 2} \cdot 4 + \cdots + \frac{p}{1} \cdot 2^{p-1} + 2^p.$$

He subtracts $1 + 2^p$ from both sides, leaving a right-hand side where every term contains a factor of p, and giving $3^p - 2^p - 1$ on the left, and p must divide that formula.

But, $3^p - 2^p - 1 = 3^p - 3 - 2^p + 2$, and we know, from the previous theorem, that p divides $2^p - 2$, leaving that p must divide $3^p - 3$. Since p is prime and not 3, this implies that p divides $3^{p-1} - 1$. Q. E. D.

Now, Euler's induction step follows the same procedure as the case $a = 3$, though he says it in a way that is a little strange to the modern reader:

Theorem. *Denoting p a prime number, if $a^p - a$, can be divided by p, then the formula $(a+1)^p - a - 1$ can also be divided by that number p.*

Euler's proof of this is almost identical to his proof in the case $a = 3$.

This theorem is, of course, the induction step of the induction proof. Euler leaves the last paragraph of this paper to explain how these theorems, the base step and the induction step, together prove the general case. It seems that the mechanics of an inductive proof still needed to be explained more explicitly than we would explain them now.

This ends Euler's first proof of Fermat's Little Theorem. We will get to see him prove it several more times in his career, and he will use the result many more times. The theorem will return in E-98, when Euler uses it to prove Fermat's Last Theorem, in the case $n = 4$, as well as other theorems about the representation of numbers as sums of powers.

References

[M] Miller, Jeff, First Use of Symbols in Mathematics, `http://members.aol.com/jeff570/mathsym.html`, as updated September 7, 2001.

28

E-55: Methodus universalis series summandi ulterius promota*

Further universal methods for summing series

Euler returns to his beloved series. In each of volumes 5 and 7 of the *Commentarii*, Euler has published two papers on infinite series. He only wrote one paper on series for volume 6, but this is his third on the subject for volume 8. It serves as a sequel to E-46 and E-47. His interest in the subject will continue, with three more papers in volume 9, none in volume 10, but four in volume 11 and one in volume 12. Then he will abandon the subject for most of the 1740s and 1750s, publishing only four papers on series in that time, before publishing 27 papers on series after 1760. He additionally left 33 papers on series for posthumous publication, mostly written in the 1770s and 1780s. So, we see that series were a major research interest early and late in Euler's career. In the middle of his career, though, they appeared prominently in his books on calculus, algebra, and in the *Introductio* and the *Institutiones calculi differentialis*, and he incorporated them in his work on other subjects, but they seldom stood alone as the focus of a research paper.

This is a short and technical paper, only 12 pages long. It apparently had little impact in its own time, for none of Euler's letters (and there are over 3000 letters) mention it. This work builds immediately on E-46 and E-47. Euler acknowledges that this is a sequel when he begins "The universal methods of summing series, which I have explained earlier this year ... " and he restates the formula from E-46 and E-47, though in a slightly different form, for the sum of the first x terms of a sequence with general term given as X:

$$\int X\,dx + \frac{X}{1\cdot 2} + \frac{dX}{1\cdot 2\cdot 3\cdot 2dx} - \frac{d^3X}{1\cdot 2\cdot 3\cdot 4\cdot 5\cdot 6\,dx^3} + \text{etc.}$$

Note that if X is a polynomial, then

$$\frac{d^nX}{dx^n} = 0$$

for sufficiently large n and the series is finite. In this case, the formula is exact. He writes that in this paper he intends to demonstrate that this formula works even if X involves quantities that are "exponential, or even transcendental," though he will start with algebraic quantities. He begins with the finite sum of a progression, which he represents as

$$\overset{a}{A} + \overset{a+b}{B} + \overset{a+2b}{C} + \cdots + \overset{x}{X} = S.$$

Here, upper case letters represent the values to be summed, and the lower case letters above represent the indices of each term. He is thinking of the capital letters as the values we get when we substitute the lower case letters into some function. He wants another expression for X that does not require

Comm. Acad. Sci. Imp. Petropol. 8 (1736) 1741 147–158; *Opera Omnia* I.14 124–137

that x be an integer, so he replaces x with $x - b$ and makes the appropriate subtraction to produce

$$X = \frac{b\,dS}{1\,dx} - \frac{b^2\,ddS}{1\cdot 2\,dx^2} + \frac{b^3\,d^3S}{1\cdot 2\cdot 3\,dx^3} - \frac{b^4\,d^4S}{1\cdot 2\cdot 3\cdot 4\,dx^4} + \text{etc.}$$

Now, he substitutes this into the formula he carried over from E-46 and E-47 to get that

$$\begin{aligned} S = \int \frac{X\,dx}{b} + \frac{X}{1\cdot 2} + \frac{b\,dX}{1\cdot 2\cdot 3\cdot 2\,dx} - \frac{b^3\,d^3X}{1\cdot 2\cdot 3\cdot 4\cdot 5\cdot 6\,dx^3} + \frac{b^5\,d^5X}{1\cdot 2\cdot 3\cdot\cdots\cdot 7\cdot 6\,dx^5} \\ - \frac{3b^7\,d^7X}{1\cdot 2\cdot 3\cdot\cdots\cdot 9\cdot 10\,dx^7} + \frac{5b^9\,d^9X}{1\cdot 2\cdot 3\cdot\cdots\cdot 11\cdot 6\,dx^9} - \frac{691b^{11}\,d^{11}X}{1\cdot 2\cdot 3\cdot\cdots\cdot 13\cdot 210\,dx^{11}} \\ + \frac{35b^{13}\,d^{13}X}{1\cdot 2\cdot 3\cdot\cdots\cdot 15\cdot 2\,dx^{13}} - \frac{3617b^{15}\,d^{15}X}{1\cdot 2\cdot 3\cdot\cdots\cdot 17\cdot 30\,dx^{15}} + \frac{43867b^{17}\,d^{17}X}{1\cdot 2\cdot 3\cdot\cdots\cdot 19\cdot 42\,dx^{17}} - \text{etc.} \end{aligned}$$

Euler reminds us that we must add a constant to this so it satisfies the initial condition $S = A$ when $x = a$. Euler wants terms as far as the 17th derivative so that he can get quite accurate approximations later in this article.

This is the main result of this paper, so Euler turns to an example. He takes $X = x^n$ so that the sum he seeks is

$$S = a^n + (a+b)^n + (a+2b)^n + \cdots + x^n.$$

He gathers his tools,

$$\int X\,dx = \frac{x^{n+1}}{n+1}, \quad \frac{dX}{dx} = nx^{n-1}, \quad \frac{d^3X}{dx^3} = n(n-1)(x-2)x^{n-3}, \text{ etc.,}$$

substitutes and calculates to get, in the case $n = 1$,

$$\begin{aligned} a + (a+b) + (a+2b) + \cdots + x &= \frac{x^2}{2b} + \frac{x}{2} + \frac{b}{12} - \frac{a^2}{2b} + \frac{a}{2} - \frac{b}{12} \\ &= \frac{x^2 - a^2 + bx + ab}{2b} \end{aligned}$$

and in the case $n = 2$,

$$\begin{aligned} a^2 + (a+b)^2 + (a+2b)^2 + \cdots + x^2 &= \frac{x^3}{3b} + \frac{x^2}{2} + \frac{bx}{6} - \frac{a^3}{3b} + \frac{a^2}{2} - \frac{ab}{6} \\ &= \frac{2x^3 - 2a^3 + 3bx^2 + 3a^2b + b^2x - ab^2}{6b}. \end{aligned}$$

Both of these forms are equivalent to those in E-47. This example is just to demonstrate that the main result works as promised, at least for its first few terms.

Now, Euler considers the tail of the infinite series, beginning with the term of index x. He writes

$$\overset{x}{X} + \overset{x+b}{Y} + \overset{x+2b}{Z} + \text{etc. } \textit{in infinitum} = S.$$

Again, he substitutes $x + b$ for x to get $S - X$, then subtracts to get an expression for X, the negative of the one above because there he substituted $x - b$ and here he substituted $x + b$. Then he uses the equation from E-47 and gets an expression for S, the negative of the one above that goes to the term of degree 17. Here, he tells us to take the constant of integration so that $S = 0$ when $x = \infty$.

It is time for another example. Returning to the Basel Problem, he takes $X = \frac{1}{x^2}$, so that the series is

$$\frac{1}{x^2} + \frac{1}{(x+b)^2} + \frac{1}{(x+2b)^2} + \text{etc.}$$

Euler calculates that

$$\int X\,dx = -\frac{1}{x}, \quad \frac{dX}{dx} = \frac{-2}{x^3}, \quad \frac{d^3X}{dx^3} = \frac{-2\cdot 3\cdot 4}{x^5}, \quad \text{etc.,}$$

then substitutes to get

$$S = \frac{1}{bx} + \frac{1}{2x^2} + \frac{b}{6x^3} - \frac{b^3}{30x^5} + \frac{b^5}{42x^7} - \frac{b^7}{30x^9} + \frac{5b^9}{66x^{11}} - \frac{691b^{11}}{2730x^{13}} + \frac{7b^{13}}{6x^{13}} - \frac{6317b^{15}}{510x^{17}} + \frac{43867b^{17}}{35910x^{19}} - \text{etc.}$$

He notes that the constant of integration here "is not required" because it is zero. Euler ends this example here. It is not really clear why he bothered to do this, for he had done almost the same example, in the case $b = 1$, in E-47, and got the same result, without the b and stopping a few terms earlier.

Euler repeats these derivations for finite alternating series, which he writes as

$$\overset{a}{A} - \overset{a+b}{B} + \overset{a+2b}{C} - \overset{a+3b}{D} + \cdots + \overset{x}{X} = S$$

and gets similar formulas. He does an example to derive a formula for the alternating sum of squares, which has the general form

$$a^2 - (a+b)^2 + (a+2b)^2 - (a+3b)^2 + \cdots + x^2 = S$$

and finds that

$$S = \text{Const.} + \frac{x^2}{2} + \frac{bx}{2}.$$

He solves for the constant by taking $x = a$, and gets

$$S = \frac{a^2 - ab + x^2 + bx}{2}.$$

He checks this with $a = b = 1$ and finds that

$$1 - 4 + 9 - 16 + 25 - \cdots + 121 = 66.$$

The result on finite alternating series extends easily to infinite alternating series. He uses that to sum

$$1 - \frac{1}{3} + \frac{1}{5} - \frac{1}{7} + \frac{1}{9} - \frac{1}{11} + \text{etc.}$$

which he knows to sum to $\frac{\pi}{4}$. He begins by summing up to $-\frac{1}{23}$ by hand, then calculating the remainder term as 0,020797471915. This gives a respectable 12 decimal approximation of π.

Euler continues his generalization to consider sums of the form

$$\overset{a}{An^a} + \overset{a+b}{Bn^{a+b}} + \overset{a+2b}{Cn^{a+2b}} + \cdots + \overset{x}{Xn^x} = Sn^x.$$

Note how he plans to simplify some of his calculations by calling the sum Sn^x instead of S. This time his calculations are more complicated. He substitutes $x - b$ for x, subtracts and solves for S, but he has to introduce a number of coefficients, α, β, χ, δ, ε and ζ as he did in E-47. Then he spends a couple of pages finding those coefficients. The calculations very closely parallel the work in E-47. Eventually, he gets that the sum is

$$n^x \left\{ \begin{array}{c} \dfrac{n^b X}{n^b - 1} - \dfrac{n^b b\,dX}{1(n^b-1)^2 dx} + \dfrac{(n^{2b}+n^b)b^2\,ddX}{1\cdot 2(n^b-1)^3 dx^2} - \dfrac{(n^{3b}+4n^{2b}+n^b)b^3 d^3X}{1\cdot 2\cdot 3(n^b-1)^4 dx^3} \\ + \dfrac{(n^{4b}+11n^{3b}+11n^{2b}+n^b)b^4 d^4X}{1\cdot 2\cdot 3\cdot 4(n^b-1)^5 dx^4} - \text{etc.} \end{array} \right\} + \text{Const.}$$

For those of us who actually want to apply this equation, he tells us to find the constant by setting $x = a$ and the sum equal to Ana. There is a similar sum for infinite series.

As an example, he finds that

$$\begin{aligned} &a^2n^a + (a+b)^2n^{a+b} + (a+2b)^2n^{a+2b} + \cdots + x^2n^x \\ &= n^a\left(a^2 - \frac{n^b a^2}{n^b-1} + \frac{2n^b ab}{(n^b-1)^2} - \frac{(n^{2b}+n^b)b^2}{(n^b-1)^3}\right). \end{aligned}$$

As an example of an infinite series, Euler finds an expression for

$$S = \frac{n^x}{x} + \frac{n^{x+2}}{x+2} + \frac{n^{x+4}}{x+4} + \frac{n^{x+6}}{x+6} + \text{etc.}$$

and gets

$$S = n^x \left\{ \begin{array}{c} \dfrac{-1}{(n^2-1)x} - \dfrac{2n^2}{(n^2-1)^2x^2} - \dfrac{4(n^4+n^2)}{(n^2-1)^3x^3} - \dfrac{8(n^6+4n^4+n^2)}{(n^2-1)^4x^4} \\ - \dfrac{16(n^8+11n^6+11n^4+n^2)}{(n^2-1)^5x^5} - \text{etc.} \end{array} \right\}.$$

Euler will recycle this result at a key point in E-74. Taking $x = 25$ and $n = \frac{1}{\sqrt{-3}}$ gives a series that converges to $\frac{\pi}{6}$, or, as he describes it, the arc of 30 degrees in a circle of radius $= 1$.

Finally, Euler turns to series formed by products, which he writes as

$$\overset{a}{A} + \overset{a+b}{AB} + \overset{a+2b}{ABC} + \cdots + \overset{x}{ABC\cdots VX}.$$

The calculations are by now familiar. He subtracts consecutive partial sums to get an expression for $SX - X$ in terms of S and some derivatives. Then he inverts that expression to find S. For an example, he sums the reciprocals of the factorial numbers to find that

$$\frac{1}{1\cdot 2\cdot 3\cdot\cdots\cdot x}+\frac{1}{1\cdot 2\cdot 3\cdot\cdots\cdot x(x+1)}+\text{etc.}$$
$$=\frac{1}{1\cdot 2\cdot 3\cdot\cdots\cdot x}\left(1+\frac{1}{x-1}-\frac{1}{(x-1)^3}+\frac{x(x+2)}{(x-1)^5}-\text{etc.}\right).$$

Euler expects us to calculate the first several terms of the series by hand, and then use this easy form, "by which the sum is found with a minimum of labor."

This brings us to the end of E-55. Its purpose has been to complete details omitted in E-46 and E-47. It shows how reluctant Euler was to leave any loose ends. This is not a great paper. We have omitted many of its calculations. The interested reader will be able to follow them in the *Opera Omnia* without knowing too much Latin, for there are almost as many formulas as there are words.

29

E-56: Curvarum maximi minimive proprietate gaudentium inventio nova et facilis*

A new and easy way of finding curves enjoying properties of maximum or minimum

Four years earlier, in 1732, Euler wrote a difficult but wonderful paper, E-27, about generalized isoperimetric problems. He gave long tables describing how to solve problems of various forms and demonstrated the use of the tables on a dozen different examples, some familiar and some new. The analysis in the paper, though, was quite tedious. This was partly because the material is difficult and because the technique of considering individual arc elements requires so many different symbols, but it was also partly because Euler lacked a good system of notation and because he had not completely organized his thoughts and his presentation. The present paper, E-56, is a kind of do-over of E-27. He uses almost identical illustrations and the same kinds of calculations, but he devises a more efficient notation (still cumbersome by our standards.) He also greatly reduces the number of distinct cases, from 24 formulas in the tables in E-27 to a single general theory. The analysis in the two papers corresponds rather closely, but the present paper includes fewer examples.

Euler begins with a paragraph about how important these problems are, attributing their formulation to Jakob Bernoulli. He reminds us of his previous paper, E-27, and its 24 formulas, admitting that some of them are complex. Then he introduces his new notation. If y is a value on a curve, then Euler will denote a nearby value one step away by $\overset{I}{y}$. The next one, another step away, he will write $\overset{II}{y}$, then $\overset{III}{y}$, and so forth. Now we recognize this as a kind of index notation and we might use subscripts instead of this superscript notation. In Euler's day, though, indices, subscripts and superscripts were not commonly used. When people encountered a need for them, they would sometimes make up a notation. Leibniz, for example, in 1678, used numbers to denote indexed variables. He wrote, 12,2 to denote the product of the second coefficient in the first equation of a system of linear equations, denoted by 12, and the second unknown, denoted by 2. The use of numbers to denote both numbers and variables was not a notation that would catch on quickly. Cramer, in 1750, when he was writing about the algorithm now known as Cramer's rule, denoted whole equations by superscripted capital letters. He wrote $A^1 = Z^1z + Y^1y + X^1x + V^1v + \&c.$, where the symbol A^1 denotes the equation, z, y, x, v etc. denote variables, and the superscripted capital letters denote coefficients. Even Gauss still used a system of primes and double primes where we would use subscripts. Modern subscripts only developed about 1820, so we can well see why Euler had so much trouble finding a convenient notation for his ideas. [C]

Euler explains a bit how his new notation works. For the values of the function itself, he tells us that

$$\overset{I}{y} = y + dy$$

Comm. Acad. Sci. Imp. Petropol. 8 (1736) 1741, 159–190; *Opera Omnia* I.25, 54–80

$$\overset{II}{y} = y + 2\,dy + ddy$$
$$\overset{III}{y} = y + 3\,dy + 3\,ddy + d^3y.$$

Similarly,

$$d\overset{I}{y} = dy + ddy$$
$$d\overset{II}{y} = dy + 2\,ddy + d^3y$$
$$d\overset{III}{y} = dy + 3\,ddy + 3\,d^3y + d^4y.$$

He gives similar rules for second differentials.

He turns to his analysis, studying what he calls the first class of problems, finding a curve that, among all the curves with the same given endpoints, enjoys a given property of maximum or minimum. He denotes the curve in Figure 1 by oa, and, using OA as his x-axis. It runs vertically this time. He lets $OA = x$, $Aa = y$ and the arc length $oa = s$. He takes two differential elements $AB = BC = dx$ and draws their applications, that is, ordinates, Bb and Cc. Then, as he did in E-27, he takes a nearby curve that passes through the point β instead of the point b. This part of Figure 1 is rather cluttered, so we have enlarged and cleaned up the essential parts in Figure 1a. He plans to compare the curve element abc with the element $a\beta c$.

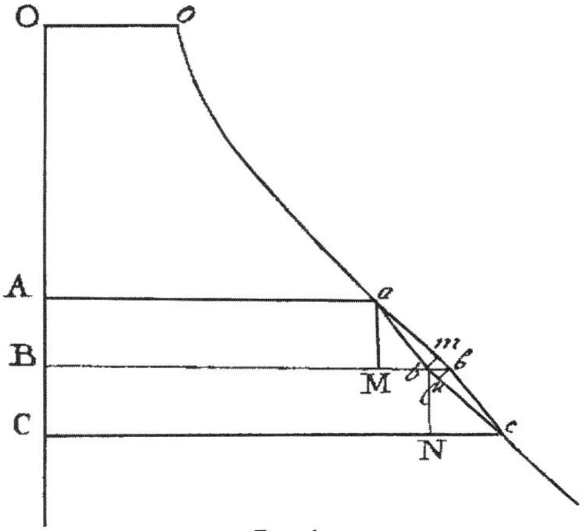

Fig. 1

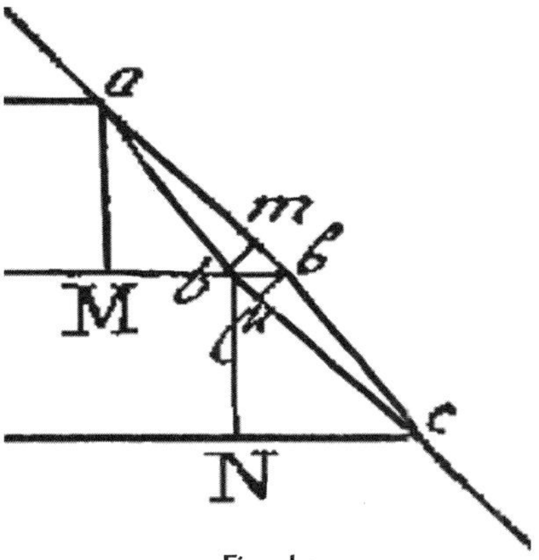

Fig. 1a

As in E-27, the difference between the two curves measured by the property that is to be maximized or minimized, can always be expressed in the form $P \cdot b\beta$, where P depends only on x, y, s and their differentials. For the property to be maximized or minimized, we must have $P = 0$. As we have seen, Euler seems to think that this *necessary* condition is actually *sufficient*.

Euler asks that the property to be maximized or minimized be expressed as $\int Q\,dx$, where Q depends on x, y, s and their differentials dx, dy and ds. He sets $dy = p\,dx$ and $ds = dx\sqrt{1+pp}$. In this way, he is able to avoid many of the cases involving integration with respect to ds and dy that he had to consider in E-27. Then the differential of Q will have the form

$$dQ = L\,ds + M\,dy + N\,dx + V\,dp.$$

By allowing some of these terms to be absent, he can avoid considering 16 different forms as he did in E-27.

Euler sets out to derive a form for that function P. In Figure 1, consider ab and bc to be elements of the "genuine" curve, and take $a\beta$ and βb to be elements of a nearby alternative. Then the contribution of the two elements ab and bc to the integral $\int Q\,dx$ can be expressed using Euler's notation as $\left(Q + \overset{I}{Q}\right) dx$. Using the same ideas of a particular differential being zero that we have seen in E-9 and in E-27, for ab and bc to be part of a maximum or minimum the difference between this form and the corresponding form for the alternate elements must be zero. This is the difference he says can always be expressed as $P \cdot b\beta$.

Euler continues his analysis. If $Q\,dx$ represents the contribution of the element ab to the integral $\int Q\,dx$, then to find the contribution of the element $a\beta$ the expressions p and $\frac{bM}{aM}$ should be replaced by $p + \frac{b\beta}{dx}$ and $\frac{\beta M}{aM}$ respectively. The other quantities, x, y and s remain the same because they depend on quantities the two curves have in common.

Euler takes ds, dy and dx to be zero, and in place of dp he substitutes $\frac{b\beta}{dx}$, so that the differential of $Q\,dx$ is

$$d\,Qdx = (L\,ds + M\,dy + N\,dx + V\,dp)\,dx = V \cdot b\beta.$$

Next, $\overset{I}{Q}\,dx$ represents the contribution of the element bc to the integral $\int Q\,dx$, where $\overset{I}{Q}$ is the same function of the variables $\overset{I}{s}$, $\overset{I}{y}$, $\overset{I}{x}$ and $\overset{I}{p}$ as Q is a function of s, y, x and p. From this, we get the expression corresponding to the contribution of the element βc, if we make a few substitutions. In place of $\overset{I}{y} = Bb$ we put $\overset{I}{y} + b\beta$; in place of $\overset{I}{s} = oa + ab$ substitute

$$\overset{I}{s} + \beta m = \overset{I}{s} + \frac{dy \cdot b\beta}{ds}$$

and in place of

$$\overset{I}{p} = \frac{cN}{bN}$$

put

$$\overset{I}{p} - \frac{b\beta}{dx}.$$

Further, we can find the difference between expressions corresponding to the elements bc and βc if we differentiate $\overset{I}{Q}\,dx$ and then substitute b? for $d\overset{I}{y}$, substitute $\frac{dy \cdot b\beta}{ds}$ in place of $d\overset{I}{s}$, and substitute $-\frac{b\beta}{dx}$ for $d\overset{I}{p}$. Euler does all this for the differential that describes arc element $b\beta$,

$$d\overset{I}{Q}\,dx = \left(\overset{I}{L}\,d\overset{I}{s} + \overset{I}{M}\,d\overset{I}{y} + \overset{I}{N}\,dx + \overset{I}{V}\,d\overset{I}{p}\right) dx.$$

Now the difference between the effects of the two arc elements gives

$$\frac{\overset{I}{L}\,dx\,dy \cdot b\beta}{ds} + \overset{I}{M}\,dx \cdot b\beta - \overset{I}{V} \cdot b\beta.$$

The term $\overset{I}{N}\,dx$ is neglected because it involves dx^2, and throughout Euler's analysis he takes dx to be constant. This last formula tells us that the difference between the effects of the elements $ab+bc$ and the effects of the elements $a\beta + \beta c$ is

$$\left(V + \frac{\overset{\mathrm{I}}{L}\,dx\,dy}{ds} + \overset{\mathrm{I}}{M}\,dx - \overset{\mathrm{I}}{V}\right)b\beta = \left(\frac{L\,dx\,dy}{ds} + M\,dx - dV\right)b\beta = P \cdot b\beta.$$

From this, we can read our function P and get the necessary conditions

$$P = \frac{L\,dx\,dy}{ds} + M\,dx - dV = 0 \quad \text{or} \quad L\,dx\,dy + M\,dx\,ds = ds\,dV.$$

Euler offers his first example. Unlike those in E-27, this one seems contrived, and he does not give us a clue to the origins of the problem. He proposes finding a curve that minimizes the expression

$$\int \frac{ds^2}{u\,dy + \sqrt{c^2\,ds^2 - u^2\,dx^2}}$$

where u is taken to be a function of x. He puts $dy = p\,dx$ and substitutes to eliminate ds, and gets

$$Q = \frac{1 + pp}{pu + \sqrt{c^2(1 + pp) - u^2}}.$$

Then $L = 0$, $M = 0$ and

$$V = \frac{c^2(1 + p^2)p - 2pu^2 + u(pp - 1)\sqrt{c^2(1 + p^2) - u^2}}{\left(pu + \sqrt{c^2(1 + p^2) - u^2}\right)^2 \sqrt{c^2(1 + p^2) - u^2}}.$$

Now, because $dV = P = 0$, V must be a constant, which Euler takes to be $\frac{1}{a}$. Then a bit of algebra and the substitution $dy = p\,dx$ leads to the solution

$$dy = \frac{(u^2 - au - cc)dx}{c\sqrt{(u - a)^2 - c^2}}.$$

"Because this rule is so easy," Euler shows how to apply it to the form $\int Q\,dx \int R\,dx$ (where, as before, the integral of R is to be interpreted as part of the integrand involving Q). When Euler worked with similar forms in E-27, he found them considerably more difficult and complicated. Here he takes $dQ = L\,ds + M\,dy + N\,dx + V\,dp$, as before, and also $dR = E\,ds + F\,dy + G\,dx + I\,dp$, and, as always, lets $dy = p\,dx$. Then he gives an analysis that closely parallels the one detailed above, though it does require us to neglect certain second-order differentials as being insignificant. He concludes that

$$\begin{aligned} P &= \left(\frac{L\,dx\,dy}{ds} + M\,dx\right)\int R\,dx - dV \int R\,dx - RV\,dx + QI\,dx \\ &= -d \cdot V \int R\,dx + dx\left(\frac{L\,dy}{ds} + M\right)\int R\,dx + QI\,dx \\ &= 0. \end{aligned}$$

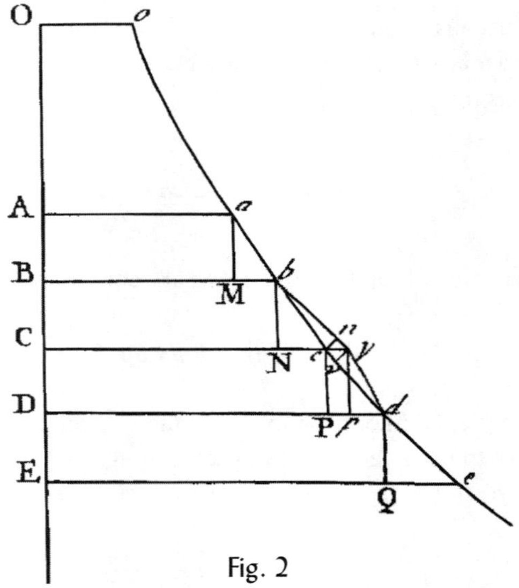

Fig. 2

As a second example of the method, Euler proposes to find the curve for which $\frac{\int s\,dx}{\int y\,dx}$ is maximized or minimized. Because this is a form we have not seen before, he has to go through the analysis of the curve elements to derive his function P. Eventually, he shows that

$$P = \frac{dx\,dy \int y\,dx - ds\,dx \int s\,dx}{ds\left(\int y\,dx\right)^2}.$$

Setting this equal to zero and substituting $dy = q\,ds$ leads first to

$$\int y\,dx = \frac{s\,dx - qy\,dx}{dq},$$

which, in turn reduces to Euler's idea of a solution to this problem,

$$qs\,ds^2\,dds^2 + 2s\,dy^2\,dds^2 = 3y\,ds\,dy\,dds^2 + s\,ds\,dy^2\,d^3s - y\,dy^3\,d^3s.$$

Euler turns to problems in which $\int Q\,dx$ involves the second derivative of the curve. As before, he takes $dy = p\,dx$, so that p is the first derivative of the curve. Here, he further takes $dp = r\,dx$, so that r is the second derivative. Then he has to consider the differential of Q to have the form

$$dQ = L\,ds + M\,dy + N\,dx + V\,dp + W\,dr,$$

and an analysis, again similar to the one we saw above, but this time relying on Figure 2, leads to

$$P = \frac{ddW}{dx} - dV + \frac{L\,dx\,dy}{ds} + M\,dx$$

so that the solution will satisfy the differential equation

$$ddW - dV\,dx + \frac{L\,dx^2\,dy}{ds} + M\,dx^2 = 0.$$

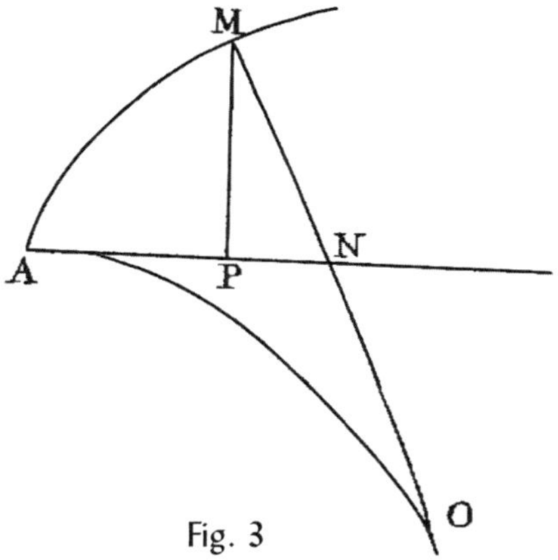

Fig. 3

We use this in another example to find the curve from among all the curves between A and M (Figure 3) whose evolute, AO, will sweep out the smallest possible area, AOM. The analysis depends on knowing that the radius of curvature, MO, can be written as

$$MO = -\frac{ds^3}{dx\,ddy} = \frac{dx(1+pp)^{3/2}}{dp}.$$

Then he finds that the area AMO can be given by the integral

$$-\frac{1}{2}\int \frac{(1+pp)^{3/2}\,ds}{r} = -\frac{1}{2}\int \frac{(1+pp)^2\,dx}{r}.$$

From this, he knows Q, so he can find L, M, N, V and W, only the last two of which are not zero. It turns out that $ddW = dV\,dx$, and he introduces a new quantity A so that $dW = V\,dx + A\,dx$. This enables him to derive the differential equation

$$\frac{ds^4\,d^3y}{ddy^3} = \frac{4\,dy\,ds^2}{ddy} + A\,dx.$$

Further calculation introduces two constants of integration, a and b, and leads Euler to conclude that $s\,ds = a\,dy + b\,dx$, which he recognizes as the differential equation of a cycloid.

For a different example involving second derivatives, Euler seeks a curve AM for which $\int \frac{ddy}{ds}$ is maximized or minimized. Easy substitutions turn this into an integral involving dx, and he finds that

$$\int \frac{ddy}{ds} = \int \frac{r\,dx}{\sqrt{1+pp}}.$$

This tells him Q, which, in turn, lets him find the necessary differential equation to be

$$\frac{2\,dp + pp\,dp}{p+p^3} = \frac{2\,dy}{y}.$$

He solves this, introducing a as a constant of integration, and gets

$$dx\sqrt{2} = \frac{dy}{y}\sqrt{-y^2 \pm \sqrt{y^4+4a^4}}.$$

The solution of this, he says, "depends on the quadratures of the circle and the hyperbola." This means it will involve logarithms and trigonometric functions.

For his fifth example in this paper, Euler returns to brachistochrone problems (Figure 4). Here, we are to find the curve from A to M that allows a body to descend as quickly as possible, when there are forces resisting its descent. He takes his variables to be $AP = x$, $PM = y$, $AM = s$ and $dy = p\,dx$. Now, his presentation becomes a bit confusing, as he uses the symbols P and p in two different senses and the symbol Q in three meanings. He supposes that forces acting on the body at M include a constant force P in the direction MP and another constant force Q in the direction MQ, which is parallel to the axis AP. He resolves these forces into tangential and normal components, and shows that the tangential component is

$$\frac{Q - Pp}{\sqrt{1+pp}} = \frac{Q\,dx - P\,dy}{ds},$$

and the normal component is

$$\frac{Qp + P}{\sqrt{1+pp}} = \frac{Q\,dy + P\,dx}{ds}.$$

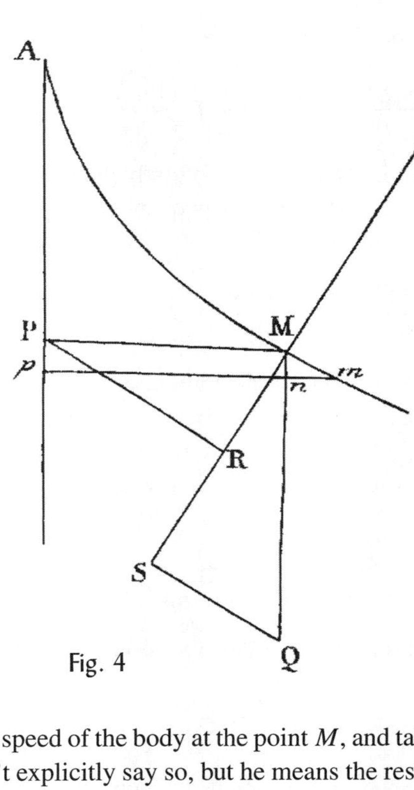

Fig. 4

Now he lets v be the speed of the body at the point M, and takes the resistance to be R, a function of that speed. He doesn't explicitly say so, but he means the resistance to be tangent to the direction of motion. Then

$$dv = Q\,dx - P\,dy - R\,ds,$$

and the time of descent along the arc AM will be

$$\int \frac{ds}{\sqrt{v}} = \int \frac{dx\sqrt{1+pp}}{\sqrt{v}}.$$

This is the quantity to be minimized.

Once he gets this equation, he can identify Q and determine his differential equation. Then, he introduces the variable z to measure MO, the radius of curvature, by setting

$$z = \frac{ds^2}{dx\,ddy} = \frac{dx(1+pp)^{3/2}}{dp}$$

and the path to his solution is straightforward. He finds that

$$\frac{2v}{z} = \frac{P+Qp}{\sqrt{1+pp}}.$$

This solution, though, is only a partial one, as it does not involve the resistance function R in any way. This is because the quantity v depends on the solution of the problem that Euler's technique cannot quite accommodate. He puts this example down and will return to it later.

Euler now turns to constrained optimization problems, which he formerly called problems of the second class. He does not use that term in this paper. Instead he calls them isoperimetric problems. The first such problems had constraints on arc length, Euler explains, hence Bernoulli, Taylor and Hermann gave these problems the name "isoperimetric." When they generalized the problems to any kind of constraints given by integration, they kept the name.

Euler has to do another derivation, this time based on Figure 5 and involving an alternate path determined by three curve elements, $a\beta$, $\beta\gamma$ and γc, so that he has enough flexibility to satisfy the given constraint as he seeks to optimize the quantity.

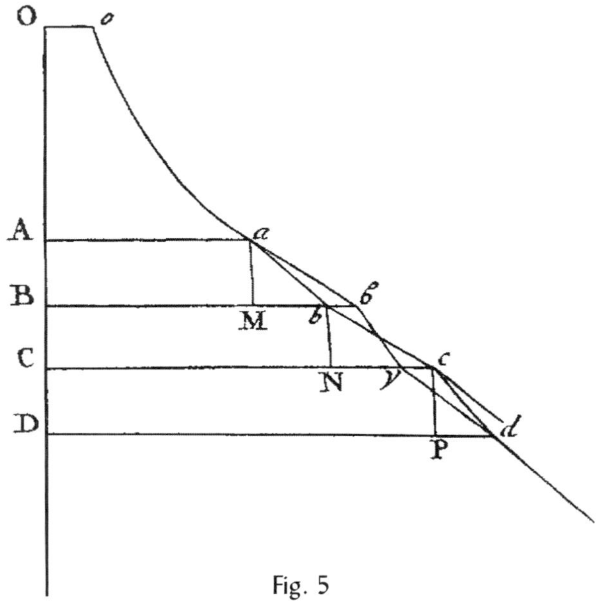

Fig. 5

Euler supposes that $\int Q\,dx$ is either the quantity to be maximized or minimized or the quantity that describes the constraint. A subtle symmetry in the analysis makes it not matter which it is. Then, as before, he gets two differential equations of the form $P = 0$ that the curve must satisfy. For the integral that is to be maximized or minimized, the equation $P = 0$ describes the condition at a maximum or minimum that the differential is zero. For the constraint, it describes the requirement that the new curve passing through $a\beta\gamma d$ and the old curve through $abcd$ give the same value for the integral. Euler proposes to combine the two differential equations by insisting that one plus any multiple of the other be zero. This is an improvement on similar steps Euler made in E-27, when he asked that the sum of the two equations be zero.

The analysis proceeds as we would expect, though with more curve elements. There are a great many more symbols, and he has to introduce the new variable $q = \frac{dy}{ds}$. As before, he takes dQ to have the form

$$dQ = L\,ds + M\,dy + N\,dx + V\,dp$$

and he finds that

$$\frac{dP}{P} = \frac{-ddV + \overset{I}{L}\,dx\,dq + dx\,d(\overset{I}{L}\,q + \overset{I}{M})}{-dV + dx(\overset{I}{L}\,q + \overset{I}{M})}.$$

Integration produces

$$P = e^{\int \frac{L\,dx\,dq}{-dV + dx(Lq+M)}}(-dV + dx(Lq + M))$$

where he still feels compelled to remind us that e is the number whose logarithm is 1.

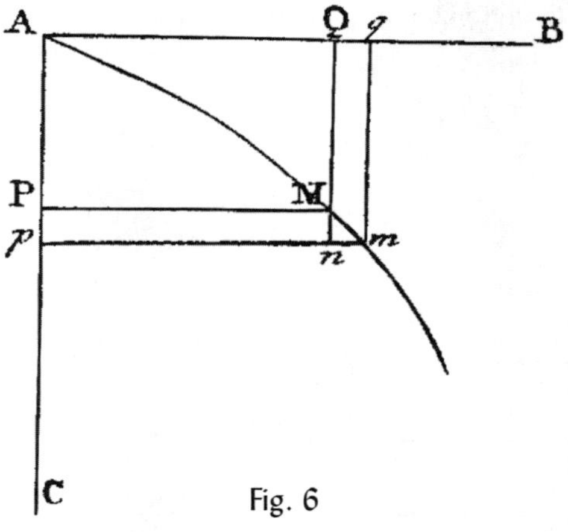

Fig. 6

Euler claims that all 16 distinct cases that he described in E-27 follow directly from this analysis.

He turns to another example, his sixth in this paper. Referring to Figure 6, he wants to find from among all curves *AM* that describe a given area *AMQ*, the curve which gives minimum resistance to fluid flow parallel to *AB*. He selects his variables to be $AP = x = QM$, $PM = y = AQ$, $dy = p\,dx$ and $AM = s$. The area $AQM = \int x\,dy$ needs to be rewritten so that the integral is with respect to x, so he substitutes $dy = p\,dx$ and gets $AQM = \int xp\,dx$.

The resistance to the fluid is given by

$$\int \frac{dx^3}{ds^2},$$

because of principles derived elsewhere.

First we consider the area constraint $\int xp\,dx$. This makes $Q = xp$ so that $dQ = p\,dx + x\,dp$. It follows that $L = 0$, $M = 0$, $N = p$ and $V = x$, so that $P = -dx$.

The resistance integral is a little more complicated. First, some easy substitutions yield

$$\int \frac{dx^3}{ds^2} = \int \frac{dx}{1+pp}$$

so that

$$Q = \frac{1}{1+pp} \quad \text{and} \quad dQ = -\frac{2p\,dp}{(1+pp)^2}.$$

From this, $L = 0$, $M = 0$, $N = 0$ and

$$V = -\frac{p}{(1+pp)^2}.$$

Substituting these into the formula for P gives

$$P = \frac{dp}{(1+pp)^2} - \frac{4p^2\,dp}{(1+pp)^3} = \frac{dp - 3pp\,dp}{(1+pp)^3}.$$

Euler now sets this value of P equal to a multiple of the other, which amounts to taking one plus a multiple of the other and setting that equal to zero. This gives

$$dx = \frac{a\,dp - 3ap^2\,dp}{(1+pp)^3} \quad \text{and} \quad x = \frac{ap}{(1+pp)^3}.$$

Now, because $dy = p\,dx$, this also is a differential equation, which he solves easily to get

$$y^4 - 2y^2x^2 + x^4 = 18cx^2y + 2cy^2 - 27c^2x^2$$

where he has collected constants involving a and replaced them with c.

Euler is ready to return to the example involving Figure 4 that he began, then abandoned earlier. Recall that we were to minimize

$$\int \frac{ds}{\sqrt{v}} = \int \frac{dx\sqrt{1+pp}}{\sqrt{v}},$$

that $dv = Q\,dx - P\,dy - R\,ds$, and that P and Q represent constant forces whereas R represents a variable resistance that depends on the speed v.

He makes v into a constraint integral by noting that $\int dv = \int \left[Q - Pp - R\sqrt{1+pp}\right] dx$. What follows is not pretty, its intense complication further aggravated by the fact that Euler uses both P and Q to be points, functions *and* forces. We will describe the variables Euler introduces and give his results, but we will leave out the nasty calculations in between.

In the integral involving dv, Euler sees that $Q = Q - Pp - R\sqrt{1+pp}$, where the Q on the left is the function in the integrand and the Q on the right is the force acting at M parallel to the axis AP. The P is the force in the direction MP. Then he introduces new functions by taking $dQ = S\,dy + X\,dx$ and $dP = T\,dy + Y\,dx$ and $dR = Z\,dv = ZQ\,dx - ZP\,dy - ZR\,ds$. With these variables in hand, he calculates P as

$$P = e^{\int \frac{ZR\,dx\,dp}{(S+Y)\,dx\,(1+pp)+Z(P\,dx+Q\,dy)\sqrt{1+pp}+\frac{R\,dp}{\sqrt{1+pp}}}} \cdot \left((S+Y)\,dx\,(1+pp) + Z(P\,dx + Q\,dy)\sqrt{1+pp} + \frac{R\,dp}{\sqrt{1+pp}}\right).$$

The integral to be minimized,

$$\int \frac{ds}{\sqrt{v}} = \int \frac{dx\sqrt{1+pp}}{\sqrt{v}},$$

submits to analysis more easily, but yields a result that is almost as complicated.

$$P = e^{\int \frac{R\,dx\,dp}{(P\,dx+Q\,dy)\sqrt{1+pp}-\frac{2v\,dp}{\sqrt{1+pp}}}} \left((P\,dx + Q\,dy)\sqrt{1+pp} - \frac{2v\,dp}{\sqrt{1+pp}}\right).$$

As before, one of these functions P is set equal to a multiple of the other, and any solution to the problem must satisfy the resulting differential equation.

Euler now regards that problem as solved and moves on to other topics. He turns to problems of the form $\int Q\,dx \int R\,dx$, a form we have seen before. He takes

$$dQ = L\,ds + M\,dy + N\,dx + V\,dp$$

and

$$dR = E\,ds + F\,dy + G\,dx + I\,dp.$$

His analysis depends on Figure 2 and leads to the result

$$P = e^{\int \frac{Q\,dI\,dx+L\,dx\,dq \int R\,dx - Q(Eq+F)\,dx^2}{-d\cdot V\int R\,dx+(Lq+M)\,dx\int R\,dx+QI\,dx}} \left(-d \cdot V \int R\,dx + (Lq + M)\,dx \int R\,dx + QI\,dx\right).$$

This form, he says, encompasses not only the 24 cases given in E-27 but also a wide variety of other forms.

It is time for more analysis. Using Figure 7, Euler considers problems in which $\int Q\,dx$ depends not only on x, y, s and p, but also on r, the second derivative of the curve. The analysis depends on a segment of the curve, *abcdef*, consisting of five curve elements, *ab*, *bc*, *cd*, etc. It is surprising, then, that only two points on the alternative segment of the curve, $ab\gamma\delta ef$, need to be different. The analysis leads to the conclusion that

$$P = e^{\int \frac{L\,dx^2\,dq}{ddW - dV\,dx + dx^2(Lq+M)}} \left[ddW - dV\,dx + d^2(Lq + M)\right].$$

We note that if $L = 0$, which happens if the integral does not involve arc length s, then P has a considerably simpler form

$$P = ddW - dV\,dx + M\,dx^2.$$

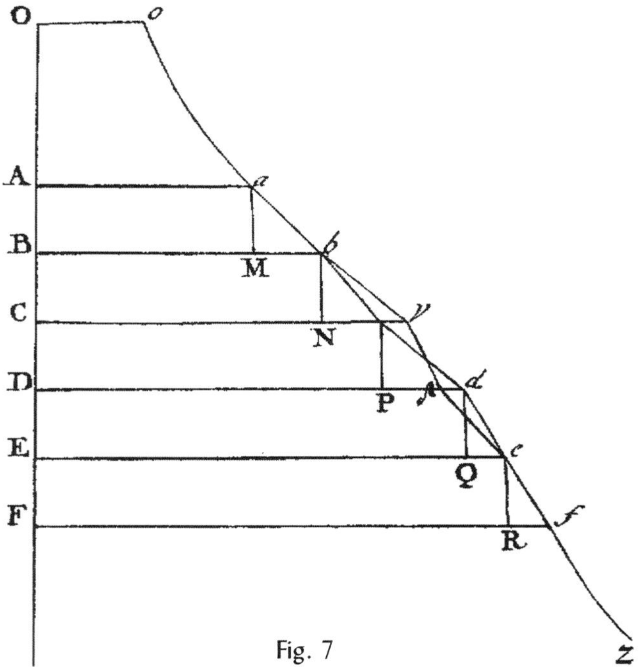

Fig. 7

This observation can be generalized. If Q depends only on x, y and their derivatives, Euler designates the higher derivatives by $dy = p\,dx$, $dp = r\,dx$, $dr = t\,dx$, $dt = u\,dx$, etc. He determines the relation between Q and these derivatives by

$$dQ = N\,dx + M\,dy + V\,dp + W\,dr + X\,dt + Y\,du, \text{ etc.}$$

Then P can be written as

$$P = M - \frac{dV}{dx} + \frac{ddW}{dx^2} - \frac{d^3x}{dx^3} + \frac{d^4y}{dx^4} - \text{etc.}$$

We get another example, the seventh of this paper. We are to find the curve that maximizes or minimizes

$$\int \frac{d^3y}{dx\,dy}.$$

As with most other examples here, Euler does not tell us what property of the curve or physical condition this quantity describes. He takes p and r to be the first and second derivatives as above, and recasts the integral to be $\int \frac{t\,dx}{p}$. This makes $Q = \frac{t}{p}$ so that

$$dQ = -\frac{t\,dp}{p^2} + \frac{dt}{p}.$$

Then

$$V = -\frac{t}{p^2} \quad \text{and} \quad X = \frac{1}{p},$$

while the other differential functions, N, M, W, Y, etc. are all zero, leaving

$$P = -\frac{dV}{dx} - \frac{d^3X}{dx^3}.$$

Before he substitutes for V and X, Euler sets $P = 0$ and integrates to get $V\,dx^2 + ddX = A\,dx^2$, where A is a constant of integration. Then he substitutes for V and X, then again for t to get

$$-\frac{2\,ddp}{p^2} + \frac{2\,dp^2}{p^3} = Adx^2.$$

Because $dp = r\,dx$ and taking $ddx = 0$, as we have seen him do before, he gets

$$ddp = \frac{dpdr}{r},$$

so

$$\frac{-2pr\,dr + 2r^2\,dp}{p^2} = A\,dp.$$

Integrating introduces another constant of integration and gives

$$B - \frac{r^2}{p^2} = Ap.$$

Now, a bit of algebra involving $dp = r\,dx$, leads to

$$dx = \frac{dp}{\sqrt{Bp^2 - Ap^3}}.$$

In case $B = 0$, this gives a hyperbola, and if $A = 0$, this is a logarithmic curve.

Euler is almost ready to wrap up this paper. Before he does, he wants to make some remarks on why some of his formulas for P are so much more complicated than others. He notices that the formulas when Q depends on s and when Q involves a second integrated quantity like $\int R\,dx$ are often particularly complicated, involving long exponentials. He notes that s is itself an integrated quantity, being, as it is, $\int \sqrt{1 + pp}\,dx$, so the question becomes an issue of explaining why integration makes Q so complicated. He addresses it with Figure 8. He notes that changing the curve from $oabc \ldots z$ to $oa\beta c \ldots z$ does not change the values of functions of x, y, p, etc. at the points d, e, f, etc., but it does change the value of s at those points, as well as the value of any other integrated quantity.

With this, a long and difficult article comes to its end. His "new and easy method" seems to be the technique of calculating with sequences of incremental quantities and not the list of formulas that ensue. He has been able to solve a variety of different kinds of examples with these new methods.

After completing this paper, Euler does not write nearly as many papers on the calculus of variations for a long time because he begins work in earnest on his great book on the subject, the *Methodus Inveniendi*, which will appear in 1744. There, he will elaborate extensively on the ideas that he began here in E-56 and earlier in E-27. Some people believe that Euler wrote and almost

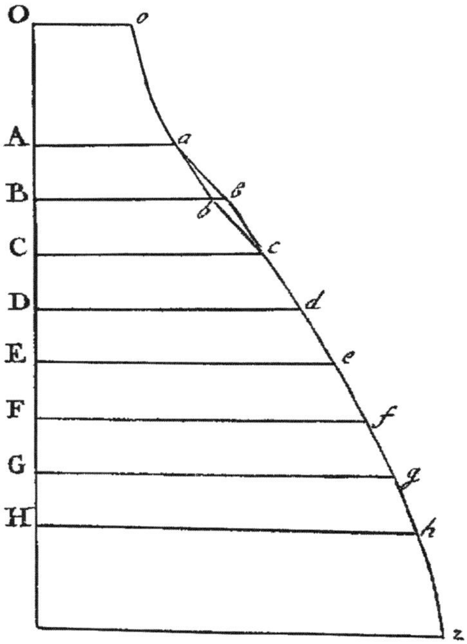

Fig. 8

completed his manuscript the *Methodus Inveniendi* while he was still in St. Petersburg and that he carried the manuscript to Berlin to be published.

References

[C] Chabert, Jean-Luc, et al., *A History of Algorithms from the Pebble to the Microchip*, Springer, Berlin, 1998. English translation by Chris Weeks of the original French edition *Histoire d'algortihms. Du caillou à la puce*, Éditions Belin, Paris, 1994.

Interlude: 1737

World events

In 1737, Benjamin Franklin founded the Philadelphia police force. The great university at Göttingen opened. It had been founded in 1734 by England's King George II, but took three years to organize.

On May 28, 1737, the planet Venus passed directly between Earth and Mercury. It is the only time in historical times that such an occultation of one planet by another has been directly observed.

On December 18, 1737, the great stringed instrument maker Antonio Stradivari (1644–1737) died.

In Euler's life

The Academy of Sciences in St. Petersburg did not publish a volume of their *Commentarii* in 1737. With no other papers or books coming out that year, Euler went the entire year without publishing anything, though his writing continued uninterrupted. Euler also had no books or papers appear in 1731, 1732, 1734 and 1742, but otherwise he would not miss a year of publishing a book or article until 1791, eight years after he died when, again, the Academy's journal did not appear.

Euler's other work

Euler's only paper outside mathematics in 1737 was E-34, Dissertatio de igne, "Dissertation on fire." It won him the Paris Prize for 1738. A Paris Prize entailed a cash award worth about as much as a modern Nobel Prize, approximately $1 million US. After winning this prize, Euler was wealthy for the rest of his life. It helped that he won the prize a total of twelve times.

Euler's mathematics

Euler wrote four mathematical papers in 1737. E-70, De constructione aequationum, "On the solution of equations" is a mostly unremarkable paper about differential equations. E-71, De fractionibus continuis dissertatio, "An essay on continued fractions" presents most of the important results about continued fractions, and some people believe that it contains a proof that the constant e is irrational [S]. E-72, Variae observationes circa series infinitas, "Various observations about infinite series" is a most remarkable paper, including, among other things, the sum-product formula for the Riemann zeta function. E-73, Solutio problematis geometrici circa lunulas a circulis formatas, "Solution to a geometric problem about lunes formed by circles" is Euler's first of several papers about an ancient geometry problem.

References

[S] Sandifer, Ed, Who proved e is irrational?, *How Euler Did It*, MAA Online (www.maa.org), February 2006.

30

E-70: De constructione aequationum*

On the solution of equations

We move to the articles Euler wrote for the *Commentarii* of 1737, published in 1744. Between the time Euler prepared E-70 and the time it was published, there was social unrest in Russia, a revolving door of rulers, and a popular sentiment against foreigners, even those like Euler who spoke good Russian. He left manuscripts to be published in St. Petersburg and set out for Berlin to work at the new Academy of Frederick the Great. Meanwhile, publication delays for the *Commentarii* grew longer and longer. The volume for 1736 had been published in 1741. The 1737 volume took until 1744. Eventually, the delay reached eleven years as the volumes for 1740 and 1741 only appeared in 1751.

This accounts for the gap in Eneström numbers as we leap from E-56 to E-70. The works in between were mostly published in Berlin in a special 1743 collection of Euler's works called *Miscellanea Berolinensis*. The gap also includes E-65, Euler's great work on the calculus of variations, the *Methodus inveniendi*.

For now, though, we turn to E-70, a largely forgotten paper on differential equations, in which Euler applies the methods of "moduli" and partial derivatives developed in E-44 and E-45 to continue his study of the Riccati equation and related problems. It is a step in the development of a technique Euler calls "constructio per quadratura curvarum," or "solution by the integration (quadrature) of curves." We will see Euler study for the first time an integral of the form $z = \int e^{ax} X\, dx$. Today we recognize how this integral is related to the Laplace transform of the function X. Laplace himself will not wrestle with these ideas until 1782. [G] At the time he wrote it, Euler's paper attracted little notice, and it goes unmentioned in his correspondence.

Euler plans to derive differential equations from integrals. He begins asking us to consider $z = \int P\, dx$, where P is what he, following Hermann, has been calling a modular function, that is, a function of a variable x and a modulus a, sometimes treated as a variable and sometimes as a constant. His first step is to find Q by taking the differential of P with respect to a, and treating x as a constant. This differential divided by da gives Q. This odd wording is his way of saying that $Q = \frac{\partial P}{\partial a}$. In the same way, he defines higher derivatives, R, S, T, etc. Next he integrates Q with respect to x to get

$$\int Q\, dx = \alpha \int P\, dx + K,$$

where α will be a function of a and K will be a function of a and of x. This function K is a bit like a constant of integration that intrudes when we differentiate P with respect to one variable and then integrate with respect to another. We saw this back in E-44. The reader who finds this confusing might work out an example, say $P = a + x$. The reader should find that $\alpha = \frac{1}{a}$ and that

Comm. Acad. Sci. Imp. Petropol. 9 (1737) 1744, 85–97; *Opera Omnia* I.22, 150–161

$$K(a,x) = \frac{-x^2}{2a} + C,$$

where C is an ordinary constant of integration.

A little thought, which Euler does not bother to explain, shows that

$$\int Q\,dx = \frac{dz - P\,dx}{da},$$

and we already know that $z = \int P\,dx$. Substituting these into the equation giving the integral of Q gives the so-called modular equation

$$\frac{dz - P\,dx}{da} = \alpha z + K,$$

where the constant in K is taken so that the expression vanishes when $x = 0$. Moreover, the integral of the expression $P\,dx$ also ought to be chosen so that it vanishes when $x = 0$. Euler writes $K - C$ instead of the constant K, so that he can separate the part C that makes $K = 0$ when $x = 0$.

It may be that $\int Q\,dx$ and $\int P\,dx$ are not related and that an equation of the form $\int Q\,dx = \alpha \int P\,dx + K$ cannot be found. In this case, Euler says that $\int Q\,dx$ does not depend on $\int P\,dx$. In this case, one can find an expression of the form

$$\int R\,dx = \alpha \int Q\,dx + \beta \int P\,dx + K$$

where α and β are constants with respect to a and where K is a function of x and a. It is unfortunate that Euler has chosen to use a as a modulus at the same time he uses α as a constant, for in the *Opera Omnia* edition of his text, they are virtually identical. In the original *Commentarii*, things are slightly clearer.

Making the same substitutions as before gives

$$\int R\,dx = \frac{d\left(\frac{dz - P\,dx}{da}\right) - Q\,dx}{da}.$$

This process may be continued, introducing successively $\int S\,dx$, $\int T\,dx$, etc., and getting successively more complicated relations, like

$$\int S\,dx = \frac{d\left(\frac{d\left(\frac{dz - P\,dx}{da} - Q\,dx\right)}{da}\right) - R\,dx}{da}.$$

The pattern is clear.

This does not seem particularly useful, but we will see what Euler makes of it. He asks us to look at what he calls a "special equation"

$$z = \int e^{ax} X\,dx$$

where X is a function of x and is independent of a, and where "e is that number whose logarithm is equal to unity." Additionally, Euler wants to take the integral so that $z = 0$ when $x = 0$. Then, in the formulas above, $P = e^{ax}X$ and, because X does not depend on a, when x is held constant, $dP = e^{ax}Xx\,da$. Today, of course, we say this more simply by writing $\frac{\partial P}{\partial a} = e^{ax}Xx$.

Now, take $Q = e^{ax}Xx$, and from the work we did earlier, we get a modular equation that is a differential equation of the first degree:

$$\int e^{ax}Xx\,dx = \alpha \int e^{ax}X\,dx + K - C.$$

Now, define a new function p by letting $K = e^{ax}Xp$. Then the differential inside the integral, holding a constant, is

$$e^{ax}Xx\,dx = \alpha e^{ax}X\,dx + e^{ax}X\,dp + e^{ax}p\,dX + e^{ax}aXp\,dx.$$

Two steps of algebra lead to

$$\frac{dX}{X} = \frac{x\,dx - \alpha\,dx - dp - ap\,dx}{p}.$$

Some time ago we had the formula

$$\frac{dz - P\,dx}{da} = \alpha z + K - C.$$

Because we now have other expressions for P and K, substituting those values gives us

$$dz - e^{ax}X\,dx = \alpha z\,da + \left(e^{ax}Xp - C\right)da.$$

Now, Euler holds p constant, setting it equal to m. He switches back to the previous differential and finds that this makes

$$\frac{dX}{X} = \frac{x\,dx - (\alpha + ma)\,dx}{m}.$$

Euler seems to realize that a and α are looking a good deal alike, so he sets $b = \alpha + ma$, and gets that

$$\frac{dX}{X} = \frac{x\,dx - b\,dx}{m}.$$

Integrating this gives a logarithm, and exponentiating gives

$$X = e^{\frac{x^2 - 2bx}{2m}}.$$

When $x = 0$, $e^{ax}Xp - C$ must be zero, so this tells us that $C = m$.

Now that we know X, we can substitute back in to the original "special equation" that defines z and we see that

$$z = \int e^{\frac{x^2 - 2bx + 2amx}{2m}}\,dx.$$

We can substitute this into the modular differential equation above to get

$$dz = (b - ma)z\,da - m\,da + e^{\frac{x^2 - 2bx + 2amx}{2m}}(dx + m\,da).$$

This satisfies Euler for the case when $p = m$ and when m and b are constants and do not depend on a. He moves on to allow p to vary, as

$$p = \beta + \gamma x.$$

This makes

$$\frac{dX}{X} = \frac{x\,dx - \alpha\,dx - \gamma\,dx - \beta a\,dx - \gamma ax\,dx}{\beta + \gamma x}.$$

Euler supposes that this expression does not actually depend on the variable a and so introduces new variables, f, g, m and n, not involving a, defining them by taking

$$\frac{dX}{X} = \frac{fx\,dx - g\,dx}{mx + n}.$$

A few lines of algebra give α, β, χ and p in terms of f, g, m and n, and lead to expressions for X, K and C. Then taking $f = 0$ "without detriment to the universality" gives

$$z = \int e^{ax}(mx + n)^{-g/m}\,dx$$

and a new form for the modular differential equation

$$dz = \frac{(g - m - na)z\,da}{ma} + \frac{e^{ax}\,(mx + n)^{-g/m}\,(ma\,dx + n\,da + mx\,da)}{ma} - \frac{n^{(m-g)/m}da}{ma}.$$

Now, introducing A by taking

$$A\,da = e^{ax}(mx + n)^{-g/m}(ma\,dx + n\,da + mx\,da)$$

recasts the expression for dz almost into the form $\frac{dz - P\,dx}{da} = \alpha z + K$ we had back at the beginning, to get

$$dz + \frac{n^{(m-g)/m}da - (g - m - na)z\,da}{ma} = \frac{A\,da}{ma}.$$

Euler tells us that the equation defining A can be solved "easily" by substituting $x = \frac{y - na}{ma}$ and then separating variables.

All of these have led to modular differential equations of the first degree. Now we move on to some that may lead to equations of higher degree. Again, take $z = \int e^{ax}X\,dx$. Then

$$P = e^{ax}X, \quad Q = e^{ax}Xx \quad \text{and} \quad R = e^{ax}Xx^2.$$

This makes

$$\int e^{ax}Xx^2\,dx = \alpha\int e^{ax}Xx\,dx + \beta\int e^{ax}X\,dx + K - C.$$

As before, introduce p by writing $K = e^{ax}Xp$ and we get, after a little algebra,

$$\frac{dX}{X} = \frac{x^2dx - \alpha x\,dx - \beta\,dx - dp - ap\,dx}{p}.$$

Euler now makes a number of substitutions. He takes

$$p = \frac{(x - \gamma)(x - \delta)}{a},$$

then $f = a\gamma + a\delta - a\alpha$ and $\beta = \frac{g}{a} - \gamma\delta$, where these new quantities γ, δ, f and g do not depend on a. With this, a few steps later, he finds that

$$X = c(x-\gamma)^{\frac{\gamma f - g - \gamma + \delta}{\gamma - \delta}}(x-\delta)^{\frac{\delta f - g - \delta + \gamma}{\delta - \gamma}}$$

where c is a constant of integration that arises when integrating $\frac{dX}{X}$.

Euler shows no mercy to tired readers and further sets

$$\lambda = \frac{\gamma f - g - \gamma + \delta}{\gamma - \delta} \quad \text{and} \quad \mu = \frac{\delta f - g - \delta + \gamma}{\delta - \gamma}.$$

This gives him new expressions for X, α, β, K and C, including

$$X = c(x-\gamma)^{\lambda}(x-\delta)^{\mu}, \quad K = \frac{ce^{ax}(x-\gamma)^{\lambda+1}(x-\delta)^{\mu+1}}{a} \quad \text{and} \quad C = \frac{(-\gamma)^{\lambda+1}(-\delta)^{\mu+1}}{a}.$$

For examples, Euler first considers $z = \int e^{ax}(x-\gamma)^{\lambda}(x-\delta)^{m}c\,dx$, then the slightly more general form $z = \int e^{ax}(\varepsilon x + \eta)^{\lambda}(\zeta x + \theta)^{\mu}dx$, in which all the Greek constants are assumed to be independent of a.

As before, this leads to a very long and complicated differential equation the parts of which can be matched to the form

$$\frac{dz - P\,dx}{da} = \alpha z + K.$$

Using that form leads to the equation

$$\frac{ddz}{da} + \left(b + \frac{c}{a}\right)dz + \left(f + \frac{g}{a}\right)z\,da = A\,da,$$

where, as before, A is introduced to absorb a great many terms of the complicated differential equation.

Euler tells us that this last second-degree differential equation can be reduced to one of the first degree, then solved to give z.

Euler continues and supposes that what he calls the "fundamental equation" has the form

$$z = E\int e^{ax}(\eta + \varepsilon x)^{\lambda}(\theta + \zeta x)^{\mu}\,dx + F\int e^{-ax}(\eta - \varepsilon x)^{\lambda}(\theta - \zeta x)^{\mu}\,dx$$

where E, F and all the Greek parameters are assumed to be independent of a. Again, Euler can find a second degree differential equation that describes z and that he can solve.

As this paper draws to a close, we see that the excitement over the first use of the Laplace transform was unwarranted. In fact, Euler only considers the form of the Laplace transform, and does not discover or use any of its special properties. All he actually does in this paper is convert certain intractable integrals into differential equations. The integrals thus become the solutions to those differential equations. Such solutions do not seem very interesting. Fortunately, more interesting things are to come.

References

[G] Gillispie, Charles Coulston, *Pierre-Simon Laplace 1749–1827: A Life in Exact Science*, Princeton University Press, 1997.

31

E-71: De fractionibus continuis dissertatio*

An Essay on Continued Fractions[1]

For most of the articles from this early period, we have been working from the Latin in the *Opera Omnia* and, because of generous access to the original journals at Yale University, from the original journals themselves. For this particular article, though, we get additional help from an English translation that appeared in 1985 in the research journal *Mathematical Systems Theory*. [E] "Why," the reader must ask, "does a translation of an Euler paper appear in a journal on systems theory?" Most of us who know anything about continued fractions probably learned about them in the context of number theory and not control theory. The editor and the translators give slightly different reasons. Christopher Byrnes, then editor of *Mathematical Systems Theory*, writes, "The relevance of this paper to the 'partial realization' problem and its relation to the Riccati equation were apparently first discovered during a seminar on algebraic system theory at the Universität Bremen during the Fall of 1982." The translators, one a Latin teacher in South Carolina and the other a mathematics professor in Ohio, tell us, "Readers interested in mathematical system theory will be concerned primarily with Euler's treatment of the Riccati equation, Sections 28-33. This aspect of the paper served as the chief motivation for the translation project."

In our discussion of E-71, when we quote Euler, we will be quoting from the Wyman and Wyman translation. [E] A different partial translation also appears in Chabert's *A History of Algorithms*. [C]

In fact we will see that in this paper, Euler does not use continued fractions to do anything we usually call "number theory" today.

Up to 1737, when Euler writes this paper, almost nothing has been done with continued fractions, but the idea is by no means original to him. Though some people trace the idea to Euclid and his algorithm for finding greatest common divisors, it is more often credited to John Wallis in 1655. [M]

Euler eases us into the idea of continued fractions. First, he tells us of quantities "which are easy to express but difficult to evaluate" like arc lengths, areas and logarithms. He notes that the area of a circle of diameter 1 may be expressed as a series,

$$1 - \frac{1}{3} + \frac{1}{5} - \frac{1}{7} + \frac{1}{9} - \text{etc.}$$

and also as an infinite product,

$$\frac{2 \cdot 4 \cdot 4 \cdot 6 \cdot 6 \cdot 8 \cdot 8 \cdot 10 \cdot 10}{3 \cdot 3 \cdot 5 \cdot 5 \cdot 7 \cdot 7 \cdot 9 \cdot 9 \cdot 11} \text{etc.}$$

He calls both of these forms "series" and asks us to consider a third kind of series "whose terms are connected by continuing division" and decides to call such series continued fractions. He tells us that he has "been studying continued fractions for a long time, and I have observed many important facts pertaining both to their use and their derivation."

**Comm. Acad. Sci. Imp. Petropol.* 9 (1737) 1744, 98–137; *Opera Omnia* I.14, 187–216

[1]Translated by Wyman and Wyman.

The most general form of a continued fraction is

$$a+\cfrac{\alpha}{b+\cfrac{\beta}{c+\cfrac{\gamma}{d+\cfrac{\delta}{e+\cfrac{\varepsilon}{f+\text{etc.}}}}}}$$

All the symbols, both Latin and Greek, are taken to be whole numbers. Euler does not say here, though he will later require, that all the numbers be positive. The Greek letters Euler calls *numerators* and the Latin are the *denominators*. This is a little unfortunate, because it will be easy to confuse these numerators and denominators of the continued fraction with the numerators and denominators of the fractions that approximate it. We will try to be clear.

Euler credits the first use of continued fractions to Viscount Brouncker. He says that Brouncker saw Wallis's formula, the infinite product given above, and replied to Wallis that it could be re-cast as

$$1+\cfrac{1}{2+\cfrac{9}{2+\cfrac{25}{2+\cfrac{49}{2+\cfrac{81}{2+\text{etc.}}}}}}$$

where the numerators are the odd squares and the denominators are all 2. Euler reports that, sadly, Brouncker's calculations do not survive. It was common in the 17th century for mathematicians to announce their results without proofs and leave the discovery of the proofs as challenges to their correspondents. Brouncker and Wallis were both part of that tradition, and we have seen in E-26 that the tradition lingered to Euler's day, at least in number theory.

Euler gives a brief account of Wallis's own derivation of the result, for Wallis was a worthy correspondent and met Brouncker's challenge, but we omit that here. Euler describes Wallis's method as unnatural and depending on "induction," that is, on observation. Euler will give his own method, quite different from Wallis's, later in the paper.

Euler starts to work. He tells us that "it appears" that a continued fraction always converges. He shows why the approximations obtained by stopping a continued fraction at some point alternate between being less than the value of the fraction and being greater. These approximations, which he variously calls *approximations*, *terms*, *fractions* or *truncations*, we now call *convergents*.

Calculating individual convergents is tedious, but Euler offers a shortcut. He writes that the fractions are given by

$$\overset{a}{\underset{\alpha}{\frac{1}{0}}},\ \overset{b}{\underset{\beta}{\frac{a}{1}}},\ \overset{c}{\underset{\gamma}{\frac{ab+\alpha}{b}}},\ \overset{d}{\underset{\delta}{\frac{abc+\alpha c+\beta a}{bc+\beta}}},\ \overset{e}{\underset{\varepsilon}{\frac{abcd+\alpha cd+\beta ad+\gamma ab+\alpha\gamma}{bcd+\beta d+\gamma b}}},\ \text{etc.}$$

and that from this, "The law of progression of these fractions is seen clearly."

Note that the indices he writes along the bottom are the Greek symbols he calls *numerators*, whereas he writes the denominators along the top.

For those of us who do not see things quite so clearly, Euler explains the numerators. He starts with the fraction $\frac{1}{0}$. The next numerator will always be "the product of a numerator with the corresponding upper indicator [the Latin letter] added to the product of the preceding numerator with the lower indicator." Moreover, "A similar rule applies for the denominators."

This rule is a bit more complicated than the ones we may be familiar with, in part because it is more general. We usually see continued fractions in which all the numerators are 1. As Euler will note soon, for such continued fractions, the rule is a good deal simpler.

Euler does a few explicit calculations to find the difference between consecutive terms, showing that the terms alternate positive and negative, and also that they converge to zero rather rapidly. From this, he concludes that continued fractions themselves always converge rapidly. Moreover, he notes that the continued fraction will converge faster when its numerators are small and its denominators are large, so the "best" continued fractions will be the ones in which the numerators are all 1. Also, Euler notes that should any numerator be zero, then "the continued fraction is broken off there and becomes a finite fraction."

This gives Euler the motivation to consider continued fractions of the special form

$$a+\cfrac{1}{b+\cfrac{1}{c+\cfrac{1}{d+\cfrac{1}{e+\cfrac{1}{f+\text{etc.}}}}}}$$

He repeats some of the more general work he did above in this special case. He tells us that the sequence of fractions is given by

$$\overset{a}{\frac{1}{0}},\ \overset{b}{\frac{a}{1}},\ \overset{c}{\frac{ab+1}{b}},\ \overset{d}{\frac{abc+c+a}{bc+1}},\ \overset{e}{\frac{abcd+cd+ad+ab+1}{bcd+b+d}},\ \text{etc.}$$

Here, because the numerators of the continued fraction are all 1, he does not write the indices along the bottom. The indices at the top are, as before, the denominators of the continued fraction. To explain the pattern here, Euler tells us that "clearly," one multiplies the numerator (or denominator) of the fraction in the sequence by the corresponding index and then adds the previous numerator (or denominator). Thus, for example, the numerator of the fraction under the index d is given by $abc+c+a=(ab+1)\cdot c+a$.

Euler turns for a moment to continued fractions in which the numerators are 1 and the denominators fractions, and shows how they can be converted to and from continued fractions with whole numbers for numerators and denominators. One way, the translation is given by

$$a+\cfrac{1}{\frac{b}{B}+\cfrac{1}{\frac{c}{C}+\cfrac{1}{\frac{d}{D}+\cfrac{1}{\frac{e}{E}+\text{etc.}}}}}=a+\cfrac{B}{b+\cfrac{BC}{c+\cfrac{CD}{d+\cfrac{DE}{e+\text{etc.}}}}}$$

He gives a similar rule for translating the other direction.

With this, Euler is satisfied that he has explained how to convert continued fractions into ordinary fractions that approximate them, and he turns to converting ordinary fractions into continued fractions with numerators all 1 and with whole numbers for denominators. He distinguishes two cases. First are finite fractions whose numerators and denominators are finite, numbers we call rational. Second are fractions with infinite numerators and denominators, which he calls irrational or transcendental.

To convert a finite fraction, that is, a rational number, $\frac{A}{B}$ into a continued fraction with numerators equal to 1, Euler follows Euclid's algorithm. He constructs the table shown below by dividing A by B with quotient a and remainder C.

B	A	a				
	C	B	b			
		D	C	c		
			E	D	d	
				F	E	e
					G	etc.

Then he divides B by C with quotient b and remainder D. Next he divides C by D with quotient c and remainder E. He continues "until a zero remainder and an infinitely large quotient is obtained." Then the quotients provide the denominators for the continued fraction, so that

$$\frac{A}{B} = a + \cfrac{1}{b + \cfrac{1}{c + \cfrac{1}{d + \cfrac{1}{e + \text{etc.}}}}}$$

Euler argues briefly why this works, starting by supposing some remainder $G = 0$, then substituting backwards. He does not seem to be aware that rational numbers have a second representation as a continued fraction with the last denominator being 1.

Euler makes brief mention that if $A < B$, then $a = 0$. Also if $A : B$ is either irrational[2] or transcendental, then A/B will be represented by an infinite continued fraction.

It is time for an example. Euler chooses to find the continued fraction for $\frac{355}{113}$. He chooses this particular fraction because in the 16th century, Adrian Antoniscoon, writing under the pseudonym "Metius," gave this as the value of π. The approximation was also known to the Ancients, both Greek and Chinese. Euler uses his method to calculate that

$$\frac{355}{113} = 3 + \cfrac{1}{7 + \cfrac{1}{16}}$$

from which he forms the series of fractions

$$\overset{3}{\frac{1}{0}}, \overset{7}{\frac{3}{1}}, \overset{16}{\frac{22}{7}}, \frac{355}{113}.$$

Euler claims that the intermediate fractions, $\frac{3}{1}$ and $\frac{22}{7}$, approximate the fraction $\frac{355}{113}$ better than "any others made from smaller numbers." While this is true, Euler gives no evidence. He also notes that $\frac{3}{1}$ is smaller than $\frac{355}{113}$, and that $\frac{22}{7}$ is larger, as he showed earlier.

[2]One may ask "How can a ratio be irrational?" A literal translation of what Euler writes is "However, the continued fraction runs on to infinity if the ratio of A to B is not that of a number to numbers, but is irrational or transcendental." (E-71 §13)

Euler introduces a new term here. He calls the fractions that arise in this sequence *principal*, and other fractions that approximate the given number as *less principal*. Contrary to what is suggested in the Wyman and Wyman translation, he does not use a word that resembles the modern term *convergent* to describe these fractions. Euler gives a rather obscure description of what he means by "less principal." A bit later, we will use an example to explain the idea.

Euler continues by finding the sequence of quotients for π itself, which he still calls "the ratio of circumference to diameter." He finds the quotients to be 3, 7, 15, 1, 292, 1, 1, 1, 2, 1, 3, 1, 14 etc. Using these, he calculates the series of principal fractions as follows:

$$\overset{3}{\frac{1}{0}},\ \overset{7}{\frac{3}{1}},\ \overset{15}{\frac{22}{7}},\ \overset{1}{\frac{333}{106}},\ \overset{292}{\frac{355}{113}},\ \overset{1}{\frac{103993}{33102}}.$$

He also gives 17 different less principal fractions, for example $\frac{19}{6}$ and $\frac{16}{5}$. These are obtained from the principal fractions $\frac{22}{7}$ and $\frac{3}{1}$ by subtracting numerators and denominators. That is, $\frac{19}{6} = \frac{22-3}{7-1}$ and $\frac{16}{5} = \frac{22-3-3}{7-1-1}$. Other less principal fractions he gives, $311/99$, $201/64$, $103638/32929$, are obtained in the same way.

Euler reminds us that the principal fractions alternate between being larger than π and being smaller, with the ones under the indices 3, 15 and 292 being the larger.

For another example, Euler turns to the scheduling of leap years, which he calls *bissextile* years, for reasons that are fascinating, but irrelevant here. He gives the actual length of the solar year to be 365 days 5 hours 49 minutes 8 seconds, which he writes as $365^d\, 5^h\, 49'8''$. He calculates the ratio of the length of a day, 24 hours, to the excess over 365 days, that is 5 hours 49 minutes 8 seconds, to be $\frac{21600}{5237}$.

This tells us that every 21600 years there ought to be 5237 leap years. However, this is not yet practical information, because it does nothing to suggest a useful pattern for distributing those leap years. He slightly modifies his method as he calculates the quotients for the continued fraction expansion of $\frac{21600}{5237}$ using the following table.

5237	21600	4			
	20948				
	652	5237	8		
		5216			
		21	652	31	
			651		
			1	21	21

The quotients are thus 4, 8, 31 and 21, which he uses to form the denominators of the continued fraction. He gets the series of principal fractions

$$\overset{4}{\frac{1}{0}},\ \overset{8}{\frac{4}{1}},\ \overset{31}{\frac{33}{8}},\ \overset{21}{\frac{1027}{249}},\ \frac{21600}{5237}.$$

The principal fraction, $\frac{4}{1}$, gives the pattern of leap years in the Julian calendar, one leap year every four years. The next one, $\frac{33}{8}$, suggests that a pattern better than $\frac{4}{1}$ would be to have eight leap years every 33 years. Euler suggests that this might be confusing, and so he turns to the less principal fractions derived from $\frac{1027}{249}$ which have numerators divisible by 4. He gives a list of these, without explaining exactly how he found them, though it is not difficult. His list is

$$\frac{136}{33}, \frac{268}{55}, \frac{400}{97}, \frac{532}{129}, \frac{664}{161}, \text{ etc.}$$

He picks the third of these, $\frac{400}{97}$, which says that every 400 years, there should be 97 leap years. Indeed, this is the pattern in the Gregorian calendar we use today. Euler notes that in 21600 years, this method will provide one too many leap years, that is 5238 leap years instead of 5237.

We return to irrational numbers. Note that

$$\sqrt{2} = 1.41421356 = \frac{141421356}{100000000}.$$

Euler uses the symbol "=" where we might use "≈." The "approximately equal" symbol had not yet been invented.

The algorithm gives the quotients 1, 2, 2, 2, 2, 2, 2, 2, etc., and that, in turn, gives principal fractions $\frac{1}{0}, \frac{1}{1}, \frac{3}{2}, \frac{7}{5}, \frac{17}{12}, \frac{41}{29}, \frac{99}{70}, \frac{239}{169}$, etc. This series, as expected, alternates being larger than $\sqrt{2}$ and being smaller. Euler also gives six less principal fractions. Thus,

$$\sqrt{2} = 1 + \cfrac{1}{2 + \cfrac{1}{2 + \cfrac{1}{2 + \cfrac{1}{2 + \text{etc.}}}}}$$

Similarly, $\sqrt{3}$ gives quotients 1, 1, 2, 1, 2, 1, 2, 1, 2, 1, etc. Euler also gives explicitly the continued fraction for this. A pattern is evident here, so Euler decides to look at continued fractions that repeat. He first looks at

$$x = a + \cfrac{1}{b + \cfrac{1}{b + \cfrac{1}{b + \text{etc.}}}}$$

By algebra, he finds that

$$x - a = \cfrac{1}{b + \cfrac{1}{b + \cfrac{1}{b + \text{etc.}}}} = \frac{1}{b + x - a}.$$

This gives a quadratic equation,

$$x^2 - 2ax + bx + a^2 - ab = 1,$$

so that

$$x = a - \frac{b}{2} + \sqrt{1 + \frac{b^2}{4}}.$$

Euler gives only one root. Euler checks his work, as he so often does, by taking $b = 2$ and $a = 1$, and finding that this form returns $\sqrt{2}$, as it should.

Next, Euler considers the case $b = 2a$ and finds that $x = \sqrt{a^2 + 1}$, thus approximating the roots of numbers of the form $a^2 + 1$, one more than the perfect squares. He dwells on this a moment and takes $a = 2$, to find the quotients 2, 4, 4, 4, 4, 4, 4 etc. and to get a number of approximations for $\sqrt{5}$.

Perhaps Euler senses solutions to the Pell equation lurking here, but he does not mention them in this paper.

Next he considers continued fractions that repeat with a period of 2, that is with quotients following the pattern $a, b, c, b, c, b, c, b, c$ etc. Then, following a calculation similar to the one he did before, he gets

$$x - a = \cfrac{1}{b + \cfrac{1}{c + \cfrac{1}{b + \cfrac{1}{c + \cfrac{1}{b + \text{etc.}}}}}} = \cfrac{1}{b + \cfrac{1}{c + x - a}}.$$

From this, he gets the quadratic equation

$$bx^2 + bcx - 2abx = abc - a^2b + c.$$

In the case $c = 2a$, this makes

$$bx^2 = a^2b + 2a \quad \text{and} \quad x = \sqrt{a^2 + \frac{2a}{b}}.$$

Euler does a similar analysis on continued fractions in which the pattern of quotients repeats in cycles of three, that is of the form a, b, c, d, b, c, d etc. and assures us that it works for cycles of any length and always results in a value x that is a root of a quadratic equation. This last claim he does not prove.

Euler is going to look at continued fractions for various expressions involving e, "the number whose natural logarithm is 1." First, he writes

$$e = 2,71828182845904$$

using, as always, a comma where Americans use decimal points. He finds that this gives the continued fraction

$$e = 2 + \cfrac{1}{1 + \cfrac{1}{2 + \cfrac{1}{1 + \cfrac{1}{1 + \cfrac{1}{4 + \cfrac{1}{1 + \cfrac{1}{1 + \cfrac{1}{6 + \cfrac{1}{1 + \text{etc.}}}}}}}}}}$$

Here, the sequence formed by every third denominator is the sequence of even numbers, and the other denominators are all 1's. He does not prove this observation just yet, but promises to prove it later.

Next, Euler does $\sqrt{e}$ and finds the series of quotients to be 1, 1, 1, 1, 5, 1, 1, 9, 1, 1, 13, etc. He tells us that the pattern here is similar to the pattern for e, and leaves it to us to describe it explicitly.

Similarly, and again without giving his calculations, Euler notes that

$$\frac{\sqrt[3]{e}-1}{2} = 0.1978062125 = \cfrac{1}{5+\cfrac{1}{18+\cfrac{1}{30+\cfrac{1}{42+\cfrac{1}{54+\text{etc.}}}}}}$$

where, all the quotients "except the first form an arithmetic progression."

Moreover,

$$\frac{e^2-1}{2} = 3.19452804951,$$

giving an arithmetic sequence of quotients, 3, 5, 7, 9, 11, 13, 15, etc., and $\frac{e+1}{e-1}$ gives the quotients 2, 6, 10, 14, 18, 22, 26, etc.

Euler decides it is useful to classify these continued fractions involving arithmetic sequences into two types. The first type includes those with denominators that are *interrupted*, like the sequence for e and $\sqrt{e}$, where the arithmetic sequence is interrupted by, in these cases, pairs of 1's. The second type comprises those without such interruptions, as the last three expansions. Euler asks (already knowing the answer) whether it is possible to convert an interrupted continued fraction into one that is uninterrupted. He proposes an arbitrary progression, a, b, c, d, e etc., whether or not it be an arithmetic progression. He interrupts it with a pair of numbers, m, n. He writes out the resulting continued fraction,

$$a+\cfrac{1}{m+\cfrac{1}{n+\cfrac{1}{b+\cfrac{1}{m+\cfrac{1}{n+\cfrac{1}{c+\cfrac{1}{m+\cfrac{1}{n+\cfrac{1}{d+\text{etc.}}}}}}}}}}$$

He tells us that this equals

$$\frac{1}{mn+1}\left((mn+1)a+n+\cfrac{1}{(mn+1)b+m+n+\cfrac{1}{(mn+1)c+m+n+\cfrac{1}{(mn+1)d+m+n+\text{etc.}}}}\right).$$

To justify this he says "The proof of the equality consists in this: the ordinary fractions which approach the limiting values agree among themselves, so the limits must coincide."

Reversing the roles of m and n changes only the first term of this last expression and yields what Euler calls "the following rather elegant theorem:"

$$a+\cfrac{1}{m+\cfrac{1}{n+\cfrac{1}{b+\cfrac{1}{m+\cfrac{1}{n+\cfrac{1}{c+\text{etc.}}}}}}}-a+\cfrac{1}{n+\cfrac{1}{m+\cfrac{1}{b+\cfrac{1}{n+\cfrac{1}{m+\cfrac{1}{c+\text{etc.}}}}}}}=\frac{n-m}{mn+1}.$$

This is one of the few results known today involving the sum or difference of two continued fractions. Moreover, as Euler notes, the difference is independent of the quantities a, b, c, d etc.

Euler returns to his problem of converting interrupted continued fractions into uninterrupted ones. Starting with the equality he got before, then "dividing 1 by each fraction and adding the quantity A to each result" he gets

$$A+\cfrac{1}{a+\cfrac{1}{m+\cfrac{1}{n+\cfrac{1}{b+\cfrac{1}{m+\cfrac{1}{n+\cfrac{1}{c+\text{etc.}}}}}}}}$$

$$=A+\cfrac{mn+1}{(mn+1)a+n+\cfrac{1}{(mn+1)b+m+n+\cfrac{1}{(mn+1)c+m+n+\text{etc.}}}}$$

Euler writes out this formula in the special case $m=n=1$ and gets a form fitting the pattern of the continued fraction for e given a few sections earlier. He applies this result to the continued fraction

$$\frac{1}{e-2}=1+\cfrac{1}{2+\cfrac{1}{1+\cfrac{1}{1+\cfrac{1}{4+\text{etc.}}}}}$$

It follows easily from the continued fraction for e, though, subtracting 2 from both sides and then taking reciprocals. Matching parts with the formula for converting between interrupted and uninterrupted fractions, he takes $A = 1$, $a = 2$, $b = 4$, and, of course, $m = n = 1$, and obtains

$$\frac{1}{e-2} = 1 + \cfrac{2}{5 + \cfrac{1}{10 + \cfrac{1}{14 + \cfrac{1}{18 + \cfrac{1}{22 + \text{etc.}}}}}}$$

Then, by taking reciprocals, Euler gets

$$e = 2 + \cfrac{1}{1 + \cfrac{2}{5 + \cfrac{1}{10 + \cfrac{1}{14 + \cfrac{1}{18 + \cfrac{1}{22 + \cfrac{1}{26 + \text{etc.}}}}}}}}$$

He uses the same technique to convert the interrupted expression for $\sqrt{e}$ and tells us that both of these expressions converge "so fast that it is an easy matter to find the values of e and $\sqrt{e}$ as closely as you please."

Euler goes through the algebra of converting uninterrupted continued fractions into interrupted ones, the inverse operation to the one he described above, putting in pairs of quotients, m and n, but he does not do any examples using these new formulas. He tells us that similar techniques would allow him to interpose any even number of terms, but that it is not possible to do it with odd numbers. He works through the formula for interposing four terms, m, n, p and q. Setting $mnpq + mn + mq + pq = P$ and $mnp + npq + m + n + p + q = Q$, he finds that

$$a + \cfrac{1}{m + \cfrac{1}{n + \cfrac{1}{p + \cfrac{1}{q + \cfrac{1}{b + \cfrac{1}{m + \text{etc.}}}}}}}$$

$$= \frac{1}{P}\left(Pa + npq + n + q + \cfrac{1}{Pb + Q + \cfrac{1}{Pc + Q + \cfrac{1}{Pd + Q + \text{etc.}}}} \right)$$

He also writes out the form for the special case $m = n = p = q = 1$.

Now we have reached Section 28 and the part of the paper that is of particular interest to the systems theorists. Things get a little more difficult. First, Euler reminds us that he only knows the pattern for the continued fractions for e and its related values by observation, and that he has not yet proved those observations to be correct. He tells us that he will be attacking the problem by looking at the differential equation

$$a\,dy + y^2\,dx = x^{-4n/(2n+1)}\,dx.$$

This is a Ricatti equation, and we have seen Euler wrestle with it before, in E-28 and in E-31. He seems to skip some steps here, but he claims he can reduce this equation to a simpler Ricatti equation,

$$a\,dq + q^2\,dp = dp$$

by setting $p = (2n+1)x^{1/(2n+1)}$. To justify this claim, Euler further claims that

$$q = \frac{a}{p} + \cfrac{1}{\cfrac{3a}{p} + \cfrac{1}{\cfrac{5a}{p} + \cfrac{1}{\cfrac{7a}{p} + \cdots}}}.$$

$$+ \cfrac{1}{\cfrac{(2n-1)a}{p} + \cfrac{1}{x^{2n/(2n+1)}y}}$$

Because this continued fraction ends after finitely many terms, for any n, it gives a finite equation between x and y that is a solution to the Ricatti equation above.

Now, Euler takes n to be an infinite number, and the continued fraction for q becomes one with an arithmetic progression of denominators,

$$q = \frac{a}{p} + \cfrac{1}{\cfrac{3a}{p} + \cfrac{1}{\cfrac{5a}{p} + \cfrac{1}{\cfrac{7a}{p} + \cfrac{1}{\cfrac{9a}{p} + \text{etc.}}}}}$$

and the second version of the Ricatti equation, for which p and q give the solution, again becomes

$$a\,dq + q^2\,dp = dp.$$

Here, the variables separate to give

$$\frac{a\,dq}{1-q^2} = dp$$

which, in turn, integrates as

$$\frac{a}{2}\,l\,\frac{1+q}{1-q} = p + C.$$

The constant of integration can be taken to be zero with the initial conditions $q = \infty$ and $p = 0$. From this, a bit of algebra gives

$$q = \frac{e^{2p/a}+1}{e^{2p/a}-1},$$

and then

$$e^{2p/a} = 1 + \frac{2}{q-1}.$$

Now, Euler has a continued fraction form for q. Substituting that form into this last equation gives

$$e^{2p/a} = 1 + \cfrac{2}{\cfrac{a-p}{p} + \cfrac{1}{\cfrac{3a}{p} + \cfrac{1}{\cfrac{5a}{p} + \cfrac{1}{\cfrac{7a}{p} + \text{etc.}}}}}$$

Various values of p, s and a give continued fractions for various expressions involving e. Euler begins with $s = \frac{a}{2p}$, so $a = 2ps$, and easily gets an expression for $e^{1/s}$, which we omit here. Euler tells us that "from the equation found earlier, it follows that"

$$\frac{e^{1/s}+1}{e^{1/s}-1} = 2s + \cfrac{1}{6s + \cfrac{1}{10s + \cfrac{1}{14s + \cfrac{1}{18s + \text{etc.}}}}}$$

Euler does not tell us which equation he means, but the simplest way to verify this seems to be to use

$$q = \frac{e^{2p/a}+1}{e^{2p/a}-1},$$

set this equal to the continued fraction form for q, and then to make the substitutions $s = \frac{a}{2p}$, as he did before.

Now, Euler applies the technique for interposing a pair of 1s into a continued fraction. That technique did not seem very important when he developed it, but now it turns out to be quite useful. He gets

$$\frac{e^{1/s}+1}{e^{1/s}-1} = 2s - 1 + \cfrac{2}{1 + \cfrac{1}{1 + \cfrac{1}{3s - 1 + \cfrac{1}{1 + \cfrac{1}{1 + \cfrac{1}{5s - 1 + \text{etc.}}}}}}}$$

Euler says that from this it follows that

$$e^{1/s} = 1 + \cfrac{1}{s - 1 + \cfrac{1}{1 + \cfrac{1}{1 + \cfrac{1}{3s - 1 + \cfrac{1}{1 + \cfrac{1}{1 + \cfrac{1}{5s - 1 + \cfrac{1}{1 + \text{etc.}}}}}}}}}$$

The calculation is easy, but not obvious. The key step is to note that

$$\frac{e^{1/s}+1}{e^{1/s}-1} = \frac{e^{1/s}-1}{e^{1/s}-1} + \frac{2}{e^{1/s}-1}.$$

Thus, taking $s = 1$, the observation about the continued fraction expansion of e is proved, and proofs of all the other observations about the continued fractions involving e follow easily.

Euler does not make the point explicitly, but he has earlier noted that rational numbers have continued fractions that terminate and that quadratic irrationals have continued fractions that repeat. Thus this result proves that e is an irrational number and is not the root of any quadratic equation.

Euler is generally credited as being the first to prove that e is irrational, though the work usually cited is his *Introductio in analysin infinitorum*, E-101 and E-102, published in 1748. See, for example [StA]. The proof given here, though, came at least ten years earlier.

Now we seek to generalize this to other continued fractions with denominators that form an arithmetic progression. Euler sets

$$s = a + \cfrac{1}{(1+n)a + \cfrac{1}{(1+2n)a + \cfrac{1}{(1+3n)a + \cfrac{1}{(1+4n)a + \text{etc.}}}}}$$

Then he uses his earlier method to approximate s, with

$$\overset{a}{\frac{1}{0}},\quad \overset{(1+n)a}{\frac{a}{1}},\quad \overset{(1+2n)a}{\frac{(1+n)a^2+1}{(1+n)a}},\quad \overset{(1+3n)a}{\frac{(1+n)(1+2n)a^3+(2+2n)a}{(1+n)(1+2n)a^2+1}},\ \text{etc.}$$

He divides the general term of this by the first term of the denominator, then substitutes $a = \frac{1}{z\sqrt{n}}$ to get a rather formidable expression. Don't panic. It will simplify.

$$s = \frac{1}{z\sqrt{n}} \cdot \frac{1 + \frac{z}{1\cdot 1} + \frac{z^2}{1\cdot 2\cdot 1(1+n)} + \frac{z^3}{1\cdot 2\cdot 3\cdot 1(1+n)(1+2n)} + \text{etc.}}{1 + \frac{z}{1(1+n)} + \frac{z^2}{1\cdot 2(1+n)(1+2n)} + \frac{z^3}{1\cdot 2\cdot 3(1+n)(1+2n)(1+3n)} + \text{etc.}}$$

Now, let

$$t = 1 + \frac{z}{1\cdot 1} + \frac{z^2}{1\cdot 2\cdot 1(1+n)} + \frac{z^3}{1\cdot 2\cdot 3\cdot 1(1+n)(1+2n)} + \text{etc.}$$

and

$$u = 1 + \frac{z}{1(1+n)} + \frac{z^2}{1\cdot 2(1+n)(1+2n)} + \frac{z^3}{1\cdot 2\cdot 3(1+n)(1+2n)(1+3n)} + \text{etc.}$$

Then $s = \frac{t}{uz\sqrt{n}}$ and $dt = u\,dz$.

In the same way, Euler notes that $u\,dz + nz\,du = t\,dz$. He substitutes $t = vu$ so that $s = \frac{v}{z\sqrt{n}}$ and also $v\,du + u\,dv = u\,dz$, as well as $u\,dz + nz\,du = uv\,dz$. Continuing his calculations, he gets

$$\frac{du}{u} = \frac{dz - dv}{v} = \frac{v\,dz - dz}{nz}$$

and $nz\,dv - v\,dz + v^2dz = nz\,dz$.

Now, he makes some more substitutions, $v = z^{1/n}q$ and $z = r^n$. This changes the last differential equation to $dq + q^2\,dr = nr^{n-2}\,dr$. He sets $r = n^{-1/n}a^{-2/n}$, which makes $s = arq$.

All this tells us that s, given by the continued fraction with denominators forming an arithmetic progression, can be found by solving the differential equation $dq + q^2\,dr = nr^{n-2}\,dr$ to find q and r. Then, knowing $s = arq$, and making appropriate substitutions, we can find s.

One also needs a constant of integration, but this is found by knowing that $a = \infty$ implies that $s = \infty$ and also $a = 0$ makes $s = 1$.

Euler notes that $dq + q^2\,dr = nr^{n-2}\,dr$ is a Riccati equation and that he is able to solve this equation only when n is of the form $\frac{1}{2m+1}$, where m is a positive integer. Substituting this value for n tells Euler that he can always give a definite value, or, what he calls a "finite expression" for a continued fraction of the form

$$a + \cfrac{1}{\frac{(2m+3)a}{2m+1} + \cfrac{1}{\frac{(2m+5)a}{2m+1} + \cfrac{1}{\frac{(2m+7)a}{2m+1} + \text{etc.}}}}$$

Euler notes that, in the case $m = 0$, this gives

$$a + \cfrac{1}{3a + \cfrac{1}{5a + \cfrac{1}{7a + \cfrac{1}{9a + \text{etc.}}}}}$$

He says that he has already found the value of this continued fraction, but we would say that he has found it only in terms of the solution of a rather difficult differential equation. Euler notes that if he sets $a = (2m+1)b$, then this gives the same form as he got for s a little while earlier, and so he knows a solution whenever m is a positive integer.

Now, Euler observes that, if m is a whole number and if $n = \frac{2}{2m+1}$, then the solution of the differential equation $dq + q^2\,dr = nr^{n-2}\,dr$ is reduced to the solution of the simpler equation, $dq + q^2\,dr = 2\,dr$.

As a final example, Euler takes $n = 2$, that is $m = 0$, so the continued fraction is

$$s = a + \cfrac{1}{3a + \cfrac{1}{5a + \cfrac{1}{7a + \text{etc.}}}}$$

Then he solves the differential equation $dq + q^2\,dr = 2\,dr$ to get

$$r = \frac{1}{2\sqrt{2}}\, l\, \frac{q+\sqrt{2}}{q-\sqrt{2}},$$

though he does not give any details of this step. From this he gets

$$q = \frac{(e^{2r\sqrt{2}}+1)\sqrt{2}}{e^{2r\sqrt{2}}-1},$$

then substitutes $r = \frac{1}{a\sqrt{2}}$ and $s = arq = \frac{q}{\sqrt{2}}$ to repeat the result he had found earlier, namely

$$s = \frac{e^{2/a}+1}{e^{2/a}-1}.$$

This has been a remarkable paper. Not only does Euler present, for the first time, the foundations of the theory of continued fractions, but also he proves that the constant e is irrational, and links continued fractions with the theory of differential equations. It may not be clear to those of us outside the specialty of systems theory, so we may not fully appreciate the subtleties and richness they see in this paper, but we can still find plenty to appreciate here.

References

[C] Chabert, Jean-Luc, et. al., *A History of Algorithms: From the Pebble to the Microchip*, Translated from the French edition *Histoire d'algorithmes. Du caillou à la puce*. Editions Belin, Paris, 1994, by Chris Weeks, Springer, Berlin, 1998.

[E] Euler, Leonhard, "An Essay on Continued Fractions" translated by Myra F. Wyman and Bostwick F. Wyman, *Math Systems Theory* **18** (1985) 295–328.

[M] Miller, Jeff, *Earliest Known Uses of Some of the Words of Mathematics*, `http://members.aol.com/jeff570/mathword.html`, as revised April 24, 2002.

[StA] "The number *e*" at `http://www-groups.dcs.st-and.ac.uk/~history/HistTopics/e.html`, MacTutor History of Mathematics Archive, University of St. Andrews, Scotland, as revised September, 2001.

32

E-72: Variae observationes circa series infinitas*

Various observations about infinite series

This article about series is really quite remarkable. It is perhaps this author's favorite among all of Euler's papers. Euler takes a series mentioned briefly at the end of E-25 and applies Goldbach's clever technique for summing the series to build a wonderful chain of conclusions. Along the way, he proves new results about what we now call the Riemann zeta function, including the so-called sum-product formula. He also suggests a new proof that there are infinitely many primes, proves that the sum of the reciprocals of the primes diverges, and shows that the rate at which it diverges is related to $\log \log n$.

Besides all that, for the first time, Euler writes a paper in the now-classic Theorem–Proof style. All of his previous papers have been in a more narrative style. This also seems to be the first time Euler uses the symbols e and π in the same paper.

Here we also see Euler commit some of his famous sins with infinity. As usual, though, everything works out in the end, and Euler's results here are always correct, though their proofs are often a bit suspect.

This is a rather long paper, 29 pages both in the original *Commentarii* and in the *Opera Omnia*. It consists of 19 theorems, their proofs, and a few corollaries. There is almost no discussion other than an introductory paragraph in which he attributes Theorem 1 to Goldbach. Goldbach's letter of April 1729, though, had been addressed not to Euler but to Daniel Bernoulli. It seems that Euler did not know the result when he wrote E-25 in 1732 or 1733, but he does know the result and Goldbach's solution in 1737. Euler tells us that he learned the result from Goldbach, and not through Daniel Bernoulli. According to Carl Boehm, one of the editors of Volume 14 of Series I of the *Opera Omnia*, the volume in which this article appears, Goldbach posed both Theorem 1 and Theorem 6 to Euler in a letter that is now lost.

After the introductory paragraph, the paper begins with Theorem 1.

Theorem 1. *This infinite series, continued to infinity,*

$$\frac{1}{3}+\frac{1}{7}+\frac{1}{8}+\frac{1}{15}+\frac{1}{24}+\frac{1}{26}+\frac{1}{31}+\frac{1}{35}+\text{etc.}$$

the denominators of which are all numbers which are one less than powers of degree two or higher of whole numbers, that is, terms which can be expressed with the formula $\frac{1}{m^n-1}$, *where* m *and* n *are integers greater than one, then the series sums to 1.*

Proof. Let

$$x = 1+\frac{1}{2}+\frac{1}{3}+\frac{1}{4}+\frac{1}{5}+\frac{1}{6}+\frac{1}{7}+\frac{1}{8}+\frac{1}{9}+\text{etc.}$$

Comm. Acad. Sci. Imp. Petropol. 9 (1737) 1744, 160–188; *Opera Omnia* I.14, 216–245

Today, this step would be regarded with suspicion, at best. Now subtract the geometric series

$$1 = \frac{1}{2} + \frac{1}{4} + \frac{1}{8} + \frac{1}{16} + \frac{1}{32} + \text{etc.}$$

leaving

$$x - 1 = 1 + \frac{1}{3} + \frac{1}{5} + \frac{1}{6} + \frac{1}{7} + \frac{1}{9} + \frac{1}{10} + \text{etc.}$$

Subtract another geometric series

$$\frac{1}{2} = \frac{1}{3} + \frac{1}{9} + \frac{1}{27} + \frac{1}{81} + \frac{1}{243} + \text{etc.}$$

leaving

$$x - 1 - \frac{1}{2} = 1 + \frac{1}{5} + \frac{1}{6} + \frac{1}{7} + \frac{1}{10} + \frac{1}{11} + \text{etc.}$$

and another geometric series

$$\frac{1}{4} = \frac{1}{5} + \frac{1}{25} + \frac{1}{125} + \text{etc.}$$

leaving

$$x - 1 - \frac{1}{2} - \frac{1}{4} = 1 + \frac{1}{6} + \frac{1}{7} + \frac{1}{10} + \text{etc.}$$

Note that Euler had to skip subtracting the geometric series

$$\frac{1}{3} = \frac{1}{4} + \frac{1}{16} + \frac{1}{64} + \frac{1}{256} + \text{etc.}$$

because the series on the right is already a subseries of the series of powers of $\frac{1}{2}$, so those terms have already been subtracted. This happens because 3 is one less than a power, 4. It happens again every time we reach a term one less than a power, so he will have to skip 7, because that is one less than the cube 8, 8 because it is one less than the square 9, 15 because it is one less than the square 16, etc.

Continuing in this way, we see that all of the fractional terms on the right-hand side can be eliminated, leaving

$$x - 1 - \frac{1}{2} - \frac{1}{4} - \frac{1}{5} - \frac{1}{6} - \frac{1}{9} - \text{etc.} = 1$$

so

$$x - 1 = 1 + \frac{1}{2} + \frac{1}{4} + \frac{1}{5} + \frac{1}{6} + \frac{1}{9} + \frac{1}{10} + \text{etc.}$$

The 1 on the left must equal the terms of the harmonic series that are missing on the right. Those missing terms are exactly the ones with denominators one less than powers, so Euler concludes that

$$1 = \frac{1}{3} + \frac{1}{7} + \frac{1}{8} + \frac{1}{15} + \frac{1}{24} + \frac{1}{26} + \text{etc.}$$

where the terms on the right have denominators one less than powers. Q. E. D.

This is a beautiful proof, and a true result. Unfortunately, the beautiful proof is of the form "infinity minus infinity equals 1" and so is not valid. Bruce Burdick, of Roger Williams University, has discovered, but not yet published, a direct proof, using the principle of inclusion and exclusion, that meets modern standards of rigor. But it somehow lacks the spirit of Goldbach's proof, as presented by Euler.

This seems to be as far as Goldbach took this problem. For the rest of this paper, we will see how much farther Euler was able to push Goldbach's idea. Euler's theorems 2 to 6 all use either Theorem 1 or a variation on its proof. We will omit many details that Euler includes. The interested reader is encouraged to consult the complete Latin version as given in the *Opera Omnia* and on The Euler Archive [EA] because the ideas of the proofs are evident from the formulas themselves.

Theorem 2. $\frac{1}{3} + \frac{1}{7} + \frac{1}{15} + \frac{1}{31} + \frac{1}{35} + \frac{1}{63} + \text{etc.} = l\,2$, *where the denominators are all one less than the powers of even numbers, and as always, l denotes the natural logarithm, and also* $\frac{1}{8} + \frac{1}{24} + \frac{1}{26} + \frac{1}{48} + \frac{1}{80} + \text{etc.} = 1 - l\,2$, *where the denominators are all one less than the powers (greater than* 1*) of odd numbers.*

Obviously, if either one of these statements is true, the other follows immediately from Theorem 1. Euler's proof begins by taking $x = \frac{1}{2} + \frac{1}{4} + \frac{1}{6} + \frac{1}{8} + \frac{1}{10} + \frac{1}{12} + \frac{1}{14} + \text{etc.}$ Then he subtracts geometric series, as before, to get

$$x = 1 + \frac{1}{5} + \frac{1}{9} + \frac{1}{11} + \frac{1}{13} + \frac{1}{17} + \frac{1}{19} + \text{etc.}$$

From this, the first result follows by knowing $l\,2 = 1 - \frac{1}{2} + \frac{1}{3} - \frac{1}{4} + \frac{1}{5} - \frac{1}{6} + \frac{1}{7} - \frac{1}{8} + \text{etc.}$ We leave the details here and for the next several proofs to the reader.

Euler ends the first part of the proof with the unusual notation "Q. E. Unum," which translates as "Which is the first one." He ends the second part with "Q. E. Alterum" or "Which is the other one." This seems to be the only time Euler (or anyone else!) does this.

Theorem 3.

$$\frac{\pi}{4} = 1 - \frac{1}{8} - \frac{1}{24} + \frac{1}{28} - \frac{1}{48} - \frac{1}{80} - \frac{1}{120} - \frac{1}{124} - \frac{1}{168} - \frac{1}{224} + \frac{1}{244} - \frac{1}{288} - \text{etc.}$$

Here, the denominators are "evenly even numbers," that is, numbers divisible by 4, that are one more or one less than powers of odd numbers. The denominators that are one more than powers have a + sign, and the others have a − sign.

Euler begins the proof knowing $\frac{\pi}{4} = 1 - \frac{1}{3} + \frac{1}{5} - \frac{1}{7} + \frac{1}{9} - \frac{1}{11} + \frac{1}{13} - \text{etc.}$ He adds alternating geometric series, starting with $\frac{1}{4} = \frac{1}{3} - \frac{1}{9} + \frac{1}{27} - \frac{1}{81} + \text{etc.}$ and also subtracts series that do not alternate, starting with $\frac{1}{4} = \frac{1}{5} + \frac{1}{25} + \frac{1}{125} + \text{etc.}$ He only adds and does not subtract those geometric series that give terms that appear on the right-hand side of the statement of the theorem, and the result follows. Note that this proof is valid by modern standards and does not require any untoward series manipulations.

Theorem 4. $\frac{\pi}{4} - \frac{3}{4} = \frac{1}{28} - \frac{1}{124} + \frac{1}{244} + \frac{1}{344} + \text{etc.}$ *Here, the denominators are multiples of four that are one more or one less than odd powers of odd numbers. The signs of the terms obey the same rule as before, positive for those one more than such powers, negative for those one less.*

To prove this, Euler begins with the result of Theorem 3. He cites a result that he has not proved here, but seems to know from some other source:

$$\frac{1}{8} + \frac{1}{24} + \frac{1}{48} + \frac{1}{80} + \frac{1}{120} + \frac{1}{168} + \text{etc.} = \frac{1}{4}$$

where the denominators are one less than the odd squares. Euler leaves verifying this series as an interesting exercise for the reader. Add this series to the result from Theorem 3 and the result follows easily.

Euler gives five corollaries to Theorem 4, two of which amount to saying that Theorem 4 gives a fairly fast converging approximation to π, and that the first two terms give $\pi = \frac{22}{7}$, the value Euler says Archimedes used.

Theorem 5. $\frac{\pi}{4} - l\,2 = \underbrace{\frac{1}{26} + \frac{1}{28}} + \underbrace{\frac{1}{242} + \frac{1}{244}} + \underbrace{\frac{1}{342} + \frac{1}{344}} + \text{etc.}$, *where the denominators are one more and one less than odd powers (greater than* 1*) of numbers which are themselves one less than a multiple of four.*

The proof is almost immediate from the series giving $\frac{\pi}{4}$ from Theorem 3 and from the one giving $1 - l\,2$ from Theorem 2.

Theorem 6. $\frac{1}{15} + \frac{1}{63} + \frac{1}{80} + \frac{1}{255} + \frac{1}{624} + \text{etc.} = \frac{7}{4} - \frac{\pi^2}{6}$, *where the denominators are one less than the square numbers that are also higher powers.*

To prove this, Euler begins with his renowned solution to the Basel Problem, that

$$\frac{\pi^2}{6} = 1 + \frac{1}{4} + \frac{1}{9} + \frac{1}{16} + \frac{1}{25} + \text{etc.}$$

Then he replaces geometric subseries on the right with their sums, starting with

$$\frac{1}{3} = \frac{1}{4} + \frac{1}{16} + \frac{1}{64} + \text{etc.} \quad \text{and} \quad \frac{1}{8} = \frac{1}{9} + \frac{1}{81} + \frac{1}{729} + \text{etc.}$$

This gives him

$$\frac{\pi^2}{6} = 1 + \frac{1}{3} + \frac{1}{8} + \frac{1}{24} + \frac{1}{35} + \frac{1}{48} + \frac{1}{99} + \text{etc.}$$

where the denominators are one less than numbers that are only squares and are not higher powers. Euler calls in a special series, closely related to the one he cited in his proof of Theorem 4, that

$$\frac{3}{4} = \frac{1}{3} + \frac{1}{8} + \frac{1}{15} + \frac{1}{24} + \frac{1}{35} + \frac{1}{48} + \frac{1}{63} + \frac{1}{80} + \text{etc.}$$

where denominators are one less than squares. The result follows by subtraction.

Euler tells us that Goldbach claimed this theorem was true but had not been able to prove it. This presents us with a mystery. When did Goldbach know this? If it was in 1729, when he told Theorem 1 to Daniel Bernoulli, or if it was any other time before Euler solved the Basel Problem in 1734, then it could have given Euler a big hint about the solution of the Basel Problem. The reason is that this proof can be "run backwards," starting with this result, which Goldbach knew, to find the answer to the Basel Problem, $\frac{\pi^2}{6}$. It would not have given Euler a *proof* for the Basel Problem, but it would have given him the important clue that the answer involves π.

Daniel Bernoulli is sometimes reported to have claimed that if he had realized the answer had involved π, he would have been able to solve the Basel Problem himself.

Euler's next theorem is dramatically different from the first six, though the proof is similar. In Theorem 7, Euler turns from infinite sums to infinite products.

Theorem 7. *The infinite product* $\frac{2\cdot3\cdot5\cdot7\cdot11\cdot13\cdot17\cdot19\cdot\text{etc.}}{1\cdot2\cdot4\cdot6\cdot10\cdot12\cdot16\cdot18\cdot\text{etc.}}$, *in which the numerators are all prime numbers and the denominators are one less than the numerators, equals the sum of the infinite series* $1+\frac{1}{2}+\frac{1}{3}+\frac{1}{4}+\frac{1}{5}+\frac{1}{6}+\text{etc.}$, *and they are both infinite.*

The factors of the infinite product can be rewritten as

$$\frac{p}{p-1}=\frac{1}{1-1/p},$$

so, in modern notation, this theorem can be rewritten as

$$\sum_{k=1}^{\infty}\frac{1}{k}=\prod_{p}\frac{1}{1-1/p}.$$

Readers will recognize this as the sum-product formula for the Riemann zeta function at the value $s=1$. Others may recognize it as part of the cover illustration of William Dunham's wonderful book on Euler. [D]

The statement of this theorem should remind us that Euler considers infinite numbers to be like ordinary numbers, and that they can be equal or unequal to each other, just like finite ones. This theorem claims that the infinite number that results from the product is the same one that results from the sum. Since Euler's time, we have learned that paradoxes arise if we treat infinite quantities in this way, and so, in a modern sense, Euler's theorem cannot be true. Still, we have things to learn from the proof Euler offers, so we will examine it in detail.

Euler begins the same way he started his proof of Theorem 1, taking

$$x=1+\frac{1}{2}+\frac{1}{3}+\frac{1}{4}+\frac{1}{5}+\frac{1}{6}+\text{etc.}$$

Euler plans to rewrite x as an infinite product. Dividing by 2 gives

$$\frac{1}{2}x=\frac{1}{2}+\frac{1}{4}+\frac{1}{6}+\frac{1}{8}+\text{etc.}$$

He subtracts this from the series giving x, which eliminates all terms with even denominators and leaves

$$\frac{1}{2}x=1+\frac{1}{3}+\frac{1}{5}+\frac{1}{7}+\text{etc.}$$

This, in turn, he divides by 3, giving

$$\frac{1}{2}\cdot\frac{1}{3}x=\frac{1}{3}+\frac{1}{9}+\frac{1}{15}+\frac{1}{21}+\text{etc.}$$

Subtracting again eliminates all remaining denominators that are multiples of 3, leaving

$$\frac{1}{2}\cdot\frac{2}{3}x = 1 + \frac{1}{5} + \frac{1}{7} + \frac{1}{11} + \frac{1}{13} + \text{etc.}$$

This process is like the ancient sieve of Eratosthenes because, at each stage, it eliminates a prime denominator and all remaining multiples of that prime denominator. Eventually, everything on the right will be eliminated except the first term, 1. Some readers may want to interpret this as a "proof" that 1 is not a prime number.

Euler carries us through one more iteration of his process. He divides this last equation by 5, and does a small rearrangement of the way he writes the product of fractions, to get

$$\frac{1\cdot 2}{2\cdot 3}\cdot\frac{1}{5}x = \frac{1}{5} + \frac{1}{25} + \frac{1}{35} + \text{etc.}$$

Subtracting leaves

$$\frac{1\cdot 2\cdot 4}{2\cdot 3\cdot 5}x = 1 + \frac{1}{7} + \frac{1}{11} + \frac{1}{13} + \text{etc.}$$

In the same way, terms with denominators that are multiples of 7, 11, and so forth for all prime numbers, will be eliminated, leaving

$$\frac{1\cdot 2\cdot 4\cdot 6\cdot 10\cdot 12\cdot 16\cdot 18\cdot 22\cdot \text{etc.}}{2\cdot 3\cdot 5\cdot 7\cdot 11\cdot 13\cdot 17\cdot 19\cdot 21\cdot \text{etc.}}x = 1.$$

But, as x is already known to be the sum of the harmonic series, the desired result is immediate.
Q. E. D.

This result is sometimes cited as Euler's proof that there are infinitely many prime numbers. Because the harmonic series diverges, the product diverges, and a product can diverge only if it is an infinite product. Hence there must be infinitely many prime numbers. Though the infinitude of primes is immediate, Euler does not explicitly make the connection here.

Of course, this proof does not meet modern standards of rigor because of the way it treats the value of the harmonic series as a number.

Euler offers without proof three corollaries.

Corollary 1. *The value of the expression* $\frac{2\cdot3\cdot5\cdot7\cdot11\cdot13\cdot\text{etc.}}{1\cdot2\cdot4\cdot6\cdot10\cdot12\cdot\text{etc.}}$ *is infinite and denoting absolute infinity by* ∞, *the value of this expression will be* $= l\,\infty$, *which infinity is the smallest of all the infinite values.*

Here, $l\,\infty$ means the natural logarithm of infinity. This is another illustration of how Euler thinks about infinite numbers and is the only use we have seen of this expression absolute infinity, which apparently means the number of positive integers. This last statement, about the logarithm of infinity being the smallest of all the infinite values, will be contradicted by a statement Euler will make at the very end of this paper. This is, indeed, a thought-provoking assertion.

Corollary 2. *This expression* $\frac{4\cdot9\cdot16\cdot25\cdot36\cdot49\cdot\text{etc.}}{3\cdot8\cdot15\cdot24\cdot35\cdot48\cdot\text{etc.}}$ *has a finite value, namely* 2, *and it follows that there are infinitely many more prime numbers than there are squares in the series of all numbers.*

Euler does not bother to tell us why the infinite product has value 2, but showing it to be so is similar to the series cited in the proofs of Theorems 4 and 5. Rewrite a square n^2 in the numerator as $n \cdot n$. Then, rewrite its corresponding denominator as $(n+1)(n-1)$. Then everything cancels except 2. It is a neat exercise to work through the details.

Corollary 3. *Because it is known that there are infinitely fewer prime numbers than integer numbers, and because the expression*

$$\frac{2 \cdot 3 \cdot 4 \cdot 5 \cdot 6 \cdot 7 \cdot \text{etc.}}{1 \cdot 2 \cdot 3 \cdot 4 \cdot 5 \cdot 6 \cdot \text{etc.}}$$

has as its value absolute infinity, it follows that the number of prime numbers is the logarithm of that infinity.

This seems to be the first statement and proof of a form of what we now call the Prime Number Theorem. If we use modern notation and let $\pi(n)$ be the number of primes less than n, then Euler seems to be interpreting his result as meaning that $\pi(\infty) = \ln \infty$. It is hard to make sense of this, but it might mean that

$$\lim_{n\to\infty} \frac{\pi(n)}{\ln(n)} = 1.$$

If that is what Euler means, then he isn't quite correct. Gauss was closer in 1792 when he conjectured a more general form in his diary, that, asymptotically, $\pi(n) \sim \frac{n}{\ln n}$, but never published anything about it. Rigorous proofs, that is, proofs that meet modern standards, were discovered independently by Hadamard and de la Vallée-Poussin only in 1896. The whole subject has a long and interesting history.

Euler follows his version of the Prime Number Theorem with the sum-product formula for $n > 1$:

Theorem 8. *If from the series of prime numbers, the following expression is formed*

$$\frac{2^n}{(2^n-1)} \cdot \frac{3^n}{(3^n-1)} \cdot \frac{5^n}{(5^n-1)} \cdot \frac{7^n}{(7^n-1)} \cdot \frac{11^n}{(11^n-1)} \cdot \text{etc.}$$

then its value will be equal to the sum of this series

$$1 + \frac{1}{2^n} + \frac{1}{3^n} + \frac{1}{4^n} + \frac{1}{5^n} + \frac{1}{6^n} + \frac{1}{7^n} + \text{etc.}$$

This is the sum-product formula itself, and its proof is almost identical to the proof of Theorem 7. Pretty much all you have to do is write in the exponents and follow the same steps. Euler gives complete detail, but we will omit that.

The reader may wonder why this function

$$\zeta(s) = 1 + \frac{1}{2^s} + \frac{1}{3^s} + \frac{1}{4^s} + \frac{1}{5^s} + \frac{1}{6^s} + \frac{1}{7^s} + \text{etc.}$$

is called the *Riemann* zeta function and not the *Euler* zeta function. A mathematical comedian once said that if everything that Euler did first were named after Euler, then everything important in mathematics would be named after him and we might as well change the name of the subject to Euler-matics instead. It's a small wonder there are so few famous mathematical comedians.

We ought to be fair to Riemann, though. Euler gives no clue that he regards the value given by the series as a function of its exponent. He thinks a function has to be given by a finite equation, though he is starting to realize that sometimes a function might be defined as a solution to a differential equation. He regards Theorem 8 as giving the equality of two infinite expressions, not as describing properties of a function.

Even more significantly, Euler regards the exponent in the series only as a real number. Riemann used analytic continuation to extend the definition to the whole complex plane. So, we can in good conscience keep calling it the Riemann zeta function.

Euler gives two corollaries to Theorem 8, product formulas for the exponents 2 and 4. He proved in E-41 that

$$1+\frac{1}{4}+\frac{1}{9}+\frac{1}{16}+\text{etc.}=\frac{\pi^2}{6}.$$

Therefore, also

$$\frac{\pi^2}{6}=\frac{2\cdot 2\cdot 3\cdot 3\cdot 5\cdot 5\cdot 7\cdot 7\cdot 11\cdot 11\cdot \text{etc.}}{1\cdot 3\cdot 2\cdot 4\cdot 4\cdot 6\cdot 6\cdot 8\cdot 10\cdot 12\cdot \text{etc.}}.$$

Likewise, when $n=4$, he gets

$$\frac{\pi^4}{90}=\frac{4\cdot 4\cdot 9\cdot 9\cdot 25\cdot 25\cdot 49\cdot 49\cdot 121\cdot 121\cdot \text{etc.}}{3\cdot 5\cdot 8\cdot 10\cdot 24\cdot 26\cdot 48\cdot 50\cdot 120\cdot 122\cdot \text{etc.}}.$$

As a supplement to the second corollary, Euler notes that if the second expression is divided by the first, it gives

$$\frac{\pi^2}{15}=\frac{4\cdot 9\cdot 25\cdot 49\cdot 121\cdot 169\cdot \text{etc.}}{5\cdot 10\cdot 26\cdot 50\cdot 122\cdot 170\cdot \text{etc.}}.$$

Theorem 9. $\frac{5\cdot 13\cdot 25\cdot 61\cdot 85\cdot 145\cdot\text{etc.}}{4\cdot 12\cdot 24\cdot 60\cdot 84\cdot 144\cdot\text{etc.}}=\frac{3}{2}.$

First, we have to untangle Euler's explanation of this infinite product. Start with the odd squares of primes. Add 1 and divide by 2. This gives the factors in the numerators. The factors in the denominator are one less than the corresponding factors in the numerator.

Making sense of the product was the hard part. The proof is easy. Divide the two expressions from the corollaries that involve π^2 to get $\frac{5}{2}=\frac{5\cdot 10\cdot 26\cdot 50\cdot 122\cdot 170\cdot 290\cdot\text{etc.}}{3\cdot 8\cdot 24\cdot 48\cdot 120\cdot 168\cdot 288\cdot\text{etc.}}$. Now divide by $\frac{5}{3}$ and cancel 2s in each factor to get the result.

Theorem 10. $\frac{\pi^3}{32}=\frac{80\cdot 224\cdot 440\cdot 624\cdot 728\cdot\text{etc.}}{81\cdot 225\cdot 441\cdot 625\cdot 729\cdot\text{etc.}}$ *where the denominators are the squares of the odd numbers that are not prime and the numerators are one less than the denominators.*

Euler's proof begins with Wallis's formula, written as $\frac{\pi}{4}=\frac{8\cdot 24\cdot 48\cdot 80\cdot 120\cdot 168\cdot\text{etc.}}{9\cdot 25\cdot 49\cdot 81\cdot 121\cdot 169\cdot\text{etc.}}$. He combines this with the infinite product for $\frac{\pi^2}{6}$ from the corollary to Theorem 8, and the result follows with just a little cancellation.

Theorem 11. $\frac{\pi}{4}=\frac{3\cdot 5\cdot 7\cdot 11\cdot 13\cdot 17\cdot 19\cdot 23\cdot\text{etc.}}{4\cdot 4\cdot 8\cdot 12\cdot 12\cdot 16\cdot 20\cdot 24\cdot\text{etc.}}$, *where the numerators are the prime numbers and the denominators are the multiples of four either one more or one less than the corresponding numerators.*

For his proof, Euler returns to the ideas of Theorem 7. He starts with the alternating series

$$\frac{\pi}{4}=1-\frac{1}{3}+\frac{1}{5}-\frac{1}{7}+\frac{1}{9}-\frac{1}{11}+\frac{1}{13}-\text{etc.}$$

Dividing by 3, then adding gives

$$\frac{4}{3}\cdot\frac{\pi}{4} = 1 + \frac{1}{5} - \frac{1}{7} - \frac{1}{11} + \frac{1}{13} + \text{etc.}$$

This has eliminated all of the denominators on the right that are multiples of 3. To eliminate the multiples of 5, we divide by 5 and subtract, leaving

$$\frac{4}{5}\cdot\frac{4}{3}\cdot\frac{\pi}{4} = 1 - \frac{1}{7} - \frac{1}{11} + \frac{1}{13} + \text{etc.}$$

Likewise, he eliminates multiples of 7. At each stage, eliminating a prime of the form $4n-1$ requires addition and so inserts a factor of the form $\frac{4n-1}{4n}$, and a prime of the form $4n+1$ requires subtraction, giving a factor of the form $\frac{4n+1}{4n}$. He writes the resulting infinite product as

$$\frac{\text{etc.}\, 24\cdot 20\cdot 16\cdot 12\cdot 12\cdot 8\cdot 4\cdot 4}{\text{etc.}\, 23\cdot 19\cdot 17\cdot 13\cdot 11\cdot 7\cdot 5\cdot 3}\cdot\frac{\pi}{4} = 1$$

and the theorem follows by dividing both sides by the infinite product.

We note that this proof fails modern standards of rigor because of its shameless rearranging of terms in a conditionally convergent series.

Theorem 12. *Start with the sequence of odd prime numbers. Take the pair of numbers one more and one less than each prime, and divide each one by* 2. *If the result is even, that result is to be a factor in the numerator. If it is odd, the factor goes in the denominator. The resulting infinite product,* $\frac{2\cdot 2\cdot 4\cdot 6\cdot 6\cdot 7\cdot 10\cdot 12\cdot \text{etc.}}{1\cdot 3\cdot 3\cdot 5\cdot 7\cdot 9\cdot 9\cdot 11\cdot \text{etc.}} = 2.$

We have adapted the wording of this theorem rather extensively because what makes sense in Latin does not seem to translate fluidly into English. The proof, though, is straightforward. Square and invert the expression for $\frac{\pi}{4}$ from Theorem 11 to get an infinite product for $\frac{16}{\pi^2}$. Then take the expression for $\frac{\pi^2}{6}$ from the corollary to Theorem 8 and multiply by $\frac{3}{4}$ to get an infinite product for $\frac{\pi^2}{8}$. The result follows by dividing the two infinite products, then canceling 2s.

Theorems 13 to 18 continue in this vein, introducing no new ideas. We will simply state the results and omit most discussion of the proofs. Euler, of course, gives all of them in complete detail.

Theorem 13. *Start with the sequence of odd numbers that are not prime (except* 1.) *Take the pair of adjacent numbers and divide them by* 2. *Use the even results in the numerator of the infinite product and the odd results in the denominator. Then the resulting infinite product is*

$$\frac{\pi}{4} = \frac{4\cdot 8\cdot 10\cdot 12\cdot 14\cdot 16\cdot 18\cdot 20\cdot 22\cdot 24\cdot \text{etc.}}{5\cdot 7\cdot 11\cdot 13\cdot 13\cdot 17\cdot 17\cdot 19\cdot 23\cdot 25\cdot \text{etc.}}.$$

Hint: Start with Wallis's infinite product for $\frac{\pi}{2}$.

Theorem 14. $\frac{\pi}{2} = \frac{3\cdot 5\cdot 7\cdot 11\cdot 13\cdot 17\cdot 19\cdot 23\cdot 29\cdot 31\cdot \text{etc.}}{2\cdot 6\cdot 6\cdot 10\cdot 14\cdot 18\cdot 18\cdot 22\cdot 30\cdot 30\cdot \text{etc.}}$ *where the numerators are the odd prime numbers and the denominators are the even numbers not divisible by* 4 *that are one more or one less than the corresponding numerators.*

Hint: Use Theorem 11 and the corollary to Theorem 8.

Theorem 15.

$$\frac{\pi}{2} = 1+\frac{1}{3}-\frac{1}{5}+\frac{1}{7}+\frac{1}{11}-\frac{1}{13}-\frac{1}{15}-\frac{1}{17}+\frac{1}{19}+\frac{1}{21}+\frac{1}{23}+\frac{1}{25}+\frac{1}{27}-\frac{1}{29}+\frac{1}{31}+\frac{1}{33}-\frac{1}{35}-\frac{1}{37}-\text{etc.}$$

where the denominators are all of the odd numbers. The signs are determined by a multiplication rule. Primes of the form $4n-1$ *are assigned a* $+$ *sign. Those of the form* $4n+1$ *are assigned a* $-$ *sign. The signs for a composite odd number is found by multiplying the signs corresponding to its prime factors.*

Theorem 16. $\frac{\pi}{2} = 1+\frac{1}{2}-\frac{1}{6}+\frac{1}{6}+\frac{1}{10}-\frac{1}{14}-\frac{1}{16}-\frac{1}{18}+\frac{1}{18}+\frac{1}{20}+\text{etc.}$, *where the denominators are either one more or one less than odd numbers that are not powers. If an odd number has a* $+$ *sign in Theorem 15, then the denominator is one less, and the sign is* $+$. *If the odd number has a* $-$ *sign in Theorem 15, then the denominator is one more, and the sign is* $-$.

Hint: Euler does not prove this, but instead tells us to apply the techniques of Theorems 1, 2 and 3 to the series in Theorem 15.

Theorem 17. *If an odd prime number of the form* $4n-1$ *takes a* $+$ *sign and the remaining ones, being of the form* $4n+1$, *take a* $-$ *sign, and if composite numbers take signs as the product of their factors, then*

$$\frac{3\pi}{8} = 1+\frac{1}{9}-\frac{1}{15}+\frac{1}{21}+\frac{1}{25}+\frac{1}{33}-\frac{1}{35}-\frac{1}{39}+\frac{1}{49}-\frac{1}{51}-\text{etc.}$$

where the denominators are the product of an even number of prime numbers.

Hint: Add the series for $\frac{\pi}{2}$ to the alternating series of reciprocals of odd numbers, which gives $\frac{\pi}{4}$. Then divide by 2.

Theorem 18. *If all prime numbers are given the* $-$ *sign, and if composite numbers take signs as the product of their factors, then the sum of the series*

$$1-\frac{1}{2}-\frac{1}{3}+\frac{1}{4}-\frac{1}{5}+\frac{1}{6}-\frac{1}{7}-\frac{1}{8}+\frac{1}{9}+\frac{1}{10}-\frac{1}{11}-\frac{1}{12}-\text{etc.}$$

equals 0.

Hint: Set the series equal to x, then use the process from Theorems 7 and 8.

Those versed in more modern ideas in number theory might see the shadows of the Möbius function in these last few theorems.

Admittedly, there has not been much excitement in Theorems 9 to 18. But he finishes with a flourish, though.

Theorem 19. *The sum of the series of reciprocals of prime numbers*

$$\frac{1}{2}+\frac{1}{3}+\frac{1}{5}+\frac{1}{7}+\frac{1}{11}+\frac{1}{13}+\text{etc.}$$

is infinitely large, but it is infinitely less than the sum of the harmonic series. In fact, the sum of the reciprocals of the primes is almost like the logarithm of the harmonic series.

Proof. Put

$$A = \frac{1}{2} + \frac{1}{3} + \frac{1}{5} + \frac{1}{7} + \frac{1}{11} + \text{etc.}$$
$$B = \frac{1}{2^2} + \frac{1}{3^2} + \frac{1}{5^2} + \frac{1}{7^2} + \frac{1}{11^2} + \text{etc.}$$
$$C = \frac{1}{2^3} + \frac{1}{3^3} + \frac{1}{5^3} + \frac{1}{7^3} + \frac{1}{11^3} + \text{etc.}$$

Euler tells us that

$$e^{A+\frac{1}{2}B+\frac{1}{3}C+\frac{1}{4}D+\text{etc.}} = 1 + \frac{1}{2} + \frac{1}{3} + \frac{1}{4} + \frac{1}{5} + \frac{1}{6} + \frac{1}{7} + \text{etc.}$$

because

$$A + \frac{1}{2}B + \frac{1}{3}C + \frac{1}{4}D + \text{etc.} = l\,\frac{2}{1} + l\,\frac{3}{2} + l\,\frac{5}{4} + l\,\frac{7}{6} + \text{etc.}$$

Then the claim follows from applying Theorem 7 to the right-hand side. However, it is a little tricky to see why this last statement is true. It follows by performing the sum on the left, one column at a time. For example, the first column is given by

$$\frac{1}{2} + \frac{1}{2}\cdot\frac{1}{2^2} + \frac{1}{3}\cdot\frac{1}{2^3} + \frac{1}{4}\cdot\frac{1}{2^4} + \cdots + \frac{1}{k}\cdot\frac{1}{2^k} + \cdots$$

Such series can be summed by integrating geometric series. For a prime p, the sum is

$$\ln\frac{1}{1-\frac{1}{p}} = \ln\frac{p}{p-1},$$

as required. Then Euler's claim follows by applying a couple of elementary properties of logarithms.

Euler notes that, not only are B, C, D, etc. finite, but also $\frac{1}{2}B+\frac{1}{3}C+\frac{1}{4}D+\text{etc.}$ is finite. Because of this, Euler says, and because A is infinitely large, the terms $\frac{1}{2}B + \frac{1}{3}C + \frac{1}{4}D + \text{etc.}$ vanish with respect to A, and so

$$e^A = e^{A+\frac{1}{2}B+\frac{1}{3}C+\frac{1}{4}D+\text{etc.}} = e^{\frac{1}{2}+\frac{1}{3}+\frac{1}{5}+\frac{1}{7}+\frac{1}{11}+\text{etc.}}$$

Thus,

$$e^{\frac{1}{2}+\frac{1}{3}+\frac{1}{5}+\frac{1}{7}+\frac{1}{11}+\text{etc.}} = 1 + \frac{1}{2} + \frac{1}{3} + \frac{1}{4} + \frac{1}{5} + \text{etc.}$$

Taking logarithms of both sides makes the sum of reciprocals of primes equal the logarithm of the harmonic series. Because the harmonic series is, itself, the logarithm of the ordinary infinity, this makes the sum of the reciprocals of the primes $= l.\,l.\,\infty$. Q. E. D.

This, too, is sometimes cited as Euler's proof that there are infinitely many prime numbers. Again, Euler does not mention this corollary.

This conclusion that $\ln\ln\infty$ is still infinity and is infinitely less than $\ln\infty$ seems to disagree with Euler's remark in the corollary to Theorem 7 that the value of the harmonic series, $\ln\infty$ is the smallest of the infinities.

Euler's treatment of infinity in this paper is quite different and considerably more sophisticated than in previous papers. In fact, his use of infinity in Theorem 19 is more subtle than it was in Theorem 1. Modern readers may complain that he did it "wrong" in view of what we now know about series and convergence. This may be true, but Euler's results are correct if we make appropriate translations into the language of limits. Also, these proofs have a beauty and a charm that rigorously correct modern proofs of the same theorems lack. Finally, calculations like the ones Euler does here will inspire others to do similar calculations. Those calculations will lead to paradoxes that will eventually lead others to figure out exactly what one can and cannot do with series.

References

[D] Dunham, William, *Euler: The Master of Us All*, Dolciani Mathematical Expositions vol. 22, Mathematical Association of America, Washington, DC, 1999.

[EA] The Euler Archive, `www.EulerArchive.com`. Consulted March 21, 2006.

33

E-73: Solutio problematis geometrici circa lunulas a circulis formatas*

Solution to a geometric problem about lunes formed by circles

A lune, or lunule, is the shape left when a circle cuts a piece out of the edge of another circle. It is also called a crescent. The study of lunes makes a long but slender thread through the history of mathematics. The earliest work that is still remembered seems to be due to Hippocrates of Chios (approx 460–380 BCE). Hippocrates was able to calculate exactly the areas of certain lunes. At the time, this led him and his contemporaries to hope that the work could be extended to calculate exact areas for circles as well. Their optimism was reasonable, though wrong. Because lunes are more complicated objects than circles, it makes sense to think that finding their areas might be harder than finding the area of a circle. Accounts of Hippocrates's work are widely reprinted [C], and Dunham [D] gives a summary of his main results and their significance.

Our first figure (not from Euler) shows the first lune of Hippocrates, described in detail in Dunham. The figure is less cluttered if we use semicircles instead of circles. We take the larger circle to be radius 1. Triangle ACO is a right triangle and its hypotenuse is the diameter of the smaller semicircle. The moon-shaped area $AECF$ is the lune that the larger circle cuts from the smaller circle. Hippocrates shows that this lune has the same area as the triangle ACO, that is $\frac{1}{2}$.

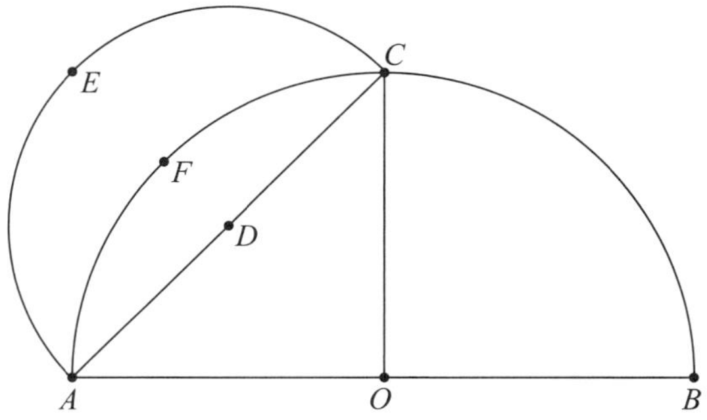

A lune can be described by giving the radii of the two circles and the distance between their centers. Of course, only the ratios of these measurements really matter. Now, we describe a lune by giving the ratio of the squares of its radii, in the form $(m : n)$. Euler uses N and n. This notation does not capture the distance between the centers. The lune of Hippocrates is the lune we now call $(1 : 2)$, because it is the lune left when a circle of radius $1/\sqrt{2}$ is cut by a circle of radius 1. Only a few particular lunes have constructible quadratures, or areas. Hippocrates discovered three such lunes, the ones we now describe as $(1 : 2)$, $(1 : 3)$ and $(2 : 3)$. Around 1600, Viète apparently worked on the lune $(1 : 4)$.

*$Comm. Acad. Sci. Imp. Petropol.$ 9 (1737) 1744, 207–221; *Opera Omnia* I.26 , 1–15

Euler studies lunes twice, here in E-73 and thirty-four years later (1771) in E-423. Christian Goldbach, G. W. Krafft and both Daniel and Johann Bernoulli also worked on lune problems, so the subject is "in the air" in the 1730s.

Between his two papers, Euler finds some conditions on m and n that are necessary for the lune $(m : n)$ to be quadrable,[1] and he discovers several new quadrable lunes. In 1947, Dorodonov showed that there are no quadrable lunes beyond the ones Euler found, so it would seem that the subject of lunes is now closed. "Closed" problems in mathematics have a way of popping open again, though, so we should not be surprised if somebody discovers new and interesting problems about lunes.

Here in E-73, Euler works on a problem related to lunes posed by Christian Goldbach. This time, instead of getting the problem directly from Goldbach, he says that he learned of the problem from an article Daniel Bernoulli had written in 1724.

The problem is described in terms of Euler's Figure 1. Goldbach wants to find lunes formed by two circles *aObmS* and *AOBMS*, with centers c and C respectively and intersecting at O and S that have two special properties. First, the line segments Ab and aB must be equal. Second, the areas of the two trilateral regions *OAb* and *OaB* should be equal.

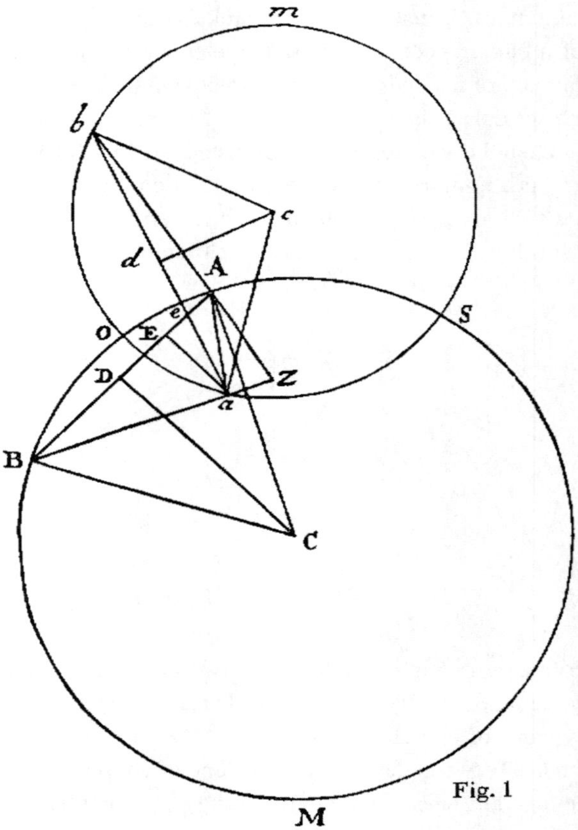

Fig. 1

[1] The word "quadrable" hardly ever occurs except followed by the word "lune." Originally, a region was quadrable if a square with the same area could be found by a ruler and compass construction. Now it means that the area is given by an algebraic expression that does not involve transcendental quantities like π.

Euler will take a very analytical approach to this problem. Euler has no regrets that his solution is so analytical. This is part of a larger trend throughout the 18th century to replace geometrical proofs with analytical ones. Euler is one of the prime movers in this transition from geometry to analysis. As the century passes, we will see him using geometry less and less. By 1765, when he discovers the "Euler line" of a triangle, his proof will be entirely and unapologetically analytic.

He supposes that his Figure 1 describes a solution to the problem posed by Goldbach. Because the trilaterals *OAb* and *OaB* are equal, there will be equality if the trilateral *AOa* is added to each. Thus, trilaterals *AbOa* and *aBOA* are also equal.

These two trilaterals will be equal if sector *CBOA* has the same area as sector *cbOa* and also the two quadrilaterals *bcaA* and *BCAa* are equal. These are only sufficient conditions. If these two conditions can be satisfied, and if also $Ab = aB$, then the problem will be solved. So, Euler's analysis will proceed in three steps, making three assumptions:

I. Sector *cbOa* = sector *CBOA*,
II. Area *bcaA* = area *BCAa*, and
III. Segment aA = Segment Aa.

He will examine the consequences of these assumptions, in this order, and then, about halfway through the article, Goldbach's problem will be solved. The rest of the article is spent exploring some of the consequences of his solution, including a problem of Daniel Bernoulli, but he does not give a list of the quadrable lunes. That will wait until 1771 and E-423.

Euler makes some unfortunate choices in his variable names. He lets $c = ca = cb$ be the radius of one circle and $C = CA = CB$ be the radius of the other. He draws the chords ab and AB, and lets their midpoints be d and D respectively, and also draws cd and CD. Then he denotes lengths

$$ad = db = b \quad \text{and} \quad AD = DB = B.$$

Readers will have to be careful to note when symbols b, B, c and C denote lengths and when they denote points. The distinction is always clear from the context.

This makes the sine of half of angle $bca = \frac{b}{c}$ and the sine of half of the angle $BCA = \frac{B}{C}$, "putting the whole = 1." This curious little remark deserves some comment.

In the early days of trigonometric tables, decimal fractions had not yet been invented. In fact, Ptolemy's tables from early in the first millennium did not give any of the trigonometric tables familiar today. Instead, he tabulated a trigonometric function called the Chord, which we would now describe as half the sine of double the angle. That is,

$$\text{Chord}(x) = \frac{1}{2}\sin 2x.$$

In the problem Euler is working on, the length d is related to the Chord of the angle bca. To avoid using fractions, tables would give the Chords for circles with a large radius: 1000, 20 000, or even 100 000. This practice continued even after the Chord was abandoned as a fundamental trigonometric function and we adopted our current practice of using sines, cosines and tangents. Tables would be given as lengths on a circle of some conveniently large radius, and people using the tables would have to multiply or divide by appropriate scaling factors. The identity $\tan x = \frac{\sin x}{\cos x}$ was true only for tables for which the radius was 1. When Euler says to put "the whole = 1," he is telling us to use a sine table where the sine of a whole angle, that is a right angle, equals 1. That amounts to using modern sine tables (if anybody is still using tables instead of using a calculator) for a circle of radius equal to 1.

Euler explains the arcsine function, denoting it by "A sin." To Euler, the arcsine gives the arc of a circle, not an angle, and it is necessary to state that the radius of that circle is 1. He uses the arcsine to determine that

$$\text{arc}\, bOa = 2c \cdot \arcsin \frac{b}{c} \quad \text{and} \quad \text{arc}\, BOA = 2C \cdot \arcsin \frac{B}{C}.$$

Similarly, the areas of the sectors are given by

$$bOac = c^2 \cdot \arcsin \frac{b}{c} \quad \text{and} \quad BOAC = C^2 \cdot \arcsin \frac{B}{C}.$$

Euler wants these two sectors to be equal, so

$$c^2 \cdot \arcsin \frac{b}{c} = C^2 \cdot \arcsin \frac{B}{C},$$

or, in other words,

$$\arcsin \frac{b}{c} : \arcsin \frac{B}{C} = C^2 : c^2.$$

The relationship between the arcs of the circles will not be what Euler calls "algebraic" unless this ratio is rational. Of course, Euler is using the word "algebraic" differently than we do. He takes

$$C^2 : c^2 = N : n$$

where N and n are integers. This makes

$$\sin\left(n \cdot \arcsin \frac{b}{c}\right) = \sin\left(N \cdot \arcsin \frac{B}{C}\right).$$

Euler knows the series expansion for the sine of a multiple of an arcsine, and he gives us the following:

$$\begin{aligned}
&\frac{nb(c^2 - b^2)^{(n-1)/2}}{1\,c^n} - \frac{n(n-1)(n-2)b^3(c^2 - b^2)^{(n-3)/2}}{1 \cdot 2 \cdot 3\,c^n} \\
&\quad + \frac{n(n-1)(n-2)(n-3)(n-4)b^5(c^2 - b^2)^{(n-5)/2}}{1 \cdot 2 \cdot 3 \cdot 4 \cdot 5\,c^n} - \text{etc.} \\
&= \frac{NB(C^2 - B^2)^{(N-1)/2}}{1\,C^N} - \frac{N(N-1)(N-2)B^3(C^2 - B^2)^{(N-3)/2}}{1 \cdot 2 \cdot 3\,C^N} \\
&\quad + \frac{N(N-1)(N-2)(N-3)(N-4)B^5(C^2 - B^2)^{(N-5)/2}}{1 \cdot 2 \cdot 3 \cdot 4 \cdot 5\,C^N} - \text{etc.}
\end{aligned}$$

Because N and n are finite whole numbers, these are both finite series, for the numerators eventually become zero. Hence, knowing $C^2 : c^2$ gives $N : n$, and this, in turn, defines a relation between the half chords B and b.

For small values of N and n, Euler finds this formula can be quite useful. He takes as an example $C^2 : c^2 = 2 : 1$, so that $C = c\sqrt{2}$, $N = 2$ and $n = 1$. In this case, the formula tells us that

$$\frac{b}{c} = \frac{2B\sqrt{C^2 - B^2}}{C^2} \quad \text{or} \quad bc = B\sqrt{C^2 - B^2}.$$

For a second example Euler looks at the case 3 : 1, so that $C = c\sqrt{3}$. This makes $N = 3$ and $n = 1$, so that

$$\frac{b}{c} = \frac{3B(C^2 - B^2)}{C^3} - \frac{B^3}{C^3} \quad \text{or} \quad 3bCc = 3BC^2 - 4B^3.$$

Third, he considers 4 : 1, so that $C = 2c$, $N = 4$ and $n = 1$. This makes

$$\frac{b}{c} = \frac{4B(C^2 - B^2)^{3/2}}{C^4} - \frac{4B^3\sqrt{C^2 - B^2}}{C^4}$$

or

$$4bcC^2 = (4BC^2 - 8B^3)\sqrt{C^2 - B^2}.$$

Fourth is the case $C^2 : c^2 = 3 : 2$, so that $N = 3$ and $n = 2$. Euler tells us that this makes

$$\frac{2b\sqrt{c^2 - b^2}}{c^2} = \frac{3BC^2 - 4B^3}{C^3}.$$

In the fifth case, $C^2 : c^2 = 4 : 3$, he gets

$$\frac{3bc^2 - 4b^3}{c^3} = \frac{(4BC^2 - 8B^3)\sqrt{C^2 - B^2}}{C^4}.$$

Euler assures us that these five examples will be enough for the applications he has in mind. The logical next example would be the case 5 : 1, which leads to nasty fifth degree equations.

Euler turns to finding conditions that guarantee that the two quadrilaterals $bAac$ and $BaAC$ are equal. He decomposes each into a difference of triangles, and gets that

$$\triangle bac - \triangle baA = \triangle BAC - \triangle BAa.$$

To rewrite this analytically, he draws two perpendiculars. Ae is through A and perpendicular to ab, and aE is through a and perpendicular to AB. Euler has neglected to draw the segment Ae, though he has given us its endpoint, e. He lets the lengths $Ae = p$ and $aE = P$. Then the equation of equal areas reduces to

$$b\sqrt{cc - bb} - bp = B\sqrt{C^2 - B^2} - BP.$$

Euler tells us that this equation is important because it gives a relation between p and P that allows us to find one, given the other.

Now we consider the third condition of Goldbach's problem, that $Ab = aB$. Euler puts $de = q$ and $DE = Q$. Then, mixing b and B, the points with b and B, the lengths, he tells us that

$$be = b + q, \quad ae = b - q, \quad BE = B + Q \quad \text{and} \quad AE = B - Q.$$

Then, Pythagoras allows him to calculate Ab in terms of p, b and q, and likewise aB in terms of P, B and Q. Because the segments must have equal length, we have

$$p^2 + b^2 + 2bq + q^2 = P^2 + B^2 + 2BQ + Q^2,$$

which gives a relation between q and Q.

Now we look at the length of the segment Aa. Euler observes that it is part of the two triangles baA and BAa, but it would have been more useful to note that it is the hypotenuse both of Aea and of AEa. By Pythagoras, he gets

$$Aa^2 = p^2 + b^2 - 2bq + q^2 = P^2 + B^2 - 2BQ + Q^2.$$

This fact, first added to, then subtracted from the previous equation, gives the two simple results

$$bq = BQ \quad \text{and} \quad p^2 + b^2 + q^2 = P^2 + B^2 + Q^2.$$

After a brief review of the meanings of his variables, b, B, c, C, p, P, q and Q in Figure 1, Euler reminds us of four equations he has derived:

I. $C^2 \arcsin \frac{B}{C} = c^2 \arcsin \frac{b}{c}$

II. $B\sqrt{C^2 - B^2} - BP = b\sqrt{c^2 - b^2} - bp$

III. $BQ = bq$

IV. $P^2 + B^2 + Q^2 = p^2 + b^2 + q^2$

Euler says that from the third and fourth of these, it follows that

$$Q = b\sqrt{1 + \frac{P^2 - p^2}{B^2 - b^2}} \quad \text{and} \quad q = B\sqrt{1 + \frac{P^2 - p^2}{B^2 - b^2}}.$$

We get a quick reminder that, for the problem to be solved in this way, the two circles must have areas that "have a ratio between them as a number to a number," and that the sectors involved must be related by

$$\sphericalangle bca : \sphericalangle BCA = AC^2 : ac^2,$$

that is

$$\arcsin \frac{b}{c} : \arcsin \frac{B}{C} = C^2 : c^2 = AC^2 : ac^2.$$

From the equation for q and from equation II above, Euler gets first an expression for P in terms of b, B, c, C and p, and then progressively more complicated equations for

$$P^2 - p^2, \ \frac{P^2 - p^2}{B^2 - b^2}, \ q^2$$

and finally $p^2 + q^2$. To try to control matters here, we will give only the last of these:

$$p^2 + q^2 = \frac{2Bbp\sqrt{C^2 - B^2} - 2b^2 p\sqrt{c^2 - b^2}}{B^2 - b^2} + \frac{B^2C^2 + b^2c^2 - B^2b^2 - b^4 - 2Bb\sqrt{(C^2 - B^2)(c^2 - b^2)}}{B^2 - b^2}.$$

This relatively innocent equation has an interesting consequence. In I through IV above, Euler has shown us how the quantities C, c, B, b, P, p, Q and q, respectively, are interrelated. Now, given our circle of radius c and center c, and chord ab, the values p and q measure the position of the point A relative to d, the midpoint of the chord ab. This formula tells us that A is not uniquely determined, but instead has a circle as its locus.

And now, in the middle of the paper, it is a bit of a surprise that Goldbach's problem is solved, as "innumerable points A are given."

As often happens with Euler, having solved the problem is no reason to end a paper. Euler still has an idea, and he uses it to solve Daniel Bernoulli's problem, to describe the special cases in which the chords ab and AB are either parallel or the same segment. He also solves a new problem of his own devising, to describe the special cases when those chords are perpendicular. We will consider these only very briefly. The results that Euler will use in E-423 are all earlier in the paper.

We introduce a new point Z at the intersection of the two lines bA and Ba and ask about the angle at Z. Euler starts by looking at triangle baA. He finds that

$$\tan bAa = \frac{2bp}{p^2 + q^2 - b^2}$$

by finding $\tan bAe$ and $\tan aAe$ and using the angle sum formula. From this,

$$\tan ZAa = \frac{2bp}{b^2 - p^2 - q^2}.$$

Similarly, he finds the tangent of ZaA, and, knowing the tangent of two angles of $\triangle ZaA$, he finds the tangent of the third to be

$$\tan aZA = -\frac{2bp(B^2 - P^2 - Q^2) + 2BP(b^2 - p^2 - q^2)}{(B^2 - P^2 - Q^2)(b^2 - p^2 - q^2) - 4BbPp}.$$

Euler continues his detailed algebraic analysis and soon finds that

$$\tan \frac{1}{2} Z = \frac{B^2 - b^2}{B\sqrt{C^2 - B^2} - b\sqrt{c^2 - b^2}}.$$

Armed with this formula, Euler is ready to consider two special cases for the angle Z; $Z = 0$ and Z a right angle.

The first case, $Z = 0$ was already considered by Daniel Bernoulli. This would mean that Ab and aB are parallel when they cut off equal pieces from the lunes. From the tangent formula he just found, we see that $B^2 - b^2 = 0$, and because both B and b are positive, $B = b$.

There are two sub-cases. It may be that the segments aB and Ab are in fact the same segment. This amounts to taking the points A and a to be the same point, and naming it α, and likewise taking B and b to be the single point β. This possibility is illustrated in Figure 2, where both circles are drawn with the same chord $\alpha\beta$. Euler finds conditions, which we omit here, that describe this situation.

The other possibility, shown in Figure 3, is that the segments Ab and aB are parallel and distinct. This makes $AbaB$ a parallelogram. Again, Euler finds conditions, omitted here, that describe this situation.

Finally, Euler takes angle Z to be a right angle. This seems to be a problem of Euler's own design, not one posed by Bernoulli or Goldbach. With Z a right angle, the tangent of the half angle

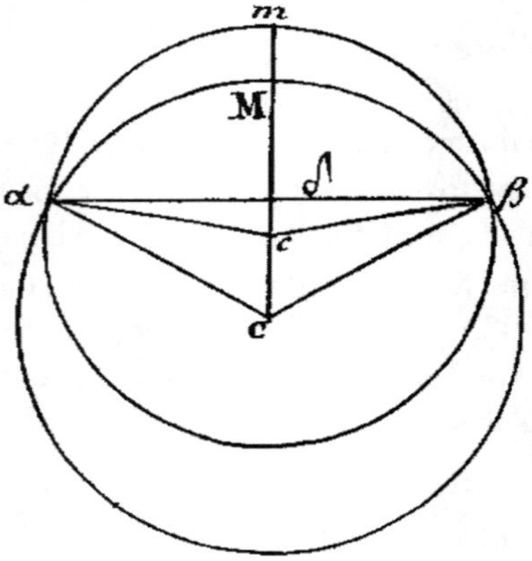

Fig. 2

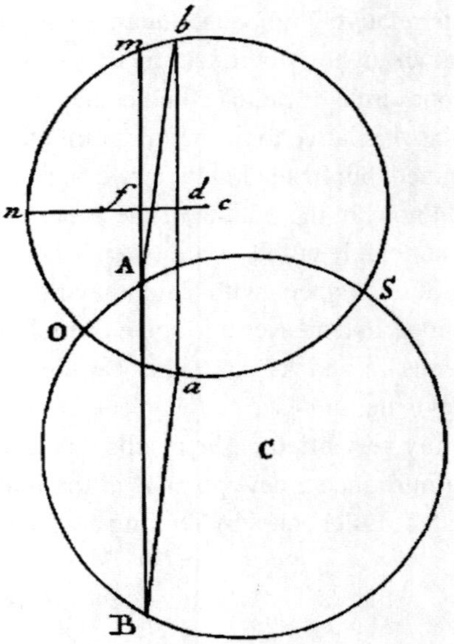

Fig. 3

is 1, so Euler's formula for tangent of half the angle gives immediately that

$$B^2 - b^2 = B\sqrt{C^2 - B^2} - b\sqrt{c^2 - b^2}.$$

It is not surprising that Euler is able to analyze this case as well and to construct a solution.

With this, we put down this paper, without quite finishing it. We leave out some parts that neither Euler nor anyone else seems to have followed up.

This has been an enticing paper, reaching back to antiquity for its basic problem, but it has really answered none of the questions we had at the outset. Euler's results are entirely analytical, and the lunes he describes are not constructible in the "ruler and compass" sense. Euler may have been somewhat dissatisfied with this paper as well, since he put the problem aside until 1771.

References

[C] Calinger, Ronald, *Classics in Mathematics*, Prentice Hall, Englewood Cliffs, NJ, 1995.

[D] Dunham, William, *Journey Through Genius: The Great Theorems of Mathematics*, Wiley, NY, 1990.

[F] Felmann, Emil, "Leonhard Euler—Ein Essay über Leben und Werk," *Leonhard Euler—Beiträge zu Leben und Werk*, Gedenkband des Kantons Basel-Stadt, Birkhäuser, Basel, 1983.

Interlude: 1738

World events

England's King George III (1738–1830) was born on June 4.

Pierre Louis Maupertuis published *Sur la figure de la terre*, "On the shape of the earth," in which he reports the results of expeditions that show that the earth is slightly flattened at the poles and bulges at the equator, as predicted by Newton.

British astronomer Sir Frederick William Herschel (1738–1822) was born in Germany this year. Legend has it that on the day Euler died in 1783, he had discussed at dinner Herschel's announcement of the discovery of the planet he had named "Herschel." Now we call it Uranus.

In Euler's life

On August 21, Euler wrote a letter to Goldbach that began "Die Geographie ist mir fatal." "Geography will be the death of me." He complained that the eyestrain of working with the Atlas of Russia was causing him to go blind. Goldbach replied the very same day and got Euler relieved of the most tedious mapmaking tasks. It was Euler's first letter to Goldbach in German; all previous letters had been in the more formal language, Latin, and this change was an acknowledgement that Euler had become more than just a colleague of the older Goldbach, but also a friend.

Euler's other work

Euler wrote his first book on the theory of ships, *Scientia Navalis*, E-110 and E-111, in 1738. It was published in 1749.

Indian and Hindu culture and artifacts were popular in Europe in 1738. Reacting to this fad, Euler wrote E-18, De Indorum anno solari astronomico, "On the Indian solar year."

He produced an article in Russian, E-32, the title of which translates as "Contribution on the shape of the earth." It seems to be a description of Maupertuis' results. (See above.)

Euler did some papers on practical physics. E-78, Dissertation sur la meilleure construction du cabestan, "Dissertation on the best construction of the capstan," would win Euler a quarter-share of the Paris Prize in 1741. He also wrote E-93, Disquisitio de bilancibus, "Dissertation on scales" and E-94, De motu cymbarum remis propulsarum in fluviis, "On the motion of oars propelling boats in rivers," as well as E-96, De machinarum tam simplicium quam compositarum usu maxime lucroso, "On the best use of simple machines compared to composite ones." Finally, E-97, De attractione corporum sphaeroidico-ellipticorum, "On the attraction of sphero-elliptical bodies" is a study of the gravitational properties of objects shaped like the earth, slightly elliptical spheroids.

Euler's mathematics

Euler did more work in science than mathematics in 1738. E-23, De curvis rectificabilibus algebraicis atque traiectoriis reciprocis algebraicis, "On rectifiable algebraic curves and algebraic reciprocal trajectories" continues his study of reciprocal trajectories that began with E-3 and E-5. E-74, De variis modis circuli quadraturam numeris proxime exprimendi, "On various ways of closely approximating numbers for the quadrature of the circle" is about approximations of π. E-95, De aequationibus differentialibus quae certis tantum casibus integrationem admittunt, "On differential equations which sometimes can be integrated" gives some new methods for solving differential equations. A paper in number theory, E-98, Theorematum quorundam arithmeticorum demonstrationes, "Proofs of some theorems of arithmetic" includes Euler's proof of Fermat's Last Theorem in the case $n = 4$. Finally, E-99, Solutio problematis cuiusdam a celeberrimo Daniele Bernoullio propositi, "Solution of some problems that were posed by the celebrated Daniel Bernoulli," describes Euler's solution to a problem in the calculus of variations involving minimizing curvature.

34

E-23: De curvis rectificabilibus algebraicis atque traiectoriis reciprocis algebraicis*

On rectifiable algebraic curves and algebraic reciprocal trajectories

Readers who learned arc length integrals will recall that they come in two basic forms:

$$\int \sqrt{1+(f'(x))^2}\,dx \quad \text{and} \quad \int \sqrt{x'(t)^2+y'(t)^2}\,dt,$$

and that almost every example seems to reduce to one or two particular integrals. Euler knew this. He observed, "if we consider ordinary curves, it occurs most rarely among these that they admit rectification." Euler wrote this paper to try to expand the population of rectifiable curves, that is, curves for which arc length integrals can actually be integrated, and he uses this to amplify his results from E-3 and E-5 about algebraic trajectories. He will revisit these results in E-48.

Euler begins with some historical perspective. In the previous century, many Geometers thought that lengths of curves were essentially different from lengths of lines, and they could not be compared. Descartes was among those who believed that "rectification of curves," that is calculation of arc lengths, was not possible. The new intellectual technology of calculus, though, showed Descartes to be wrong, and any curve known to Euler could, in theory, be rectified. In the 19th century, of course, Weierstrass, Peano and others, will discover classes of curves that do not have arc lengths.

Euler's problem is that, although he can write down an arc length integral for any curve (at least any curve that he knew), he can only evaluate that arc length integral in a few special cases. Two classes of curves, caustics and evolutes, have easy arc lengths because of the special ways they are constructed. He means to exclude these special classes when he says that rectifiable curves are very rare among "ordinary curves."

The purpose of this paper is to expand the known population of rectifiable curves. For his first class of curves, Euler takes z to be a parameter, the differential of which is to be constant. He further takes P and Q to be functions of z. Though they could be any functions whatever of z, Euler is particularly interested if they are algebraic. From these, Euler builds two functions, x and y, that parameterize a curve, and their arc length function s, as follows:

$$x = P + \frac{dQ(dP^2 - dQ^2)}{dQ\,ddP - dP\,ddQ}, \qquad y = \frac{(dP^2 - dQ^2)^{3/2}}{dQ\,ddP - dP\,ddQ}$$

and then

$$s = Q + \frac{dP(dP^2 - dQ^2)}{dQ\,ddP - dP\,ddQ}.$$

To check Euler's claim, it is necessary only to check that $ds^2 = dx^2 + dy^2$, a task that Euler describes as "a great deal of work requiring no skill." Mercifully, he omits it.

*_Comm. Acad. Sci. Imp. Petropol._ 5 (1730/1) 1738, 169–174; _Opera Omnia_ I.27, 24–28

Note that these formulas give a rectifiable curve regardless of the functions P and Q used to create them (as long as P and Q have the required differentials and the formulas make sense), and they are algebraic curves with algebraic arc lengths whenever P and Q are themselves algebraic.

Uncharacteristically, Euler offers no examples. The reader who wants to try an example that works out neatly might take $P = \frac{z^2}{2}$ and $Q = z$, chosen so that all the denominators are 1. Instead of giving examples, Euler goes farther. Let L, M and N be three functions of z, and let

$$x = L + \frac{(dL^2 + dM^2 - dN^2)(dL\,dN - dM\sqrt{dL^2 + dM^2 - dN^2})}{dL\,dN\,ddL + dM\,dN\,ddM - dL^2\,ddN - dM^2\,ddN + (dL\,ddM - dM\,ddL)\sqrt{dL^2 + dM^2 - dN^2}},$$

$$y = M + \frac{(dL^2 + dM^2 - dN^2)(dM\,dN + dL\sqrt{dL^2 + dM^2 - dN^2})}{dL\,dN\,ddL + dM\,dN\,ddM - dL^2\,ddN - dM^2\,ddN + (dL\,ddM - dM\,ddL)\sqrt{dL^2 + dM^2 - dN^2}}.$$

Euler claims that

$$s = N + \frac{(dL^2 + dM^2 - dN^2)(dL^2 + dM^2)}{dL\,dN\,ddL + dM\,dN\,ddM - dL^2\,ddN - dM^2\,ddN + (dL\,ddM - dM\,ddL)\sqrt{dL^2 + dM^2 - dN^2}}.$$

Euler again omits the calculations, which certainly require even more work, but no more skill. Euler gives no examples, but notes that if $L = P$, $M = 0$ and $N = Q$, then these formulas reduce to his first class of formulas. We get no examples other than this.

Euler could stop here, but it would leave his paper only two pages long. Euler does write a few papers that short, but they are usually corrections or comments on his own or somebody else's papers and not papers that are intended to stand alone.

Instead of stopping, Euler returns to the subject of reciprocal trajectories he dealt with several years earlier in E-3 and E-5. In particular, he brings back his first diagram in E-3, here rotated 90 degrees as Figure 1.

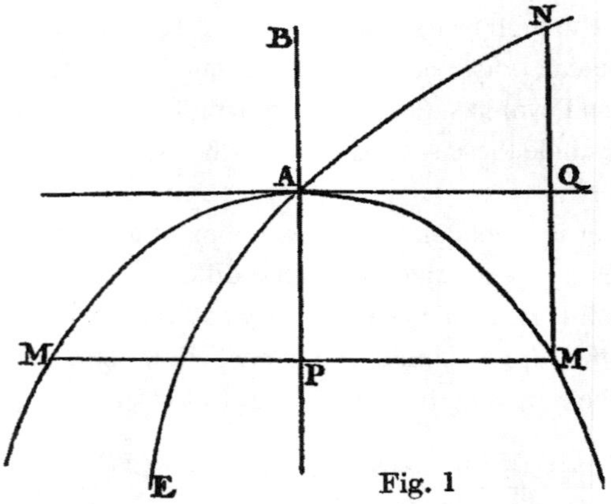

Fig. 1

Recall, the curve *NAE* is a reciprocal trajectory because, if it is reflected across the axis *BAP*, then moved up or down that axis, the resulting curve always intersects the original at a right angle.

In E-3, Euler constructed the curve *NAE* from the *MAM*. That curve was taken to be symmetric about the axis *BAP*. Then, for any point *M* on *MAM*, he found the arc length *AM*. Then *N*, the point on the reciprocal trajectory directly above the point *N* is located so that the distance *MN* equals the arc length *AM*.

Euler is grievously overworking the symbol *M* in this exposition. In a way, it, like the problem itself, is a throwback to the 17th century style of mathematics. It almost makes sense. Here, *M* is a point on the curve. Because *MAM* is symmetric, he is comfortable calling its twin point *M* as well. Then, as the curve passes through the point *A* and the two points named *M* are all on the curve, he calls the curve *MAM*. It makes sense, but it certainly isn't modern. It isn't even 18th century.

When Euler presented this construction in E-3, he gave few details and was a little ambiguous about the origins of the construction. Here in E-23, Euler is three years older and has since written twenty more papers of his own, so he does not need to appear to claim credit for the construction. He simply calls it "a theorem of Bernoulli."

We see that if *MAM* is an algebraic curve with an algebraic rectification, then *NAE* will also be algebraic. This provides the link between rectifiable algebraic curves and algebraic trajectories suggested in the title.

For the benefit of those who do not want to work through the details, Euler gives formulas, though not derivations, for *NQ* in terms of the three functions L, M and N that describe the curve *MAM*.

These results excited at least some interest in the mathematical community of the times. Two minor mathematicians, J. D. Gordak in 1745 and J. T. Waltz in 1746, wrote to Euler to ask for explanations of the formulas in this work. None of Euler's letters to either correspondent seem to survive, so we do not know if Euler replied, though he probably did, because each of them wrote several more times. We also do not know how Euler explained these formulas or how he came to discover them.

There is a second curious aspect to this paper. It appeared in the volume of the *Commentarii* for the years 1730 and 1731 but Eneström tells us he did not write the paper until 1738, the very year the volume finally got published. How did this happen? Was the paper dated incorrectly and actually written earlier than we think? Did somebody in the editorial chain of command think it was particularly important and deserved to be rushed to print? No one knows.

What we can only describe as one of Euler's minor papers ends, fittingly, with a couple of minor mysteries.

35

E-74: De variis modis circuli quadraturam numeris proxime exprimendi*

On various ways of closely approximating numbers for the quadrature of the circle

Finally, Euler is about to take aim at one of the most popular problems in mathematics, the value of π. Some people have measured progress in mathematics by how accurately we know π, from $\pi = 3$ in early Ancient times, to the billions of decimal places to which we can calculate π today. The 1998 cult movie "Pi" showed, if nothing else, how the number still captures the public imagination.

Euler has given ways to estimate π in some of his earlier papers (E-41, for example.) There, though, the estimates were more by-products of the properties of series rather than ends in themselves. Here, in E-74, Euler turns to estimating π and to comparing different methods for doing this. His standard of comparison will be the number of steps each method requires to achieve various numbers of decimal place accuracy, usually 10 or 100 decimal places.

Euler begins by reminding us of the results of Archimedes, who found that if a 96-sided regular polygon is inscribed in a circle of radius 1, then each side has length

$$\sqrt{2-\sqrt{2+\sqrt{2+\sqrt{2+\sqrt{3}}}}}.$$

Likewise, if one is circumscribed, then the sides have length

$$\frac{2\sqrt{2-\sqrt{2+\sqrt{2+\sqrt{2+\sqrt{3}}}}}}{\sqrt{2+\sqrt{2+\sqrt{2+\sqrt{2+\sqrt{3}}}}}}.$$

He multiplies each of these by 96 and tells us that 2π lies between the resulting values. He does not write out those two values as decimals. Instead, he credits to Machin (1675–1751) in 1706 and to De Lagny (1660–1734) in 1719 a 127 decimal place estimate of π, which he writes out over an expanse of three lines of text! Details of the work of Machin and De Lagny, as well as some of Euler's other efforts involving π, are described in Beckmann's book. [B] Some of Beckmann's history is a bit fantastic, his account of the burning of the Library at Alexandria, for example, but his mathematical details are for the most part accurate.

Euler himself will never join in on the "digits race," though with his legendary powers of calculation, he surely could have done so. In fact, De Lagny's record of 127 decimal places stood until 1794, eleven years after Euler's death, when the Slovenian mathematician and national hero Baron Jurij Vega added 13 decimal places to De Lagny's work, and also discovered an error in De Lagny's 113th decimal place. Euler's theoretical work, however, will be the basis for much of the progress in approximation for the next century.

Comm. Acad. Sci. Imp. Petropol. 9 (1737) 1744, 222–236; *Opera Omnia* I.14, 245–259

Euler will rely mostly on infinite series to approximate π, rather than on other infinite expressions like the ones Archimedes used. He has two criteria for a "good" procedure. First, it should converge quickly. Second, it should be easy to calculate. Archimedes' calculation relied on extracting square roots at every step, so Euler sought a simpler process.

Euler proceeds in a way that should now be familiar. He begins with the simple arctangent series,

$$\arctan x = \frac{x}{1} - \frac{x^3}{3} + \frac{x^5}{5} - \frac{x^7}{7} + \frac{x^9}{9} - \text{etc.}$$

For small values of x, this series converges fairly rapidly. Euler claims that "if $x = \frac{1}{10}$, then it is easy to find the arc tangent of $\frac{1}{10}$ as a decimal fraction to a thousand figures." Although powers of ten are easy to calculate, it is not really so easy. It takes almost a thousand terms to reach ten decimal places, and also arctan $\frac{1}{10}$ is not commensurate with the circumference of a circle, so even knowing it to 100 decimal places would not help us calculate π. If $x = 1$, the series converges to arctan 1, which Euler describes as "the arc of 45°," or $\frac{\pi}{4}$. Euler says that it takes more than 10^{50} terms for this series to give an approximation that is accurate to 100 decimal places. This is not very good. He gets some improvement by taking $x = \frac{1}{\sqrt{3}}$, which gives the arc of 30°, or $\frac{\pi}{6}$. This gives Euler's 100 decimal places in 210 terms. This is the method that Abraham Sharp (1651–1742) used in 1705 to give π to 72 decimal places (though Beckmann says in one place that he used an arcsine series, not an arctangent).

In general,

$$\arctan \frac{1}{p} = \frac{1}{p} - \frac{1}{3p^3} + \frac{1}{5p^5} - \frac{1}{7p^7} + \text{etc.}$$

Euler lets S denote the sum of the first n terms of this series, and turns to the techniques he developed in E-55. When we described that paper, we mentioned that he would be using one particular formula here in E-74. In that formula, he takes $n = \frac{\sqrt{-1}}{p}$ and $x = 2n + 1$ and gets

$$\arctan \frac{1}{p} = S + \frac{1}{p^{2n+1}} \left\{ \frac{p^2}{(1+p^2)(2n+1)} + \frac{2p^2}{(1+p^2)^2(2n+1)^2} - \frac{2^2(p^4 - p^2)}{(1+p^2)^3(2n+1)^3} + \frac{2^3(p^6 - 4p^4 + p^2)}{(1-p^2)^4(2n+1)^4} - \text{etc.} \right\}$$

Euler notes that if n is large enough, then the new series converges much faster than the previous one did.

Euler makes a substitution that does him a little bit of good here. Set

$$q = \frac{2}{(1+pp)(2n+1)}.$$

This lets him rewrite his arctangent as

$$\begin{aligned} \arctan \frac{1}{p} &= \frac{1}{p} - \frac{1}{3p^3} + \frac{1}{5p^5} - \text{etc.} \\ &= S + \frac{1}{2p^{2n-1}} \cdot \frac{1}{\frac{1}{q} - 1 + qp^2 - q^2(p^4 - 2p^2) + q^3(p^6 - 8p^4 + 4p^2) - \text{etc.}}. \end{aligned}$$

Machin's idea that we mentioned above was based on the clever fact, discovered by James Gregory (1638–1675), that

$$\frac{\pi}{4} = 4\arctan\frac{1}{5} - \arctan\frac{1}{239}.$$

Euler apparently likes this idea and seeks similar relationships. Up to now, he has been describing arctangents as "the arc whose tangent is...." Now, he invents a function notation, using $A\,t\,x$ to denote arctangent(x). He supposes

$$A\,t\,1 = A\,t\,\frac{1}{a} + A\,t\,\frac{1}{b}.$$

Note that this is not quite the form that Machin used. It follows from the formula for $\tan(\alpha+\beta)$ that

$$1 = \frac{a+b}{ab-1}, \quad \text{so that} \quad ab - 1 = a + b \quad \text{and} \quad b = \frac{a+1}{a-1}.$$

Euler seeks whole number solutions to this, and takes $a = 2$, $b = 3$ and $\alpha = \arctan 1$. Then α is given as the sum of two infinite series

$$+\frac{1}{1\cdot 2} - \frac{1}{3\cdot 2^3} + \frac{1}{5\cdot 2^5} - \frac{1}{7\cdot 2^7} + \text{etc.}$$

and

$$+\frac{1}{1\cdot 3} - \frac{1}{3\cdot 3^3} + \frac{1}{5\cdot 3^5} - \frac{1}{7\cdot 3^7} + \text{etc.}$$

For once, there are no hidden problems with conditional convergence. Both of these series converge faster than the one involving $\frac{1}{\sqrt{3}}$ that Euler used earlier and they do not involve radicals.

Euler tells us that to get 100 decimal places of accuracy, one should sum 154 terms of the first series and 96 terms of the second series. Moreover, 318 terms of the first series and 250 terms of the second will give 200 decimal places. This he compares to the series using $\frac{1}{\sqrt{3}}$, which he says requires more than 200 terms to give 100 decimal places.

The sum formula for tangents also tells us that

$$A\,t\,\frac{1}{p} = A\,t\,\frac{1}{p+q} + A\,t\,\frac{q}{p^2+pq+1}.$$

For $p = 2$ and $q = 1$, this gives

$$A\,t\,\frac{1}{2} = A\,t\,\frac{1}{3} + A\,t\,\frac{1}{7}.$$

This, with the earlier equation, gives

$$\alpha = 2\,A\,t\,\frac{1}{3} + A\,t\,\frac{1}{7}.$$

This, in turn, yields $\alpha = \frac{\pi}{4}$ as the sum of two series that converge even faster than the previous pair, but still not quite as fast as the one Machin used.

Euler offers some other arctangent series without any explanation. We leave it to the reader to discover how the first of these can be derived, and how the others follow from the first.

$$At\,\frac{p}{p^2-1} = \frac{1}{p} + \frac{2}{3p^3} + \frac{1}{5p^5} - \frac{1}{7p^7} - \frac{2}{9p^9} - \frac{1}{11p^{11}} + \text{etc.}$$
$$At\,\frac{2p}{2p^2-1} = \frac{1}{p} + \frac{1}{3\cdot 2p^3} - \frac{1}{5\cdot 2^2p^5} - \frac{1}{7\cdot 2^3p^7} + \frac{1}{9\cdot 2^4p^9} + \text{etc.}$$
$$At\,\frac{3p}{3p^2-1} = \frac{1}{p} - \frac{1}{5\cdot 3^2p^5} + \frac{1}{7\cdot 3^3p^7} - \frac{1}{11\cdot 3^5p^{11}} + \frac{1}{13\cdot 3^6p^{13}} - \text{etc.}$$
$$At\,\frac{3p(pp-1)}{p^4-4pp+1} = \frac{3}{1p} + \frac{3}{5p^5} - \frac{3}{7p^7} - \frac{3}{11p^{11}} + \frac{3}{13p^{13}} + \text{etc.}$$

Euler suggests using $p = 2$ in the last of these to get a series for arctan(18), then using

$$\frac{\pi}{2} = \arctan 18 + \arctan\frac{1}{18}.$$

Now we will look for identities of the form

$$At\,\frac{a}{b} = At\,\frac{1}{m} + At\,\frac{1}{n}.$$

This time, the tangent sum formula yields

$$\frac{m+n}{mn-1} = \frac{a}{b}, \quad \text{or} \quad (ma-b)(na-b) = a^2+b^2.$$

Euler is very good at solving equations like this, and he easily finds the solution $a = 7$, $b = 9$, $m = 2$ and $n = 5$ giving

$$At\,\frac{7}{9} = At\,\frac{1}{2} + At\,\frac{1}{5}.$$

This is not particularly useful, but it gives Euler more ideas. He considers a sequence of related forms:

$$\text{I. } At\,\frac{x}{y} = At\,\frac{ax-y}{ay+x} + At\,\frac{1}{a}$$

$$\text{II. } At\,\frac{x}{y} = At\,\frac{ax-y}{ay+x} + At\,\frac{b-a}{ab+1} + At\,\frac{1}{b}$$

$$\text{III. } At\,\frac{x}{y} = At\,\frac{ax-y}{ay+x} + At\,\frac{b-a}{ab+1} + At\,\frac{c-b}{bc+1} + At\,\frac{1}{c}$$

$$\text{IV. } At\,\frac{x}{y} = At\,\frac{ax-y}{ay+x} + At\,\frac{b-a}{ab+1} + At\,\frac{c-b}{bc+1} + At\,\frac{x}{y}$$
$$= At\,\frac{ax-y}{ay+x} + At\,\frac{b-a}{ab+1} + At\,\frac{c-b}{bc+1} + At\,\frac{d-c}{cd+1} + At\,\frac{1}{d}$$

This can be continued indefinitely by taking a, b, c, d, etc. to be any sequence that grows to infinity. Euler suggests taking $\frac{x}{y} = 1$ and taking the sequence to be the odd numbers 3, 5, 7, 9, etc.

to get the formula

$$A\,t\,1 = A\,t\,\frac{1}{2} + A\,t\,\frac{1}{8} + A\,t\,\frac{1}{18} + A\,t\,\frac{1}{32} + \text{etc.}$$

He notes in passing a theorem that is "not inelegant." If A is any arc of a circle, then

$$A = \frac{\sin A}{\cos\frac{A}{2}\cdot\cos\frac{A}{4}\cdot\cos\frac{A}{8}\cdot\cos\frac{A}{16}\cdot\cos\frac{A}{32}\cdot\text{etc.}}.$$

Euler gives only a hint about why this is true. He writes "coronidis," which can mean that he is using the theorem as a colophon, a device or flourish for ending this article. On the other hand, it means literally "by the curved line or tail," and the proof of the theorem depends on the way the sine function curves as it approaches zero. Either way, we will leave the proof of this "not inelegant" little theorem to be discovered by the reader.

Recasting the cosines as secants, then taking logarithms gives us

$$\ln A = \ln\sin A + \ln\sec\frac{A}{2} + \ln\sec\frac{A}{4} + \ln\sec\frac{A}{8} + \text{etc.}$$

Euler consults a table of base 10 logarithms of $\sin 1^\circ$ and of secants of $30'$ (30 minutes of arc or half a degree), $15'$, $7.5'$ and $3.75'$, then calculates the log of 1° of arc. Adding the log of 180 gives the log of π to be 0.4971497.

Finally, for those of us who could not figure it out for ourselves, Euler offers his version of a proof of his "not inelegant" theorem. His proof begins

$$\frac{\sin A}{\cos\frac{A}{2}} = 2\sin\frac{A}{2}$$

and ends

$$\frac{\sin A}{\cos\frac{A}{2}\cdot\cos\frac{A}{4}\cdot\cos\frac{A}{8}\cdot\cos\frac{A}{16}\cdot\text{etc.}} = \infty\sin\frac{A}{\infty}$$

which, according to Euler, equals that arc length A. A modern reader may want to re-cast this using limits.

Thus, we see that Euler contributes to the digit race for π, yet he himself remains above the fray. We can only speculate why he did not bother to set a digits record himself. Perhaps he knew his talents would be better spent elsewhere, or he knew that he had already done the hard part with his discoveries about series and convergence. Perhaps, too, he was already sensitive to criticisms like Voltaire would later level at him, that he filled more pages with calculations than anyone else in his century.

References

[B] Beckmann, Petr, *A History of π*, Dorset Press, New York, 1989.

36

E-95: De aequationibus differentialibus, quae certis tantum casibus integrationem admittunt*

On differential equations which sometimes can be integrated

Euler had a lifelong interest in differential equations. This is the tenth article on the subject that we have encountered. After this one, though, he will write no articles on the subject for twelve years, until 1750, with E-188. Twenty of his papers on differential equations will be published posthumously. Much of his work on this subject during the middle of his life will find its way into his great *Institutiones calculi integralis*, published in three volumes in 1768, 1769 and 1770. People who think that today's calculus books are getting a bit thick might be interested that Euler separated his calculus book into four volumes, one 880-page volume on differential calculus (often bound as two volumes) and three volumes on integral calculus of lengths 542, 526 and 639 pages. This sums to 2587 pages, far more than any modern text. And Euler's texts did not contain homework problems!

The paper at hand, E-95, was written in 1738 but not published until 1747, long after Euler had left St. Petersburg for Berlin. It concerns one particular differential equation and Euler's efforts to solve it by a method of undetermined coefficients. Later in the paper, Euler considers some generalizations of the given equation.

The equation that interests Euler this time is

$$(a + b^n)x^2\, ddv + (c + f^n)x\, dx\, dv + (g + hx^n)v\, dx^2 = 0.$$

Euler does not tell us why this equation interests him. He supposes that the equation has a solution of the form

$$v = Ax^m + Bx^{m+n} + Cx^{m+2n} + Dx^{m+3n} + Ex^{m+4n} + \text{etc.}$$

As always seems to happen with the method of undetermined coefficients, this choice of form for the solution is supposed to seem natural and obvious, but for people who are not very experienced with differential equations, it may retain an air of mystery.

If we take the first two differentials of v and look at the terms of degree m we get the equation

$$g + cm + am(m - 1) = 0.$$

If this condition is not satisfied, then the differential equation cannot have a solution of the form given by v.

Matching terms of degree $m + n$ gives

$$B = \frac{-A(h + fm + bm(m - 1))}{cn + an(2m + n - 1)}.$$

Comm. Acad. Sci. Imp. Petropol. 10 (1738) 1747, 40–55; *Opera Omnia* I.22, 162–180

He gets similar relationships for C in terms of B, D in terms of C, etc., and concludes that A is an arbitrary constant and the rest of the coefficients can be given in terms of A. Euler notes that if the numerator of this expression is zero, then B and all subsequent coefficients are zero, so that v has the simple form $v = Ax^m$. Similarly, Euler gets conditions, based on the numerators of the expressions for B and C, that will insure that v can be given by just two or three terms. He generalizes this slightly by noting that the numerators all have a factor of the form

$$h + f(m + in) + b(m + in)(m + in - 1).$$

If this factor is zero for some i, then the corresponding coefficient and all subsequent ones will be zero, and v will have a form that contains only $i + 1$ terms.

Euler considers a slightly different form for the solution. He supposes that the exponents count down instead of up, so that v has the form

$$v = Ax^k + Bx^{k-n} + Cx^{k-2n} + Dx^{k-3n} + Ex^{k-4n} + \text{etc.}$$

As before, for such a solution to exist, it is necessary that

$$h + fk + bk(k - 1) = 0,$$

and he gets expressions for B, C, D, E, etc., each in terms of the previous ones, and ultimately depending only on the arbitrary constant A. Again, he notices a pattern in the numerators that tells him the number of terms in v. The calculations are almost identical.

Euler moves to a more general form of the same equation

$$P\,ddv + Q\,dx\,dv + Ry\,dx^2 = 0$$

where P, Q and R are functions of x.

Suppose that $v = X$ is a particular solution to the given equation and that $v = Xz$ is a general solution. Euler finds dv and ddv and substitutes them into the given equation to get an equation he writes rather awkwardly as

$$\begin{aligned}&+ Pz\,ddX + 2p\,dX\,dz + PX\,ddz = 0\\ &+ Qz\,dX\,dx + QX\,dx\,dz\\ &+ RzX\,dx^2\end{aligned}$$

In choosing this format, popular in the 17th century, Euler is trying to make it easy to combine like terms involving the differentials of z, but to a modern reader it certainly looks confusing.

Because X satisfies the given equation, the first column of the confusing equation, as written, sums to zero. Therefore the remaining three terms also sum to zero. This leaves

$$2P\,dX\,dz + QX\,dx\,dz + PX\,ddz = 0.$$

Euler divides this by $XP\,dz$ to rewrite it as

$$\frac{2\,dX}{X} + \frac{Q\,dx}{P} + \frac{ddz}{dz} = 0.$$

Now we introduce some new functions

$$S = \int \frac{Q\,dx}{P}$$

and C defined by

$$X^2\,dz = Ce^{-S}\,dx, \quad \text{so} \quad z = C\int \frac{e^{-S}\,dx}{X^2}.$$

This lets us write the complete integral mentioned above as

$$v = CX \int \frac{e^{-\int (Q\,dx)/p}}{X^2}\,dx.$$

Euler considers this result in the special case of his original differential equation

$$(a + b^n)x^2\,ddv + (c + f^n)x\,dx\,dv + (g + hx^n)v\,dx^2 = 0.$$

There, he finds that $v = e^{\int z\,dx}$ so $z = \frac{dv}{v\,dx}$. Then the given second degree equation is transformed into the first degree equation

$$(a + bx^n)x^2\,dz + (c + fx^n)xz\,dx + (a + bx^n)x^2z^2\,dx + (g + hx^n)\,dx = 0.$$

Now, Euler notes that this first degree equation has the general form,

$$P\,dz + Qz^2\,dx + R\,dx = 0,$$

where, again, P, Q and R are to be functions of x. However, he does not choose to do anything with this general form. Instead, in his first degree equation, he substitutes $z = Ty$, where T is an as-yet unknown function of x. This gives

$$\begin{aligned}(a + bx^n)Tx^2\,dy + (a + bx^n)yx^2\,dT + (c + fx^n)Tyx\,dx \\ + (a + bx^n)T^2x^2y^2\,dx + (g + hx^n)\,dx = 0.\end{aligned} \quad (1)$$

Taking out the terms containing the factor y gives

$$\frac{(c + fx^n)\,dx}{(a + bx^n)x} + \frac{dT}{T} = 0.$$

Partial fractions, then integration and exponentiation lead to

$$T = \frac{(a + bx^n)^{(bc-af)/abn}}{x^{c/a}}.$$

Because $z = Ty$, Euler can find a similar expression for z. Now, in equation (1) above, after removing the terms containing the factor y, the remaining terms must also sum to zero. Euler substitutes for T in the diminished equation and does a little algebra to get

$$dy + \frac{(a + bx^n)^{(bc-af)/abn}y^2\,dx}{x^{c/a}} + \frac{(g + hx^n)x^{(c/a)-2}\,dx}{(a + bx^n)^{(bc-af)/(abn+1)}} = 0. \quad (2)$$

This simplifies considerably in the case $bc = af$. Euler additionally substitutes $x^{(a-c)/a} = t$ to get

$$dy + \frac{ay^2\,dt}{a-c} + \frac{a(g + ht^{na/(a-c)})dt}{(a-c)(a + bt^{na/(a-c)})tt} = 0.$$

Because this is so closely related to the equation with which we started, the same conditions on g, h and c that Euler found earlier still apply. Euler notes that there are simpler forms in the two special cases $c = 0$ and $c = a$.

Euler returns to the more general equation (2) and considers the case $c = -a(n-1)$. He makes another substitution, $(a + bx^n)^{(b-f)/bn} = t$ and translates his conditions on g and h. Then he additionally supposes that $f = b - nb$ and that $k = n$, and he finds that his equation becomes relatively simple.

Now he returns to a form we saw earlier,

$$(a + bx^n)x^2\,dz + (c + fx^n)xz\,dx + (a + bx^n)x^2z^2\,dx + (g + hx^n)\,dx = 0.$$

Last time, he supposed that $z = Ty$, for some T a function of x. This time, he allows a more general form,

$$z = Ty + S,$$

where T and S are both functions of x. The calculations parallel what he did before. He takes the differential, substitutes and looks at the terms containing the factor y, and he quickly arrives at

$$\frac{dT}{T} + 2S\,dx + \frac{(c + fx^n)\,dx}{(a + bx^n)x} = 0.$$

Here, though, he supposes that $T = x^p$ and learns that

$$S = \frac{-c - ap - (f + bp)x^n}{2x(a + bx^n)}.$$

Substituting this and dS into the equation, again omitting the terms with factor y, since they sum to zero, and applying a bit of algebra, Euler gets

$$dy + x^p y^2\,dx + \frac{p(p+2)\,dx}{4x^{p+2}} + \frac{(c + 2g)\,dx + (f + 2h)x^n\,dx}{2(a + bx^n)x^{p+2}}$$
$$= \frac{(c + fx^n)^2\,dx - 2n(bc - af)x^n\,dx}{4(a + bx^n)^2x^{p+2}}.$$

Euler reminds us again that conditions on g and h must be satisfied. Then he notes the simpler form that this takes if $bc - af = 0$.

If $b = 0$ this simplifies still further. It gets simpler and simpler as Euler assumes next that $f = 0$, and then that $p = 0$. He makes a clever substitution, unfortunate because his α is almost indistinguishable from his a:

$$a^2(p+1)^2 - (a-c)^2 + 4ag = \alpha a^2 \quad \text{and} \quad af - naf - eafk - cf = \beta\alpha f.$$

He finds that, with these assumptions, the equation can be reduced to a Riccati equation.

He examines further special cases, $\alpha = 0$, $\beta = 0$, $\alpha = -\beta^2$ and finds that they all lead to forms with a reasonable chance of being solved.

Euler briefly considers a slight generalization of his first equation,

$$(a + bx^n + cx^{2n})x^2\,ddv + (f + gx^n + hx^{2n})x\,dx\,dv + (p + qx^n + rx^{2n})v\,dx^2 = 0.$$

Again, he supposes that a solution v has the form

$$v = Ax^m + Bx^{m+n} + Cx^{m+2n} + \text{etc.}$$

As before, he finds that A is arbitrary, that B depends on A, etc., and he finds conditions that insure that $B = 0$.

Finally he returns to his very first equation of the paper,

$$(a + b^n)x^2\,ddv + (c + f^n)x\,dx\,dv + (g + hx^n)v\,dx^2 = 0.$$

He considers the special case $b = c = f = g = 0$ and finds that it, too, leads to an equation that can be solved.

In conclusion, this seems to be a specialized and unmotivated paper. Euler considers a narrow class of differential equations, and gives techniques to solve them. He does no actual examples, nor does he tell us why he is interested in them. It does highlight, though, how the Riccati equation casts a shadow across much of Euler's work in differential equations.

37

E-98: Theorematum quorundam arithmeticorum demonstrationes*

Proofs of some theorems of arithmetic

In the 17th century, Fermat left a large number of results in number theory, but hardly any proofs. When Goldbach introduced Euler to the subject in the 1720s, Euler did not know which of Fermat's claims Fermat had actually proved and among those unproven, Euler did not know which were true and which were false.

Recall that Euler's first paper on number theory, E-26, consists of a demonstration that one of Fermat's conjectures is false, as well as Euler's statement, without proof, of six conjectures of his own. In 1732, when Euler writes E-26, he seems not to have any real idea how to go about proving theorems in number theory. Now, though, it is 1738, and Euler is going to work his way systematically through Fermat's claims, proving them one after another until only one important claim will remain unresolved. This theorem, because it will be the last one to be proved (not because it was the last one Fermat posed) will become known as Fermat's Last Theorem and will defy proof until the 1990s.

Here in E-98, Euler begins his assault on Fermat's list. We shall see a number of interesting results, including one of the few for which Fermat actually left us a proof, Fermat's Last Theorem in the case $n = 4$. This paper represents important progress beyond the theorems it proves. It is written in the now-standard Theorem–Proof style we first saw in E-72 and is the first number-theoretic paper written in that style. Since 1732, Euler has figured out how to prove things in number theory. He has not quite figured out how to state theorems in number theory, though. We will find his statements of theorems rather cumbersome and long-winded.

Euler begins with two lemmas, the first of which he attributes to Euclid and does not prove himself.

Lemma 1. *The product of two or more numbers that are relatively prime can never be a square, nor a cube, nor any other power unless the factors themselves are squares or cubes or among those other powers.*

Lemma 2. *If $a^2 + b^2$ is a square, and if a and b are relatively prime, then there will exist relatively prime numbers, p and q, one even, the other odd, such that $a = pp - qq$ and $b = 2pq$.*

Diophantus knew this formula more than 1500 years before Euler, but this seems to be the first proof that all Pythagorean triples can be generated in this way. Euler's proof goes like this:

Suppose $a^2 + b^2$ is a square, and its square root is $a + \frac{bq}{p}$, where p and q are relatively prime. This makes

$$a^2 + b^2 = a^2 + \frac{2abq}{p} + \frac{bbqq}{pp}.$$

**Comm. Acad. Sci. Imp. Petropol.* 10 (1738) 1747. 125–146; *Opera Omnia* I.2 38–58

Subtracting a^2 and juggling terms a bit lead to the ratio

$$a : b = pp - qq : 2pq.$$

Now, the numbers $pp - qq$ and $2pq$ must either be relatively prime or have 2 as their common divisor. When one of p and q is odd and the other is even, the terms will be relatively prime. In this case, because a and b are also relatively prime, the ratio implies that

$$a = pp - qq \quad \text{and} \quad b = 2pq.$$

On the other hand, if $pp - qq$ and $2pq$ have 2 as their common divisor, then $a = \frac{pp-qq}{2}$ and $b = pq$. Now introduce r and s by letting $p + q = 2r$ and $p - q = 2s$. These two numbers, r and s, will also be relatively prime, and one will be odd while the other is even. Moreover, it works out that $a = 2rs$ and $b = rr - ss$. And so, reversing the roles of a and b, and using r and s instead of p and q, the claims of the lemma are met and the lemma is proved. Q. E. D.

Euler offers four corollaries and a scholion.

Corollary 1. *If the sum of two relatively prime squares is a prime, then one of the squares must be even and the other odd. The sum of two odd squares can never be a square.*

Corollary 2. *If $aa + bb$ is a square, with a odd and b even, then a will be $= pp - qq$ and $b = 2pq$.*

Corollary 3. *If, in addition to the hypotheses of Corollary 2, p and q are one odd, the other even, then b will be "evenly even," that is, divisible by* 4. *If neither p nor q is divisible by* 3, *then either $p - q$ or $p + q$ will be divisible by* 3, *and then one of the numbers, a or b will also be divisible by* 3.

Euler does not state Corollary 3 as clearly as he might. It amounts to saying that if either p or q is divisible by 3, then $b = 2pq$ is divisible by 3 and a is not, because a and b must be relatively prime. On the other hand, if neither p nor q is divisible by 3, then $a = pp - qq = (p + q)(p - q)$ must be divisible by 3. Either way, if $aa + bb$ is a square, then exactly one of the numbers a or b, must be divisible by 3.

Corollary 4. *If $a = pp - qq$ and $b = 2pq$ and if $aa + bb$ makes a square, then it is easy to find the numbers p and q. Because $a = (p + q)(p - q)$, $a > p + q$, unless $p - q = 1$, and because $b = 2pq$, b will be greater than p and q.*

In his scholion, Euler explains why, if a and b are relatively prime, and also c and d, the ratio $a : b = c : d$ implies $a = c$ and $b = d$. This explains one of the steps in the proof of Lemma 2.

This word *scholion* may be an unfamiliar one. Usually, there is no logical difference between a scholion and a corollary. They are consequences of a theorem, or, in this case, a lemma. To call such a consequence a corollary usually means that the writer thinks the consequence is useful, whereas making it a scholion suggests that it is merely interesting or illustrative. A scholion usually tells how to use a result to solve problems. Euclid, for example, shows in a theorem that the area of a rectangle is its base times its height, and, as a scholion, shows how to find the height given the area and the base. Euler seems to be using the word in yet a different sense, kind of a late lemma to help fill in a step that some readers might have found tricky.

Lemma 3. *If $aa - bb$ is a square, and a and b are relatively prime, then there are relatively prime numbers p and q, one odd, the other even, so that $a = pp+qq$ and either $b = pp-qq$ or $b = 2pq$.*

Proof. Suppose that $aa - bb$ is a square and that a and b are relatively prime. Let $a^2 - b^2 = c^2$, so that $a^2 = b^2 + c^2$ and b and c are relatively prime. Then, by Corollary 1, one of b and c is even, the other odd. Suppose first that b is odd, so that c is even. By Lemma 2, this makes $b = pp - qq$ and $c = 2pq$, where one of p and q is even, the other odd, and they are relatively prime. This makes $a = pp + qq$. On the other hand, if b were even and c were odd, then $b = 2pq$ and $c = pp - qq$, and so again $a = pp + qq$. Either way, $a = pp + qq$ and either $b = pp - qq$ or $b = 2pq$.

Q. E. D.

Euler gives three corollaries.

Corollary 1. *If the difference between two squares is a square, then it is odd, if the first two squares are relatively prime.*

Corollary 2. *Given numbers a and b, one can find smaller numbers p and q so that $a = pp + qq$ and either $b = pp - qq$ or $b = 2pq$.*

Euler is not being very precise in his statement of Corollary 2, for he omits the conditions that $aa - bb$ is a square and that a and b are relatively prime. We are supposed to assume, because it is a corollary, that the same conditions hold.

Corollary 3. *If $aa - bb = cc$, one of the numbers a, b, c will always be divisible by 5. This is because we can write $a = pp + qq$, $b = pp - qq$ and $c = 2pq$. If one of the numbers p and q is divisible by 5, then c is divisible by 5. If neither p nor q is divisible by 5, then pp and qq are numbers of the form $5n \pm 1$, so that either $pp - qq$ or $pp + qq$ will be divisible by 5.*

There are lots of interesting comments to be made about Corollary 3. First, note how Euler makes the proof of the corollary part of the statement. This is the first time he has made a statement about divisibility by 5. His earlier divisibility statements have been about 2, 3 or 4. Here, he does what we would call modular arithmetic by a manipulation of forms. Where we might say a number is congruent to 1 or 4 mod 5, he says it is of the form $5n \pm 1$. Later in his career, he will develop an arithmetic on remainders. Later than that, Gauss will develop our modern "mod" notation.

Now we come to one of the major results of this paper, Euler's proof of Fermat's Last Theorem in the case $n = 4$. We remind the reader that a *biquadratic* is a number to the fourth power. They were sometimes called *squared squares*.

Theorem 1. *The sum of two biquadratics, $a^4 + b^4$, is never a square, unless one of the biquadratics vanishes.*

Euler begins by explaining that his proof will depend on a technique that we and Fermat call the Method of Infinite Descent, though Euler does not call it that. Rather, he explains that he will suppose that there were numbers a and b so that $a^4 + b^4$ is a square. He says that he will use those to find smaller numbers that also form a square. He explains that because smaller numbers do not always exist, then it is concluded that the larger ones could not exist either. It is not the clearest explanation of the Method of Infinite Descent. This done, Euler begins the proof itself. Because biquadratics are also squares, the lemmas tell us that

$$aa = pp - qq \quad \text{and} \quad bb = 2pq,$$

where p and q are relatively prime, one odd, the other even.

Here, p must be the odd number and q the even one, or else $pp - qq$ could not be a square. Euler does not explain why, but it is easy to see using a mod 4 argument. Now, from Lemma 1, for $2pq$ to be a square, both p and $2q$ must be squares. By lemma 3, for $pp - qq$ to be a square, there must be relatively prime numbers m and n, one even, the other odd, such that

$$p = mm + nn \quad \text{and} \quad q = 2mn.$$

Because $2q$ is a square, so also is $4mn$, and so mn is, as well. As m and n are relatively prime, Lemma 1 tells us that they must both be squares. Write them as

$$m = xx \quad \text{and} \quad n = yy.$$

Now,

$$p = m^2 + n^2 = x^4 + y^4.$$

Thus, if $a^4 + b^4$ is a square, then $x^4 + y^4$ must also be square.

Now, Euler completes the argument by filling in some of the gaps in his explanation of the Method of Infinite Descent. He says that "it is manifest" that the numbers x and y must be smaller than the numbers a and b, and they are all integers. In the same way, even smaller biquadratics can be found, until one of the biquadratics must be 0, and this is a contradiction. Q. E. D.

Fermat also used the Method of Infinite Descent in his proof.

This theorem is not exactly the case $n = 4$ of Fermat's Last Theorem, but, since any biquadratic is also a square, it follows immediately. Euler makes sure we are aware of this with his first of six corollaries. Euler continues to embed his proofs in the statements of his corollaries.

Corollary 1. *Because the sum of two biquadratics can never be a square, it cannot be a biquadratic, either.*

Corollary 2 tells us that $a^4 + b^4 = c^2$ does not have any rational solutions either, for if it did, there would have to be integer solutions.

Corollary 3 tells us that there are no numbers p and q such that p, $2q$ and $pp - qq$ are all square, for if there were, we could take $a = \sqrt{pp - qq}$ and $b = \sqrt{2pq}$, and then a and b would be integers such that $a^4 + b^4$ is a square, contrary to Theorem 1.

Corollaries 4, 5 and 6 are that $x^4 - 4y^4$, $ab(a^2 + b^2)$ and $2ab(aa - bb)$ are never squares.

Theorem 2. *The difference of two biquadratics, $a^4 - b^4$, can never be a square unless $b = 0$ or $b = a$.*

We will omit the proof here, noting only that Euler uses Infinite Descent again, and he divides the proof into two cases, depending on whether b is even or odd.

Euler gives us six corollaries to Theorem 2. The first four expand on some of the details of the proof of Theorem 2. The fifth tells us that neither $ab(aa - bb)$ nor $2ab(aa + bb)$ can be squares, if a and b are relatively prime. In Corollary 6, he says that he is giving nine more forms that can never be squares, but, in a rare lapse in his counting skills, Euler lists ten, some of which we already knew about.

I. $a^4 + b^4$	VI. $a^4 - b^4$
II. $a^4 - 4b^4$	VII. $4a^4 + b^4$
III. $4a^4 - b^4$	VIII. $ab(aa - bb)$
IV. $ab(aa - bb)$	IX. $2ab(aa + bb)$
V. $2ab(aa - bb)$	X. $2a^4 \pm 2b^4$

Theorem 3. *The double of the sum of two biquadratics, $2a^4 + 2b^4$, can never be a square unless $a = b$.*

The proof supposes there is such a square, and uses it to construct another square of the form $n^4 - m^4$, which is impossible by Theorem 2.

Six more corollaries tell us that a variety of other forms, $2ab(aa+bb)$, $(a^2 + b^2)^2 + (a^2 - b^2)^2$, $a^4 \pm 6aabb + b^4$, cannot be squares.

Theorem 4. *The double of the difference of two biquadratics, $2a^4 - 2b^4$, cannot be a square, unless $a = b$.*

The proof supposes that the form were a square and that a and b are relatively prime. Euler factors it into $2(a - b)(a + b)(aa + bb)$, where a and b must both be odd. Then, the form must be divisible by 16, so he divides. Because the factors are relatively prime and the product is square, the factors must be squares. He sets

$$\frac{a-b}{2} = pp \quad \text{and} \quad \frac{a+b}{2} = qq.$$

Note that the third factor, $\frac{aa+bb}{2}$ is also square. This makes $a = pp + qq$, $b = qq - pp$ and

$$\frac{aa+bb}{2} = p^4 + q^4.$$

But this last is also a square, which is impossible, by Theorem 2. Q. E. D.

Euler gives us no corollaries to Theorem 4. It is the only theorem in this paper that is so deprived.

Theorem 5. *Neither $ma^4 - m^3b^4$ nor $2ma^4 - 2m^3b^4$ can be a square.*

Euler does not think it is necessary to mention that m ought not be zero. For the first form, he supposes that a and b are relatively prime. He explains that he can also assume that m is "neither a square nor divisible by a square," a condition we now call "square free." If m is divisible by a square, then dividing by that square gives a smaller solution. Next, Euler factors the first form to get $m(aa - mbb)(aa + mbb)$, which must be a square. The factors are relatively prime and their products are square, so $m = pp$ and $(aa - mbb)(aa + mbb)$ must be square. That would make $(aa - ppbb)(aa + ppbb)$ a square. This last expands to give $a^4 - (pb)^4$. Rather than object to the conclusion that m must be both square and non-square, Euler rejects the last expression being a square because he has shown that a difference of fourth powers cannot be a square.

The second part of the proof is similar.

Four corollaries tell us that a variety of other forms cannot be squares. Among those forms are $mn(m^2a^4 - n^2b^4)$, $2mn(m^2a^4 - n^2b^4)$ and $n(p^4 \pm 6nppqq + n^2q^4)$. Corollary 2 has a slightly different form. It tells us that if $maa + nbb$ is a square, then neither $m^2naa - mn^2bb$ nor twice that can be squares.

Theorem 6 is closely related to Theorem 5.

Theorem 6. *Neither $ma^4 + m^3b^4$ nor $2ma^4 + 2m^3b^4$ can be a square.*

The proof follows the patterns we have been seeing in the proofs of the other theorems.

Theorem 7. *No triangular number can be a biquadratic except* 1.

This is another of the small collection of theorems for which Fermat left proofs. As we have noted before, Fermat's usual practice was to announce results, usually in letters to Mersenne, but to leave the proofs as "exercises" for his correspondents.

Proof. Euler reminds us that triangular numbers have the form $\frac{x(x+1)}{2}$. Euler writes this as $\frac{x}{2}(x+1)$ in case x is even, and as $x\frac{x+1}{2}$ in case x is odd. Either way, the two factors are relatively prime. Thus, by Euclid's lemma, for the product to be a biquadratic, the factors themselves would have to be biquadratics. Neither case is possible, though. In the first case, $\frac{x}{2}$ would have to be biquadratic. Call it m^4. Then $x = 2m^4$ and $x + 1 = 2m^4 + 1$ must be biquadratic. In case x is odd, the same calculation leads to having $x = 2m^4 - 1$ be a biquadratic.

Euler turns to showing that $2m^4 \pm 1$ is never a biquadratic. Suppose it were equal to n^4. Then $4m^4 = 2n^4 \mp 2$. But $4m^4$ is a square, and we have already seen that numbers of the form $2a^4 \pm 2b^4$ cannot be squares unless $a = b$. This admits the solutions $x = 0$ and $x = 1$, but excludes all others.

Q. E. D.

Two corollaries tell us that the form $8y^4 + 1$ cannot be a square unless $y = 0$ or $y = 1$, and that $2z^2 - 2$ cannot be a biquadratic unless $z = 1$ or $z = 3$.

Theorem 8. *The sum of three biquadratics, two of which are equal to each other, that is, numbers of the form $a^4 + 2b^4$, cannot be squares unless $b = 0$.*

Euler's proof here is long. We suppose that $a^4 + 2b^4$ is a square, and that its root is $a^2 + \frac{m}{n}b^2$, where we may take a and b to be relatively prime, as well as m and n. Squaring the root and equating expressions gives

$$\frac{b^2}{a^2} = \frac{2mn}{2n^2 - m^2}.$$

The fraction on the right will either be in lowest terms or, if m is even, will be reduced to lowest terms by dividing by 2. This makes two cases.

In the first case, we get $b^2 = 2mn$ and $a^2 = 2n^2 - m^2$. This now splits into two more cases, depending on whether n is odd or even. In case n is odd, then m must also be odd or else the fractions would not be in lowest terms. But then $2mn$ could not be a square, so this case is eliminated. On the other hand, in case n is even, we must have a and m both odd. But $a^2 + m^2 = 2n^2$. The expression on the left must be divisible by 2 but not by 4, while the expression on the right must be divisible by 8. Thus the first case is impossible.

It is interesting to note that the structure of Latin allows Euler to consider this entire case in just one five-line sentence.

In the second case, $2mn$ and $2n^2 - m^2$ have a common factor of 2, so m must be even and n will be odd. Take $m = 2k$. Substituting and canceling 2s gives

$$\frac{b^2}{a^2} = \frac{4kn}{2n^2 - 4k^2} = \frac{2kn}{n^2 - 2k^2},$$

where now the right-hand side must be in lowest terms. Hence $b^2 = 2kn$ and $a^2 = n^2 - 2k^2$. Because n is odd and $2kn$ is a square, k must be even. By Euclid's lemma, n and $2k$ must be squares. Let $n = cc$ and $2k = 4dd$, where c must be odd. This makes $a^2 = c^4 - 8d^4$.

Now he will investigate whether such forms $c^4 - 8d^4$ can be squares. The investigations will closely parallel what we did above. Suppose that it is a square and that c and d are relatively prime. As before, let the root be

$$c^2 - \frac{2p}{q}d^2,$$

where p and q are relatively prime as well. Again squaring and equating leads to

$$\frac{dd}{cc} = \frac{pq}{pp + 2qq}.$$

As before, Euler distinguishes two cases, depending on whether p is odd or even. In case p is odd, pq and $pp + 2qq$ are relatively prime and so $dd = pq$ and $cc = pp + 2qq$. Once again, by Euclid's lemma, p and q must be squares, and Euler calls them $p = x^2$ and $q = y^2$. This makes $cc = x^4 + 2y^4$. By now, Euler expects that the reader is becoming familiar with the use of Fermat's Method of Infinite Descent, so he just tells us that x and y are smaller than a and b, and so this case (p odd) cannot happen.

The other case, p even, is similar. Q. E. D.

Corollaries tell us that several more forms cannot be squares, $2mn(2n^2 - m^2)$, $4y^4 - 2x^2$, $2y^4 - 4x^4$, $8y^4 - x^4$ and $4\alpha\beta^3y^4 - 2\alpha^3\beta x^4$.

Theorem 9. *If a form $a^4 + kb^4$ cannot be a square, then any form $2k\alpha\beta^3y^4 - 2\alpha^3\beta x^4$ cannot be a square, either.*

This theorem is a bit different from the others, for it deals with an infinite class of forms, but the proof is familiar. Euler starts by supposing that $a^4 + kb^4$ is a square, with root $= a^2 + \frac{m}{n}b^2$. He substitutes and equates as before and gets a relation. When the form is not a square, the relation is not satisfied. Then he substitutes $m = \alpha x^2$ and $n = \beta y^2$ and reaches his conclusion.

Six corollaries and a scholion follow. Euler closes the article with quite a different theorem, one that apparently does not involve biquadratics, though they arise almost immediately in the proof itself. It is a special case of what we now call the Carmichael Conjecture.

Theorem 10. *No cube, not even fractions, is ever one less than a square, except in the one case when the cube is* 8.

Proof. We recast the statement to claim that $\frac{a^3}{b^3}+1$ is never a square, except in the case when $\frac{a}{b}=2$. This proof will be based on showing that a^3b+b^4 cannot be a square unless $a=2b$. Euler does not explain why this is sufficient, but it follows easily if we suppose that $\frac{a^3}{b^3}+1=c^2$ and multiply both sides by b^4. We factor a^3b+b^4 into $b(a+b)(aa-ab+bb)$, which is supposed to be a square. Note that it could be square if $b(a+b)=aa-ab+bb$, but that implies that $a=2b$, which is exactly the case we have excluded. Otherwise, take $a+b=c$, and the form becomes $bc(cc-3bc+3bb)$. We want to show that this cannot be square unless $c=3b$. Take b and c to be relatively prime.

Euler distinguishes two cases, depending on whether or not c is a multiple of 3. First, we consider the case where c is not a multiple of 3. Then, again using Euclid's lemma, the three factors, b, c and $cc-3bc+3bb$, must all be squares. Let the root of $cc-3bc+3bb$ be $\frac{m}{n}b-c$, where, of course, m and n are relatively prime. As before, it turns out that

$$\frac{b}{c}=\frac{2mn-3nn}{mm-3nn},$$

and the fraction on the right is in lowest terms unless m is a multiple of 2. Again, this determines two sub-cases.

If m is not a multiple of 2, then either $c=3nn-mm$ and $b=3nn-2mn$ or the negatives of these, $c=mm-3nn$ and $b=2mn-3nn$. The first of these is not possible, because c is supposed to be a square and $3nn-mm$ is never a square. Euler does not tell us why, but we can convince ourselves using a mod 3 argument. And so it must be that $c=mm-3nn$ and $b=2mn-3nn$.

Let the root of c be $m-\frac{p}{q}n$, where it goes without saying that p and q are relatively prime. This makes

$$\frac{m}{n}=\frac{3qq+pp}{2pq} \quad\text{and}\quad \frac{b}{nn}=\frac{2m}{n}-3=\frac{3qq-3pq+pp}{pq}.$$

Here, b is a square, and so $pq(3qq-2pq+pp)$ is a square. This has the same form as a form we saw earlier, $bc(cc-3bc+3bb)$. We have been trying to show that the form cannot be a square. Euler tells us that the new form involves smaller numbers and we are supposed to fill in the gap by applying Infinite Descent. This completes the case where neither c nor m are multiples of 3.

Now, suppose that m is a multiple of 3. Call $m=3k$, and we get

$$\frac{b}{c}=\frac{nn-2kn}{nn-3kk}.$$

Steps just like those we saw before lead us to conclude that $c=nn-3kk$. As before, c is a square, so we take its root to be $n-\frac{p}{q}k$. Euler leads us through the three or four steps it takes to get

$$\frac{b}{nn}=1-\frac{2k}{n}=\frac{pp+3qq-4pq}{3qq+pp}.$$

Therefore, $(pp+3qq)(p-q)(p-3q)$ must be a square. We take $p-q=t$ and $p-3q=u$, and this last becomes $tu(3tt-3tu+uu)$, which again is of the same form as the original $bc(cc-3bc+3bb)$, but with smaller parameters. We are left to conclude for ourselves that, by Infinite Descent, the form cannot be a square. This concludes the case where c not a multiple of 2.

Euler tells us that if c is a multiple of 2, we should put $c=3d$, and the analysis will be similar.

Q. E. D.

Euler's first corollary is something he claims in his statement of Theorem 10 but does not mention in the proof. "In a similar way it can be proved that no cube can be one less than a square, even if they are fractions." His second corollary claims that it follows that neither x^6+y^6 nor x^6-y^6 can be squares, and also that no triangular number can be a cube except 1. He gives neither proof.

Euler is probably the only mathematician of his time who is very interested in these kinds of questions. Newton and the Bernoullis almost entirely ignored number theory, and the rest of the mathematical community mostly followed their lead. Here, though, we see Euler pursuing these questions out of his own interest. He has developed enough self-confidence that he can pose and pursue some of his own questions. In many fields, we will find that others will work on problems because Euler says they are important, but few will follow his example in number theory until Lagrange in the 1760s and then Gauss in the 1790s.

We will also see that Euler will use this Theorem–Proof style of writing more and more as he matures, and the style will gradually become the standard way of writing mathematics, supplanting the narrative style we saw Euler using earlier.

This is the last of Euler's number theory papers from his first St. Petersburg period. Although his first paper on amicable numbers, E-100, will be published in 1747, the same year that this paper appeared, E-98 suffered a nine-year publication delay. (E-100 was apparently written the same year it was published.) So, it seems that after he writes E-98, he moves to Berlin. There he puts aside number theory for almost ten years. When he returns to the field, there will be spectacular results, but we will have to wait for another time to enjoy them.

38

E-99: Solutio problematis cuiusdam a celeberrimo Daniele Bernoullio propositi*

Solution of some problems that were posed by the celebrated Daniel Bernoulli

We turn to a problem in the calculus of variations. Euler tells us that the problem was posed to him by Daniel Bernoulli in a letter from Basel on May 24, 1738, and is inspired by a "problem in mechanics." Bernoulli's problem is to find the curve of a given length that either maximizes or minimizes $\int r^m\, ds$. Here, s is arc length, r is the radius of curvature at a given point. Euler tells us that he has already solved this in the case $m = 1$ and found that

$$ds = \frac{2r\, dr}{\sqrt{g + 4nr - 4rr}}.$$

If we were looking at Euler's mechanics papers as well as his mathematics papers, especially E-8 and E-831, then we might expect that "problem in mechanics" to be the so-called Problem of Elastic Curves. Euler and Bernoulli usually consider stiff ribbons. As a ribbon is bent at a given point the force with which it resists the deformation is a function of the amount of curvature at that point; the more the curvature, the more the force. Because curvature is the reciprocal of r, the radius of curvature, Bernoulli's problem is related to elastic curves.

This makes a nice story, but it turns out that other letters between Euler and Bernoulli make it seem likely that they did not yet understand the relation between $\int r^m\, ds$ and elastic curves in May of 1738. We have the following exchange of letters between 1738 and 1743, when Euler was writing his comprehensive account of the calculus of variations, the *Methodus Inveniendi.*[1]

May 24, 1738—Bernoulli poses maximizing or minimizing $\int r^m\, ds$ as a "problem in mechanics."

November 8, 1738—Bernoulli claims that he can show that elastic curves maximize or minimize

$$\int \frac{ds^2}{r^2\, d\xi^2},$$

and promises to "send my reflections on this another time."

December 23, 1738—Euler is skeptical about Bernoulli's claim, writing, "For superficially, I see that these [elastic] curves have a maximum or minimum in the course of bending."

March 7, 1739—Bernoulli backs off on his claim a bit, writing, "I think that an elastic band which takes on itself a certain curvature will bend in such a way that the live force will be a minimum, because otherwise the band will move of itself."

**Comm. Acad. Sci. Imp. Petropol.* 10 (1738) 1747, 164–180; *Opera Omnia* I.25, 84–97

[1] These letters are numbers 120, 123, 124, 125, 147, 148, 150 and 151 in the catalog of Euler's correspondence [OO] prepared by the Editors of the *Opera Omnia*. They are available online at The Euler Archive [EA]. Thanks to Lawrence D'Antonio [D] for drawing them to my attention, and for providing translations.

October 20, 1742—After a gap in this conversation of more than three years, Bernoulli seems to have clarified his thoughts, and perhaps developed a bit of a sarcastic streak, when he writes "I should like to know if your Worship could not solve the curvature of the elastic band . . . Since no one has perfected the isoperimetric method as much as you, you will easily solve this problem of rendering $\int \frac{ds}{r^2}$ a minimum."

December 12, 1742—Bernoulli writes Euler that he is "glad that you are so pleased with my principle for finding the elastica by the isoperimetric method."

April 23, 1743—Bernoulli suggests that Euler add an appendix to his *Methodus Inveniendi* about elastic curves.

September 4, 1743—Bernoulli writes that he has received the "Isoperimetric Additions," the appendix he had proposed.

So, if in 1738 they knew, rather than just suspected, that elastic curves depended on maximizing or minimizing $\int r^m ds$, then why did it take until 1742 to agree on it and until 1743 to get it written down? We will probably never know.

Let us return to the problem at hand. Euler begins by defining his variables as usual. The abscissa of the solution is x and the application, or ordinate, is y. Eighteenth century mathematicians do not have subscripts or superscripts, and, being in the European tradition, Euler avoids notation that might be seen as Newtonian. So, to denote higher derivatives, Euler introduces new symbols. He sets $dy = p\,dx$, $dp = q\,dx$, $dq = r\,dx$, $dr = s\,dx$, etc., and seems unaware of the fact that s now denotes both arc length and fourth derivatives and that r denotes both curvature and third derivatives. This matters little because third and fourth derivatives do not arise in Daniel Bernoulli's problem.

Euler reminds us of some of the results in E-56. In particular, isoperimetric problems can be reduced to the form $\int Z\,dx$, where we denote by Z what was Q in E-56, any function of the quantities x, y, p, q, r, s, etc. Then, as we have seen him do in E-27 and E-54, he writes

$$dZ = M\,dx + N\,dy + P\,dp + Q\,dq + R\,dr + S\,ds + \text{etc.}$$

We form

$$V = N - \frac{dP}{dx} + \frac{ddQ}{dx^2} - \frac{d^3R}{dx^3} + \frac{d^4S}{dx^4} - \text{etc.}$$

The curve that maximizes or minimizes the integral will satisfy $V = 0$. Euler still mixes up necessary conditions with sufficient ones by thinking that $V = 0$ is sufficient as well as necessary.

Returning to Daniel Bernoulli's problem, Euler reminds us that $ds = \sqrt{dx^2 + dy^2}$ and that

$$r = \frac{ds^3}{dx\,ddy},$$

where s and r are arc length and curvature and not higher order derivatives. Substitution makes the integral to be minimized

$$\int \frac{(1 + pp)^{(3m+1)/2}\,dx}{q^m}.$$

This determines Z, from which we find the various differentials, M, N, P, Q, etc., most of which are zero, but

$$P = \frac{(3m+1)(1+pp)^{(3m-1)/2}p}{q^m} \quad \text{and} \quad Q = \frac{-m(1+pp)^{(3m+1)/2}}{q^{m+1}}.$$

So we know that

$$V = -\frac{dP}{dx} + \frac{ddQ}{dx^2}.$$

We now look at the differential equation $V = 0$, knowing what we know about P and Q. A solution must have the form $A\,dx = -P\,dx + dQ$. We learned before that $dp = q\,dx$ and $dZ = P\,dp + Q\,dq$. Substituting $dx = \frac{dp}{q}$ makes $A\,dp = -P\,dp + q\,dQ$. We substitute $P\,dp = dZ - Q\,dQ$, then integrate to get $Ap + B = -Z + Qq$. Thus we have defined these two functions of integration A and B. A bit of calculation shows that $Ay + Bx = (m+1)\int r^m\,ds$, so the integral will be minimized by the same function that minimizes $Ay + Bx$.

Now Euler uses an idea we have not seen before. He notes that the two variables that matter here, r and s, are not changed if the axes are changed. He says, without explaining exactly why, that it is possible to choose axes so that either A or B will be equal to 0, and so the integral to be minimized can be written either as

$$Ax = (m+1)\int r^m\,ds$$

or as

$$Ay = (m+1)\int r^m\,ds.$$

The variable B has disappeared. Euler often exploits the virtues of homogeneity, and he does so here as well. He sets $A = a^m$, so that

$$a^m x = (m+1)\int r^m\,ds$$

and, after differentiating,

$$a^m\,dx = (m+1)r^m\,ds.$$

Euler now uses a fact that most of us have forgotten (if we've ever seen it at all) that $\frac{dx}{ds}$ is the sine of the arc whose differential is $\frac{ds}{r}$. He writes this as $\frac{dx}{ds} = \sin\int\frac{ds}{r}$. In our problem, this means that

$$a^m \sin\int\frac{ds}{r} = (m+1)r^m.$$

Euler substitutes

$$c^m = \frac{a^m}{m+1}$$

to rewrite this as

$$\int\frac{ds}{r} = \arcsin\frac{r^m}{c^m} = \int\frac{mr^{m-1}\,dr}{\sqrt{c^{2m} - r^{2m}}}.$$

Differentiating, multiplying by r and integrating again give arc length in terms of radius of curvature, namely

$$s = m \int \frac{r^m\, dr}{\sqrt{c^{2m} - r^{2m}}}.$$

For the first time, Euler calls his coordinates "orthogonal" as he tells us that "from the equation between the orthogonal coordinates x and y," he gets an equation relating p and q:

$$a^m q^m = (1 + pp)^{(3m+1)/2}.$$

In this, Euler solves for q, substitutes $q\, dx = dp$ and integrates to get

$$y = -\frac{ma}{(m+1)(1+pp)^{(m+1)/2m}} + b.$$

He substitutes $\frac{ma}{m+1} = -c$ and holds $b = 0$, to get an explicit solution to the original isoperimetric equation,

$$x = \int \frac{y^{m/m+1}\, dy}{\sqrt{c^{2m/(m+1)} - y^{2m/(m+1)}}}.$$

Euler claims that this will be algebraic whenever m has the form $m = \frac{1}{2i}$ or $m = -\frac{1}{2i+1}$, for i an integer.

Euler briefly considers elementary cases. First, he takes $m = -1$, so the integral to be maximized is $\int \frac{ds}{r}$. This makes $m + 1 = 0$ which he says makes the equations "inconvenient" because of the divisions by zero.

On the other hand, if $m = 0$ and we are to minimize the integral, because

$$A^m q^m = (1 + pp)^{(3m+1)/2}$$

and $m = 0$, we get that p is constant and the curve is a straight line. This makes sense, because when $m = 0$, the integral is just arc length.

In the case $m = 1$, Euler shows that

$$x = \int \frac{\sqrt{y}\, dy}{\sqrt{c - y}} = \int \frac{y\, dy}{\sqrt{cy - yy}},$$

which gives the cycloid as a solution. Euler knows this is correct from work others have done on cycloids.

He also looks at the case $m = -2$ and finds that

$$x = \int \frac{y^2\, dy}{\sqrt{c^4 - y^4}}$$

is a kind of elastic curve that he already understands from another context.

In this problem, Euler finds what he calls a paradox that does not arise in other problems of this kind. He notes that, for $m = 1$, different values of c still give equations that satisfy the conditions for optimization. This corresponds to generating cycloids from circles of different radii. If we do this with a relatively small value of c, then it corresponds to going through more than one cycle of the cycloid, and it generates a curve with cusps in it. Because Euler is only barely beginning to develop the modern notion of a function, he can hardly be expected to recognize that his questions here could be answered by restricting his solutions to a suitably continuous class of functions. Instead, he suggests some other conditions that could reduce the number of solutions. He considers requiring that the solution pass through some third point, in addition to specifying the endpoints, and also considers specifying the tangents to the curve at the endpoints.

A few comments on adding extra points may be in order. Let us suppose that the original problem is to find a curve from A to B optimizing certain criteria. Now we change the problem by requiring the curve to go through a third point C. Euler would not be satisfied by finding two curves, one from A to C and another from C to B, and then connecting the two curves at the point C. He does not think that this makes a true curve because the two pieces of the curve are not given by the same equation.

Euler spends a bit of time looking at the equations that arise by adding additional points. He shows no actual calculations here, but he says that each time we add another point it increases the degree of the differential equation and thus makes the problem more difficult to solve.

He also discusses some other ideas without giving proofs. He tells us that if an integral to be maximized or minimized is like $\int (ay - xx)y\,dx$, then the curve must satisfy $2ay = xx$.

Up to this point, Euler has only considered curves that maximize or minimize the form $\int r^m\,ds$. He has not yet used his isoperimetric condition. The curve must have a fixed arc length, so it must give a particular value to the integral $\int ds$. Euler showed us how to handle such criteria in E-56. Each condition separately gives its own function V that must satisfy $V = 0$, so the sum of the two functions, both denoted by V here and by Q in E-56, must also equal 0. He follows his program from E-56 and finds that there must be constants A, B and C such that

$$A\,ds + B\,dy + C\,dx = (m+1)r^m\,ds.$$

Now, "as I have noted above," we can choose the axes so that either B or C is zero.

Euler takes $B = 0$ and recalls that

$$\frac{dx}{ds} = \arcsin \int \frac{ds}{dr}.$$

He finds that the curve maximizing or minimizing $\int r\,ds$ is given by

$$s = \int \frac{r\,dr}{\sqrt{c^2 - r^2 \pm 2ar - aa}}.$$

He says that this agrees with Bernoulli's solution,

$$ds = \frac{2r\,dr}{\sqrt{g + 4nr - 4rr}}$$

if we make the two changes of variables $\frac{g}{4} = c^2 - a^2$ and $n = \pm 2a$.

If also $C = 0$, Euler finds that the equations give the arc of a circle. He once again lets $m = 1$ and finds that the equations still give him a cycloid.

It seems that this paper adds little new to the world's understanding of the calculus of variations. The important cases $m = 0$ and $m = 1$ had been solved by Bernoulli. Although Euler gave differential equations for other values of m, he did not do much with them. This paper does test the ideas he developed in E-56, and it shows that those ideas are easily adapted to accommodate different differential quantities, this time r, the radius of curvature.

It may be that the most interesting idea here is choosing his axes so late in the problem in order to make certain formulas easier to manipulate. People had long selected convenient axes. We have seen Euler use major or minor axes of ellipses as one of his axes, choosing one or the other depending on whether he wanted a particular parameter to be greater than or less than one. Until now, though, he has chosen his axes at the beginning of his analysis. Here, he decides very late.

We have also been reminded again that Euler expects a curve to be given by a particular equation or system of equations. He does not conceive of the idea of a piecewise defined curve.

Euler's next effort in the calculus of variations will appear in 1744, six years after he writes this paper but three years before it sees print. That will be his great *Methodus inveniendi*, or "Method of finding curved lines that enjoy properties of maximum or minimum, or the solution of isoperimetric problems in the broadest accepted sense." It will be one of his greatest works.

References

[D] D'Antonio, Lawrence, "The fabric of the universe is most perfect," presented at The Euler Society Annual Conference, Roger Williams University, Bristol, RI, 2003.

[EA] The Euler Archive, `www.EulerArchive.com`.

[OO] *Leonhard Euler Briefwechsel, Opera Omnia*, Ser. IVA Vol. I, Birkhäuser, Basel, 1975.

Interlude: 1739

World events

In 1739, England went to war with Spain, a war known as the War of Jenkins' Ear. The blind Lucasian Professor Nicolas Saunderson (1682–1739) died that year.

In Euler's life

Euler's book on the theory of music, E-33, *Tentamen novae theoriae musicae ex certissimis harmoniae principiis dilucide expositae* was published this year.

Childhood was dangerous in the 18th century. In 1739, Euler lost two of his children. Marie Gertrud died five days after her second birthday on March 3, and an infant daughter, Anna Elizabeth, died on November 8, two weeks after she was born. The Eulers had lost another six-week-old daughter in 1736. Eventually, the Eulers would have 13 children, only five of whom would survive to adulthood.

Euler's other work

In 1739, Euler wrote an essay on tides that would win a share of the Paris Prize for 1740. That essay, E-57, was titled Inquisitio physica in causam fluxes ac refluxus maris, "Physical inquiry into the cause of the ebb and flow of the sea." He wrote a seldom-mentioned paper on weather, E-124, Determinatio caloris et frigoris graduum pro singulis terrae locis ac temporibus, "Determination of the degrees of heat and cold for a particular place and time on the earth." There was a mechanics paper, E-126, De novo genere oscillationum, "On a new kind of oscillation," and an optics paper, E-127, Explicatio phaenomenorum quae a motu lucis successive oriuntur, "Explanation of phenomena which arise from the successive motion of light."

Euler's mathematics

We will consider six articles that Euler wrote in 1739. He wrote a sequel to E-19 about infinite products, E-122, De productis ex infinitis factoribus ortis, "On products arising from infinitely many factors." There was another paper on continued fractions, E-123, De fractionibus continuis observationes, "Observations on continued fractions." He investigated π in E-125, Consideratio progressionis cuiusdam ad circuli quadraturam inveniendam idoneae, "Consideration of some progressions appropriate for finding the quadrature of the circle," and continued analyzing trigonometric issues in E-128, Methodus facilis computandi angulorum sinus ac tangentes tam naturales quam artificiales, "An easy method for computing sines and tangents of angles both natural and artificial." There was an analytic geometry article, E-129, Investigatio curvarum quae evolutae sui similes producunt, "Investigation of curves which produce evolutes that are similar to themselves." Finally, he explained some of the details of his solution to the Basel problem in E-130, De seriebus quibusdam considerationes, "Considerations about certain series."

39

E-122: De productis ex infinitis factoribus ortis*

On products arising from infinitely many factors

Euler has been neglecting his gamma function lately. He has had nothing to say about it since 1730/31, when he wrote E-19, the first paper on the subject. Now in 1739, almost ten years later, he returns to the interpolation of infinite products. The paper will suffer an eleven-year publication delay, until 1750, so it will seem as if he has been away from infinite products for two decades.

Euler reminds us that an infinite product can be converted into an infinite sum by taking logarithms. Then he gives a short history lesson, citing Wallis's product from 1655 as the first nontrivial infinite product to be successfully evaluated. Before he gets to new material, Euler reviews a few of the major results of E-19. Euler studies what he calls a "hypergeometric progression" by looking at the terms of an infinite series:

$$\begin{aligned}&\overset{1}{(f+g)}+\overset{2}{(f+g)(f+2g)}+\overset{3}{(f+g)(f+2g)(f+3g)}\\&\qquad+\overset{4}{(f+g)(f+2g)(f+3g)(f+4g)}+\text{etc.}\end{aligned} \tag{1}$$

Note how he labels the terms with index numbers written above each term, a kind of proxy for subscripts. The term of index n is the previous term multiplied by $f+ng$. Euler does not seem to care whether or not the series converges. In E-19, he found the term of degree n is given by the formula

$$\frac{g^{n+1}\int dx(-\ln x)^n}{(f+(n+1)g)\int x^{f:g}dx(1-x)^n}.$$

Here, we write $\ln x$ where Euler wrote $l\,x$ to denote the natural logarithm. Euler still describes the integrals as he did before. We are to choose "the integral that vanishes when $x=0$ and then make $x=1$."

This formula is defined even for fractional values of n, and these values interpolate the terms of his infinite series. To Euler, infinite products arise naturally from such interpolation problems.

Euler notes that this formula involves a difficult integral, $\int dx(-\ln x)^n$, but that for rational $n=\frac{p}{q}$, the integral can be given by a root of a product

$$\sqrt[q]{\begin{matrix}1\cdot2\cdot3\cdots p\left(\dfrac{2p}{q}+1\right)\left(\dfrac{3p}{q}+1\right)\left(\dfrac{4p}{q}+1\right)\cdots(p+1)\\ \times\int dx(x-xx)^{p/q}\int dx(x^2-x^3)^{p/q}\int dx(x^3-x^4)^{p/q}\cdots\int dx(x^{q-1}-x^q)^{p/q}\end{matrix}}\;.$$

*Comm. Acad. Sci. Imp. Petropol. 11 (1739) 1750, 3–31; Opera Omnia I.14, 260–291

This does not seem like progress, but the product is not that complicated when p and q are small. In particular, if $n = \frac{1}{2}$, $p = 1$ and $q = 2$, it becomes the relatively simple formula

$$\int dx(-\ln x)^{1/2} = \sqrt{1 \cdot 2 \int dx\sqrt{x - xx}} = \frac{\sqrt{\pi}}{2}.$$

Let z be the term of index $\frac{1}{2}$ in Euler's infinite series. Then Euler writes a new series, that he says follows "from the law that gives the terms" of his other series,

$$z + z\left(f + \frac{3}{2}g\right) + z\left(f + \frac{3}{2}g\right)\left(f + \frac{5}{2}g\right) + z\left(f + \frac{3}{2}g\right)\left(f + \frac{5}{2}g\right)\left(f + \frac{7}{2}g\right) + \text{etc.}$$

If we consider the term z to have index $\frac{1}{2}$, then the terms of this sequence obey the same recursive relationship to z as the terms of the series (1) have to $(f+g)$, if we consider that term to have index 1.

Euler notices that, in some ways this series resembles a geometric series. The similarity he sees is that in a geometric series, the ratio of consecutive terms is constant. In this series, the ratio of consecutive terms approaches the constant g. Based on this, he says that this series gives a sequence of approximations for z:

I. $z = \sqrt{f + g}$

II. $z = \sqrt{\dfrac{(f+g)(f+g)(f+2g)}{1(f+\frac{3}{2}g)(f+\frac{3}{2}g)}}$

III. $z = \sqrt{\dfrac{(f+g)}{1}\dfrac{(f+g)}{(f+\frac{3}{2}g)}\dfrac{(f+2g)}{(f+\frac{3}{2}g)}\dfrac{(f+2g)}{(f+\frac{5}{2}g)}\dfrac{(f+3g)}{(f+\frac{5}{2}g)}}.$

Euler extends these approximations to give z as the root of an infinite product. He also cites the results of E-19 to describe this same term in integral form as

$$z = \frac{g\sqrt{\pi g}}{(2f+3g)\int x^{f:g}dx\sqrt{1-x}} = \frac{\sqrt{\pi g}}{(2f+3g)\int y^{f+g-1}dy\sqrt{(1-y^g)}}.$$

Note the substitution $x = y^g$.

Euler does the calculations taking $g = 1$ and derives Wallis's formula. We have seen Euler check his work like this a great many times now. His motivation seems to be to verify his results rather than to exhibit yet another way to derive Wallis's formula. Euler seems to give us new proofs of old results in two circumstances. Sometimes, the new proofs really are better, but usually he uses them as he does here to check for errors. We have not seen him give us new proofs just for the sake of novelty. There was a curious exception to this rule in E-41, when he gave us three slightly different solutions to the Basel Problem. We can speculate that perhaps he was building suspense in his presentation, because the third solution is the most interesting of the three, or perhaps he was using the first two solutions to prepare us for the intricacies of the third and most elegant solution.

We take $g = 2$ and f any odd number. Then our integral formulas give us

$$\frac{fg}{\pi}\left(\int \frac{y^{f-1}\,dy}{\sqrt{1-y^2}}\right) = \frac{(2f+g)(2f+g)(2f+3g)(2f+3g)(2f+5g)(2f+5g)}{2f(2f+2g)(2f+2g)(2f+4g)(2f+4g)(2f+6g)}\text{etc.}$$

It bothers him a little bit that this expression contains π, so he replaces some of the f and g with h and k, then divides and rearranges factors. After taking square roots, he has a quotient of integrals and an infinite product

$$\frac{\int y^{f-1}\,dy : \sqrt{1-y^g}}{\int y^{h-1}\,dy : \sqrt{1-y^k}} \cdot \sqrt{\frac{g}{k}} = \frac{2h(2f+g)(2h+2k)(2f+3g)(2h+4k)(2f+5g)}{2f(2h+k)(2f+2g)(2h+3k)(2f+4g)(2h+5k)}\text{etc.}$$

He sets to work manipulating this expression. First, he tries grouping the factors on the right two at a time, then again separating the $2h/2f$ and associating them two at a time. He examines the case $k = g$, and also rewrites $2f = a$ and $2h = b$, and substitutes $y = x^2$. Then, by taking various values of g and a, he gets seventeen different integrals that, integrated from zero to one, give π. We give a selection.

$$\pi = 2\int \frac{dx}{\sqrt{1-x^2}} \cdot \int \frac{x\,dx}{\sqrt{1-x^2}} \qquad \pi = 24\int \frac{x^2 dx}{\sqrt{1-x^8}} \cdot \int \frac{x^6 dx}{\sqrt{1-x^8}}$$

$$\pi = 14\int \frac{dx}{\sqrt{1-x^{14}}} \cdot \int \frac{x^7 dx}{\sqrt{1-x^{14}}} \qquad \pi = 56\int \frac{x^3 dx}{\sqrt{1-x^{14}}} \cdot \int \frac{x^{10} dx}{\sqrt{1-x^{14}}}$$

All of his examples involve the product of two integrals of the forms

$$\int \frac{x^m\,dx}{\sqrt{1-x^{2g}}} \quad \text{and} \quad \int \frac{x^{m+g}\,dx}{\sqrt{1-x^{2g}}},$$

where we are supposed to know that the integrals are definite integrals, taken from zero to 1. Euler finds that these integrals are closely related. If we take

$$\int \frac{x^m dx}{\sqrt{1-x^{2g}}} = A,$$

then Euler tells us that

$$\int \frac{x^{m+g}\,dx}{\sqrt{1-x^{2g}}} = \frac{\pi}{2(m+1)gA},$$

$$\int \frac{x^{m+2g}\,dx}{\sqrt{1-x^{2g}}} = \frac{(m+1)A}{m+g+1}$$

and so on down to

$$\int \frac{x^{m+5g}\,dx}{\sqrt{1-x^{2g}}} = \frac{(m+g+1)(m+3g+1)\pi}{2(m+1)(m+2g+1)(m+4g+1)gA}.$$

Euler notes that only the odd numbered forms "require the quadrature of the circle," that is to say, involve π.

Now we return to the series above that begins with z. Before, he took z to be the value of the term of index $\frac{1}{2}$. Now, we take z to be the term of index $\frac{p}{q}$. Euler writes the series

$$\overset{\frac{p}{q}}{z} + \overset{\frac{p+q}{q}}{\frac{z(fq+(p+q)g)}{q}} + \overset{\frac{p+2q}{q}}{\frac{z(fq+(p+q)g)(fq+(p+2q)g)}{q^2}} + \text{etc.}$$

Euler repeats many of the calculations that he did above. He starts with a list of approximations for z, then gives z as an infinite product. He finds $\int dx(-\ln x)^{p/q}$ as a product of radicals and integrals, and defines a value P by taking $\sqrt[q]{P}$ equal to this integral. He will also use Q, defined by

$$Q = \int \frac{y^{f-1}\,dy}{(1-y^g)^{(q-p)/q}}$$

and gives a formula for z in terms of P and Q. He continues to introduce new values, R, S and T, and develops a long series of relationships giving expressions involving P, Q, R, S and T in terms of f, g, p and q, sometimes in terms of infinite products and sometimes not. Eventually, after ten pages of dense calculations and substitutions Euler leads us to what he wants:

$$\int x^{m-1}\,dx(1-x^n)^{p/q}$$
$$= \frac{q}{(p+q)n} \cdot \frac{1(mq+(p+q)n)}{m(p+2q)} \cdot \frac{2(mq+(p+2q)n)}{(m+n)(p+3q)} \cdot \frac{3(mq+(p+3q)n)}{(m+2n)(p+4q)} \cdot \text{etc.}$$

From this, he gets what is probably the climax of this article, a collection of "more notable" infinite products. Nobody should be surprised that the first of these is Wallis's formula again.

$$\int \frac{dx}{\sqrt{1-xx}} = 1 \cdot \frac{1\cdot 4}{1\cdot 3} \cdot \frac{2\cdot 8}{3\cdot 5} \cdot \frac{3\cdot 12}{5\cdot 7} \cdot \text{etc.} = \frac{2\cdot 2\cdot 4\cdot 4\cdot 6\cdot 6}{1\cdot 3\cdot 3\cdot 5\cdot 5\cdot 7}\text{etc.}$$
$$\int \frac{x\,dx}{\sqrt{1-xx}} = 1 \cdot \frac{1\cdot 6}{2\cdot 3} \cdot \frac{2\cdot 10}{4\cdot 5} \cdot \frac{3\cdot 14}{6\cdot 7} \cdot \text{etc.} = 1$$
$$\int \frac{x^2\,dx}{\sqrt{1-xx}} = 1 \cdot \frac{1\cdot 8}{3\cdot 3} \cdot \frac{2\cdot 12}{5\cdot 5} \cdot \frac{3\cdot 16}{7\cdot 7} \cdot \text{etc.} = \frac{2\cdot 4\cdot 4\cdot 6\cdot 6\cdot 7}{3\cdot 3\cdot 5\cdot 5\cdot 7\cdot 7}\text{etc.}$$

and on down to

$$\int \frac{dx}{\sqrt[4]{1-x^4}} = \frac{1}{3} \cdot \frac{4\cdot 4\cdot 8\cdot 8\cdot 12\cdot 12\cdot 16\cdot 16}{1\cdot 7\cdot 5\cdot 11\cdot 9\cdot 15\cdot 13\cdot 19}\text{etc.}$$

All nine of the formulas he gives have the forms

$$\int x^{m-1}\,dx(1-x^n)^{-m/n} = \frac{1}{n-m} \cdot \frac{n\cdot n\cdot 2n\cdot 2n\cdot 3n\cdot 3n}{m(2n-m)(m+n)(3n-m)(m+2n)(4n-m)}\text{etc.}$$

or

$$\int x^{m-1}\,dx(1-x)^{(m-n)/n}$$
$$= \frac{1}{m} \cdot \frac{n\cdot 2m\cdot 2n\cdot (2m+n)\cdot 3n\cdot (2m+2n)\cdot 4n\cdot (2m+3n)}{m(m+n)(m+n)(m+2n)(m+2n)(m+3n)(m+3n)(m+4n)}\text{etc.}$$

Euler has a few more general formulas and their consequences. For example, he gives a rather complicated infinite product for the quotient of two integrals,

$$\frac{\int x^{m-1}\,dx(1-x^n)^{p/q}}{\int x^{\mu-1}\,dx(1-x^\nu)^{r/s}}.$$

Euler does not use this result to get more infinite products, though. Instead, he finds equations between quotients of integrals by showing that they expand to the same infinite products. He also gets a cumbersome formula involving an infinite product of integrals, which we omit here.

Besides making us grateful for modern equation editors, this paper has shown us how exhaustively Euler can generalize a result. We have watched Euler extend Wallis's formula to give seventeen new integral formulas that all evaluate to π. Then he finds ways to write those formulas as infinite products. He certainly did not hint that he would discover such results when he first studied infinite products back in E-19.

40

E-123 : De fractionibus continuis observationes*

Observations on continued fractions

In 1737, Euler wrote his now classic article on continued fractions, E-71. It is now 1739 and he has had a few new ideas. Here, he finds and exploits connections between continued fractions and integration. This nicely parallels his results in E-20 and E-25, where he connects series with integration, and those in E-122, where he connects infinite products with integration. Viewed in this light, a connection between integration and continued fractions is not surprising.

Euler is writing longer and longer papers. E-123 took 50 pages of the volume of the *Commentarii* in which it first appeared, and because of its spacious equations, it takes 59 pages of the *Opera Omnia*. By either measure, it is the longest of Euler's papers that we shall see here. Later in his career, Euler will write papers that exceed a hundred pages.

Euler begins with a review of some of the results of E-71. First, he reminds us how a continued fraction can be rewritten as an alternating series of simple fractions. Then he reverses that process to construct a continued fraction from a given alternating series. We will run through this quickly. He starts with a continued fraction

$$A+\cfrac{B}{C+\cfrac{D}{E+\cfrac{F}{G+\cfrac{H}{I+\text{etc.}}}}}$$

This has the same value as the alternating series

$$A+\frac{B}{1P}-\frac{BD}{PQ}+\frac{BDF}{QR}-\frac{BDFH}{RS}+\text{etc.}$$

where $P = C$, $Q = EP + D$, $R = GQ + FP$, $S = IR + HQ$, etc. He did this part using different variable names in E-71. He substitutes in the continued fraction for C, E, G and I in terms of D, F, H, P, Q, R and S, and he assumes $A = 0$ to rewrite the continued fraction as

$$\cfrac{B}{P+\cfrac{DP}{Q-D+\cfrac{FPQ}{R-FP+\cfrac{HQR}{S-HQ+\cfrac{KRS}{\text{etc.}}}}}}$$

Now, he matches terms of this with the alternating series

$$\frac{a}{p}-\frac{b}{q}+\frac{c}{r}-\frac{d}{s}+\frac{e}{t}-\text{etc.}$$

**Comm. Acad. Sci. Imp. Petropol.* 11 (1739) 1750, 32–81; *Opera Omnia* I.14, 291–349

He sees that $B = a$, $D = b : a$, $F = c : b$, $H = d : c$, $K = e : d$, etc., and $P = p$, $Q = q : p$, $R = pr : q$, $S = qs : pr$, $T = prt : qs$, etc. Thus, this last alternating series can be written as a continued fraction

$$\cfrac{a}{p + \cfrac{bp^2}{aq - bp + \cfrac{acqq}{br - cq + \cfrac{bdrr}{cs - dr + \cfrac{cess}{dt - es + \text{etc.}}}}}}$$

Expressions like this make us glad that subscripts have not been invented yet. This is hard enough to type without a forest of subscripts, though subscripts do make the patterns a little easier to see.

It is time for an example. Euler starts with Leibniz's alternating harmonic series for ln 2,

$$1 - \frac{1}{2} + \frac{1}{3} - \frac{1}{4} + \frac{1}{5} - \frac{1}{6} + \text{etc.}$$

which he notes equals $\int \frac{dx}{1+x}$, where the integral is from 0 to 1. The variables match up to give $a = b = c = d = \text{etc.} = 1$ and $p = 1$, $q = 2$, $r = 3$, $s = 4$, etc. This makes

$$\int \frac{dx}{1+x} = \cfrac{1}{1 + \cfrac{1}{1 + \cfrac{4}{1 + \cfrac{9}{1 + \cfrac{16}{1 + \text{etc.}}}}}}$$

In typical form, our next example is one for which we already know the answer. We start knowing that

$$1 - \frac{1}{3} + \frac{1}{5} - \frac{1}{7} + \frac{1}{9} - \text{etc.} = \int_0^1 \frac{dx}{1+x^2} = \frac{\pi}{4}.$$

Taking $a = b = c = d = \text{etc.} = 1$, $p = 1$, $q = 3$, $r = 5$, $s = 7$, etc., we get

$$\int \frac{dx}{1+xx} = \cfrac{1}{1 + \cfrac{1}{2 + \cfrac{9}{2 + \cfrac{25}{2 + \cfrac{49}{2 + \text{etc.}}}}}}$$

We saw this formula in E-71. It is the one Viscount Brounker discovered when he did the first work on continued fractions in the 1650s.

We get four more continued fractions like this that give the values of integrals, all of the general form

$$\int \frac{dx}{1+x^m} = \cfrac{1}{1+\cfrac{1^2}{m+\cfrac{(m+1)^2}{m+\cfrac{(2m+1)^2}{m+\cfrac{(3m+1)^2}{m+\text{etc.}}}}}}$$

for $m = 3, 4, 5$ and 6. He also gives a formula for fractional values of m. Then he gives a related formula

$$\int \frac{x^{n-1}\,dx}{1+x^m} = \cfrac{1}{n+\cfrac{n^2}{m+\cfrac{(m+n)^2}{m+\cfrac{(2m+n)^2}{m+\text{etc.}}}}}$$

Next, we get a more complicated expression for

$$\int \frac{x^{n-1}\,dx}{(1+x^m)^{\mu/\nu}}.$$

Note how closely these integrals are related to the ones we saw in the evaluation of infinite products in the previous article, E-122. We will see in a moment that Euler too has noticed the similarities, but first Euler gives continued fractions for some different series. He starts with those that do not alternate,

$$\frac{a}{p} + \frac{b}{q} + \frac{c}{r} + \frac{d}{s} + \frac{e}{t} + \text{etc.}$$

Then he works with alternating series of the form

$$\frac{b}{p} - \frac{bd}{pq} + \frac{bdf}{qr} - \frac{bdfh}{rs} + \text{etc.}$$

We noted some connections between this article and E-123. In E-71, Euler mentions that Wallis had discovered that

$$a^2 = a-1+\cfrac{1}{2(a-1)+\cfrac{9}{2(a-1)+\cfrac{25}{2(a-1)+\text{etc.}}}} \times a+1$$
$$+\cfrac{1}{2(a+1)+\cfrac{9}{2(a-1)+\cfrac{25}{2(a+1)+\text{etc.}}}}$$

Substitution also gives $(a+2)^2$ as the product of two continued fractions, the first of which is the same as the second factor of a^2. He divides and repeats the expansion for $a+4$, $a+6$, etc. to get

$$a \cdot \frac{a(a+4)(a+4)(a+8)(a+8)(a+12)(a+12)}{(a+2)(a+2)(a+6)(a+6)(a+10)(a+10)(a+14)}\text{etc.}$$
$$= a - 1 + \cfrac{1}{2(a-1) + \cfrac{9}{2(a-1) + \cfrac{25}{2(a-1) + \text{etc.}}}}.$$

Euler makes the connection when he tells us that in E-122, we learned that the product on the left equals

$$a\,\frac{\int x^{a+1}\,dx : \sqrt{1-x^4}}{\int x^{a-1}\,dx : \sqrt{1-x^4}}.$$

He uses colons to represent fractions inside the integrals. He seems to do this just to make typesetting easier.

Now Euler turns his talents to expressing as continued fractions things he has previously expressed as infinite products. He takes as his basic example the interpolation of the series

$$\frac{1}{2} + \frac{1\cdot 3}{2\cdot 4} + \frac{1\cdot 3\cdot 5}{2\cdot 4\cdot 6} + \text{etc.}$$

The formulas Wallis and Brounker discovered arise from interpolating the terms of this series. Euler takes $AB = \frac{1}{2}$, $CD = \frac{3}{4}$, $EF = \frac{5}{6}$, $GH = \frac{7}{8}$, etc., so that he can write the terms in this series in the general form

$$AB + ABCD + ABCDEF + \text{etc.}$$

This leaves a great deal of latitude in the choice of A, C, E, G, etc., so he takes A to be the interpolated value of index $\frac{1}{2}$, B the term of index $\frac{3}{2}$, etc. This, of course, determines the values of B, D, F, H, etc.

Euler gives what seems to be a pretty shaky reason for his next step when he says "from the law of continuation," $BC = \frac{2}{3}$, $DE = \frac{4}{5}$, $FG = \frac{6}{7}$, etc. If we accept this, then it follows that $A = \frac{1}{2B}$, $B = \frac{2}{3C}$, $C = \frac{3}{4D}$, $D = \frac{4}{5E}$, etc. All this makes

$$A = \frac{1\cdot 3\cdot 3\cdot 5\cdot 5\cdot 7}{2\cdot 2\cdot 4\cdot 4\cdot 6\cdot 6}\text{etc.}$$

which is again Wallis's formula. This gives us some confidence in Euler's "law of continuation."

Euler mimics this development by seeking to interpolate a more general series

$$\frac{p}{p+2q} + \frac{p(p+2r)}{(p+2q)(p+2q+2r)} + \frac{p(p+2r)(p+4r)}{(p+2q)(p+2q+2r)(p+2q+4r)} + \text{etc.}$$

Again, he takes A to be the interpolated term of index $\frac{1}{2}$, and he seeks an analytical expression for A. He writes expressions for *AB*, *CD*, *EF*, etc., then gets, by his law of continuation, values for *BC*, *DE*, *FG*, etc.

Some intricate substitutions and expansive power series lead, after two and a half pages, to a value for A,

$$A = \frac{\int y^{p+q-1}\,dy\,(1-y^{2r})^{n-1}}{\int y^{p-1}\,dy\,(1-y^{2r})^{n-1}}.$$

In the course of his calculation, Euler writes

$$A = \frac{a}{p+2q-r}.$$

In this expression, a can be written as a continued fraction,

$$a = p + \cfrac{2p(q-r)}{p+r+\cfrac{(p+2q-r)(p+r)}{r+\cfrac{(p+2q)(p+2r)}{r+\cfrac{(p+2q+r)(p+3r)}{r+\text{etc.}}}}}$$

Euler gives an alternate form of the above expression, involving only positive numerators and denominators, in case $r > q$. In the course of his calculation, Euler introduces a variable $m = p+r$. He now rewrites this latest continued fraction for several special values of m and r, including $m = p+q$, $m = p + 2q$ and $r = 2q$.

Some special values, like $q = \frac{1}{2}$ and $p = r = 1$ give integrals that can be evaluated easily and so lead to special continued fractions like

$$\frac{\pi}{2} = 1 + \cfrac{1}{1+\cfrac{1\cdot 2}{1+\cfrac{2\cdot 3}{1+\cfrac{3\cdot 4}{1+\text{etc.}}}}} = 2 - \cfrac{1}{2+\cfrac{1^2}{2+\cfrac{2^2}{2+\cfrac{3^2}{2+\text{etc.}}}}}$$

We are now a third of the way through this very long article. Euler is going to try to interpolate the terms of a different example, the reciprocals of a series he looked at earlier,

$$\frac{2}{1} + \frac{2\cdot 4}{1\cdot 3} + \frac{2\cdot 4\cdot 6}{1\cdot 3\cdot 5} + \text{etc.}$$

Following the analysis of the other series, he calls A the term of index $\frac{1}{2}$. According to the notation from before, this makes $p = 2$, $r = 1$ and $q = -\frac{1}{2}$. He puts

$$A = \frac{a}{p+2q-r}.$$

Then $A = \frac{a}{0}$, a fact Euler describes as "inconvenient." The reader should not despair, though, for Euler has a trick. He sets $3A = 2Z$. Then $\frac{Z}{2}$ will be the term of index $\frac{1}{2}$ of this new series

$$\frac{4}{3} + \frac{4\cdot 6}{3\cdot 5} + \frac{4\cdot 6\cdot 8}{3\cdot 5\cdot 7} + \text{etc.}$$

This is really just the first series divided by 2, but now $p = 4$, $r = 1$ and $q = -\frac{1}{2}$. This solves the problem, and Euler easily gets Z as a quotient of integrals, then gets a new continued fraction for $\frac{3}{4}\pi$ and the one we saw above for $\frac{\pi}{2}$.

Euler assures us that if we understand this process, then we can use interpolation to find "innumerable continued fractions" whose values depend on "quadratures of curves or integral formulas." We can also handle other forms of integrals. Euler takes $p + 2q - r = f$ and $p + r = h$ in the integral formula above to get

$$r+\cfrac{fh}{r+\cfrac{(f+r)(h+r)}{r+\cfrac{(f+2r)(h+2r)}{r+\text{etc.}}}}$$

$$=\frac{h(f-r)\int y^{h+r-1}\,dy:\sqrt{1-y^{2r}}-f(h-r)\int y^{f+r-1}\,dy:\sqrt{1-y^{2r}}}{f\int y^{f+r-1}\,dy:\sqrt{1-y^{2r}}-h\int y^{h+r-1}\,dy:\sqrt{1-y^{2r}}}.$$

This calculation is not obvious, but this is all Euler gives us by way of explanation. Euler notes that "this equation is always real unless $f = h$." We have not seen Euler call equations "real" before. Here, "real" seems to mean it is not an indeterminate form, for if $h = r$ this equation has the form $0/0$. He tells us that in case $f = h$, we should set $f = h + dw$ and the right-hand side reduces to

$$\frac{1-(h-r)\int\frac{x^{h-1}\,dx}{1+x^r}}{\int\frac{x^{h-1}\,dx}{1+x^r}}.$$

This is really just an application of l'Hôpital's rule, but Euler does not yet have the modern limit notation, so l'Hôpital's rule is formulated as an algorithmic calculation involving differentials.

Euler gets a similar formula in case $p = f$ and $p + 2q - r = h$. This formula works unless $f = h + r$, and, again, he resolves that case by taking $f = h + r + dw$ to get

$$2r+\cfrac{h(h+r)}{2r+\cfrac{(h+r)(h+2r)}{2r+\cfrac{(h+2r)(h+3r)}{2r+\text{etc.}}}}=\frac{h+2h(r-h)\int\frac{x^{h-r}\,dx}{1+x^r}}{-1+2h\int\frac{x^{h-r}\,dx}{1+x^r}}.$$

This yields an interesting consequence in case $h = r = 1$,

$$2+\cfrac{1\cdot 2}{2+\cfrac{2\cdot 3}{2+\cfrac{3\cdot 4}{2+\cfrac{4\cdot 5}{2+\text{etc.}}}}}=\frac{1}{2\ln 2-1}.$$

Euler proposes a lemma, though he does not put it in the now-traditional Theorem–Proof format. He considers an infinite system of equations with the Greek symbols being the unknowns

$$\begin{aligned}\alpha\beta-m\alpha-n\beta-x&=0\\ \beta\gamma-(m+s)\beta-(n+s)\gamma-x&=0\\ \gamma\delta-(m+2s)\gamma-(n+2s)\delta-x&=0\\ \delta\varepsilon-(m+3s)\delta-(n+3s)\varepsilon-x&=0.\end{aligned}$$

Euler develops a system of continued fractions that solves this system. Usually, we skip such tedious and bulky derivations, but this one has a particularly interesting twist. For the first time, we see Euler using subscripts in a relatively modern looking way. His first step is to introduce new variables a, b, c, d, etc. by rewriting his equations as

$$\alpha = m + n - s + \frac{ss - ms + ns + x}{a}$$

$$\beta = m + n + s + \frac{ss - ms + ns + x}{b}$$

$$\gamma = m + n + 3s + \frac{ss - ms + ns + x}{c}$$

$$\delta = m + n + 5s + \frac{ss - ms + ns + x}{d}.$$

This he recasts as

$$ab - (m - s)a + (n + s)b - ss + ms - ns - x = 0$$
$$bc - mb - (n + 2s)c - ss + ms - ns - x = 0$$
$$cd - (m + s)b - (n - 3s)d - ss + ms - ns - x = 0$$
$$de - (m + 2s)b - (n + 4s)e - ss + ms - ns - x = 0$$

where the unknowns are a, b, c, d, etc. This is of the same form as the first system, so he introduces new variables, a_1, b_1, c_1, d_1 etc. by writing

$$a = m + n - s + \frac{4ss - 2ms + 2ns + x}{a_1}$$

$$b = m + n + s + \frac{4ss - 2ms + 2ns + x}{b_1}$$

$$c = m + n + 3s + \frac{4ss - 2ms + 2ns + x}{c_1}$$

$$d = m + n + 5s + \frac{4ss - 2ms + 2ns + x}{d_1}.$$

This, too, he recasts in the form of the original system, which, in turn he solves for a_1, b_1, c_1, d_1, etc. in terms of a_2, b_2, c_2, d_2, etc. Now, Euler chains all these solutions together into a system of continued fractions to get

$$\alpha = m + n - s + \cfrac{ss - ms + ns + x}{m + n - s + \cfrac{4ss - ms + 2ns + x}{m + n - s + \cfrac{9ss - 3ms + 3ns + x}{m + n - s + \cfrac{16ss - 4ms + 4ns + x}{m + n - s + \text{etc.}}}}}$$

$$\beta = m + n + s + \cfrac{ss - ms + ns + x}{m + n + s + \cfrac{4ss - ms + 2ns + x}{m + n + s + \cfrac{9ss - 3ms + 3ns + x}{m + n + s + \cfrac{16ss - 4ms + 4ns + x}{m + n + s + \text{etc.}}}}}$$

$$\gamma = m + n + 3s + \cfrac{ss - ms + ns + x}{m + n + 3s + \cfrac{4ss - ms + 2ns + x}{m + n + 3s + \cfrac{9ss - 3ms + 3ns + x}{m + n + 3s + \cfrac{16ss - 4ms + 4ns + x}{m + n + 3s + \text{etc.}}}}}$$

etc.

Euler aims this new tool at the interpolation of the series where he used his "law of continuation" before

$$\frac{p}{p+2q}+\frac{p(p+2r)}{(p+2q)(p+2q+2r)}+\frac{p(p+2r)(p+4r)}{(p+2q)(p+2q+2r)(p+2q+4r)}+\text{etc.}$$

Again he denotes the term of index $\frac{1}{2}$ by A, that of index $\frac{3}{2}$ by ABC, that of index $\frac{5}{2}$ by $ABCDE$, etc. and he puts

$$AB=\frac{p}{p+2q},\quad CD=\frac{p+2r}{p+2q+2r},\quad EF=\frac{p+4r}{p+2q+4r},\quad \text{etc.}$$

Euler changes variables, with an eye on applying his lemma, by setting

$$A=\frac{a}{p+2q-r},\quad B=\frac{b}{p+2q},\quad C=\frac{c}{p+2q+r},\quad D=\frac{d}{p+2q+2r},\quad \text{etc.}$$

This makes

$$\begin{aligned} ab &= p(p+2q-r)\\ bc &= (p+r)(p+2q)\\ cd &= (p+2r)(p+2q+r)\\ de &= (p+3r)(p+2q+2r)\\ &\text{etc.}\end{aligned}$$

Now, Euler's lemma with the subscripts tells us that

$$a=p+q-r+\cfrac{qr-qq}{2(p+q-r)+\cfrac{2rr+qr-qq}{2(p+q-r)+\cfrac{6rr+qr-qq}{2(p+q-r)+\cfrac{12rr+qr-qq}{2(p+q-r)+\text{etc.}}}}}$$

$$b=p+q+\cfrac{qr-qq}{2(p+q)+\cfrac{2rr+qr-qq}{2(p+q)+\cfrac{6rr+qr-qq}{2(p+q)+\cfrac{12rr+qr-qq}{2(p+q)+\text{etc.}}}}}$$

$$c=p+q+r+\cfrac{qr-qq}{2(p+q+r)+\cfrac{2rr+qr-qq}{2(p+q+r)+\cfrac{6rr+qr-qq}{2(p+q+r)+\cfrac{12rr+qr-qq}{2(p+q+r)+\text{etc.}}}}}$$

and so on for d, e, etc.

There is obviously a pattern here involving $2(p+q+mr)$ in the denominators. Euler is unable to write it out for us because it would require the use of subscripts. Euler has not figured out a way to relate the sequence $m=1,2,3,\ldots$ to the variables a, b, c, etc. He is able to give an integral form for these continued fractions, but we omit it here.

Euler checks his work by noting that $r = 2$ and $q = 1$ gives a result that we saw earlier:

$$s + \cfrac{1}{2s + \cfrac{9}{2s + \cfrac{25}{2s + \cfrac{49}{2s + \cfrac{81}{2s + \text{etc.}}}}}} = (s+1)\frac{\int y^{s+2}\,dy : \sqrt{1-y^4}}{\int y^s\,dy : \sqrt{1-y^4}}.$$

We move to another topic. It seems like a strange jump, but Euler has a reason. We consider the integrals $\int P\,dx$, $\int PR\,dx$, $\int PR^2\,dx$, $\int PR^3\,dx$, $\int PR^4\,dx$, etc., all integrated from 0 to 1. Euler uses a process he describes as being much like "the method of reducing one integral formula to two others." He apparently means what we now call "integration by parts." He gives no details, but claims that there are constants a, b, c and α, β, γ such that

$$a\int P\,dx = b\int PR\,dx + c\int PR^2\,dx$$
$$(a+\alpha)\int PR\,dx = (b+\beta)\int PR^2\,dx + (c+\gamma)\int PR^3\,dx$$
$$(a+2\alpha)\int PR^2\,dx = (b+2\beta)\int PR^3\,dx + (c+2\gamma)\int PR^4\,dx.$$
$$\text{etc.}$$

He rewrites each of these equations to isolate

$$\frac{\int PR^{m-1}\,dx}{\int PR^m\,dx},$$

then back-substitutes to get a continued fraction expansion

$$\frac{\int P\,dx}{\int PR\,dx} = \frac{b}{a} + \cfrac{c:a}{\frac{b+\beta}{a+\alpha} + \cfrac{(c+\gamma):(a+\alpha)}{\frac{b+2\beta}{a+2\alpha} + \cfrac{(c+2\gamma):(a+2\alpha)}{\frac{b+3\beta}{a+3\alpha} + \cfrac{(c+3\gamma):(a+3\alpha)}{\frac{b+4\beta}{a+4\alpha} + \text{etc.}}}}}$$

He simplifies this a bit and then writes a similar continued fraction for

$$\frac{\int PR\,dx}{\int P\,dx}.$$

At last, Euler tells us why he made this excursion. Looking back about 15 pages in his article, we see that he was deriving continued fractions for forms like

$$\frac{(h-r)\int x^{h-r-1}\,dx : (1+x^p)}{\int x^{h-1}\,dx : (1+x^p)}.$$

Euler's new technique applies to such forms, and it gives the same continued fraction expansions we have seen before.

Euler has yet another idea. He is going to convert a series to a power series and then try to relate the continued fraction expansion for the series to an expansion for the power series. He starts with

$$1+\frac{p}{q+s}+\frac{p(p+s)}{(q+s)(q+2s)}+\frac{p(p+s)(p+2s)}{(q+s)(q+2s)(q+3s)}+\text{etc.}$$

He assumes that p, q and s are "affirmative" (no clue why he doesn't call them "positive") and that $q>p$. Then, by a partial fraction analysis, this sums to $=\frac{q}{q-p}$. Euler asks us to consider the related power series

$$z=x^q+\frac{p}{q+s}x^{q+s}+\frac{p(p+s)}{(q+s)(q+2s)}x^{q+2s}+\text{etc.}$$

Euler differentiates, solves for $x^{p-q-s}\,dz$, integrates, does some algebra, and after a while re-writes z as

$$z=\frac{qx^q}{q-p}-\frac{pq(1-x^s)^{(q-p)/s}}{q-p}\int\frac{x^{q-1}\,dx}{(1-x^s)^{(q-p)/s}}.$$

Integrals are always, it seems, from zero to one. We can check the derivation by noting that if $x=1$, then z has the correct value.

Next we spend a couple of sections simplifying forms giving familiar continued fractions as quotients of integrals. It seems an odd time in the paper to do this, because the process does not use any of his tools on series. It would have fit better immediately after he developed the tools about integrals of $P\,dx$, $PR\,dx$, etc. The calculations become extraordinarily detailed and involve wave after wave of substitutions, so that Euler's train of thought is obscured. Eventually, he gets some continued fractions for integral quotients more complicated than any we have seen before, such as

$$\frac{\int x^{g+r-1}\,dx(1-x^r)^{(c-b/r}(p+qx^r)^{(c-a)/r}}{\int x^{g-1}\,dx(1-x^r)^{(c-b)/r}(p+qx^r)^{(c-a)/r}}$$
$$=\cfrac{pg}{ap-bq+\cfrac{pq(c+r)(g+r)}{(a+r)p-(b+r)q+\cfrac{pq(c+2r)(g+2r)}{(a+2r)p-(b+2r)q+\text{etc.}}}}$$

In the course of the calculation, Euler uses an interesting word. He writes, "If the quantities c and g are commuted between themselves, ..." The word "commutative" will not evolve into its modern meaning until well into the next century. In common Latin, the word means "switch places." Euler's readers are familiar with the word in a legal sense, where justice is called commutative if it is fair to both parties and if it would still be balanced if the parties switched roles. [M] Euler's use of the word is mostly in this traditional sense, but, in applying it to switching variables in algebraic expressions, he is doing his part to help the modern meaning evolve.

Euler gives no new values for continued fractions, but he shows that some of these forms lead to expansions involving e and π that we saw back in E-71.

Now, five pages from the end of this long article, he turns to the differential equation of Count Riccati. Euler's continuing interest in the Riccati equation could make a book of its own. This equation has many forms, and this time he considers it as

$$ax^m\,dx+bx^{m-1}y\,dx+cy^2\,dx+dy=0.$$

He puts $y^{m+2} = t$ and $y = \frac{1}{cx} + \frac{1}{xxz}$ and gets

$$\frac{-c}{m+3} t^{(-m-4)/(m+3)}\, dt - \frac{b}{m+3} t^{-1/(m+3)} z\, dt - \frac{ac+b}{(m+3)c} z^2 dt + dz = 0.$$

This has the same general form as the previous equation. Euler further substitutes $t^{(2m+5)/(m+3)} = u$ and

$$z = \frac{-(m+3)c}{(ac+b)t} + \frac{1}{ttv}.$$

Now, he supposes that y can be given as a continued fraction by

$$y = Ax^{-1} + \cfrac{1}{-Bx^{-m-1} + \cfrac{1}{Cx^{-1} + \cfrac{1}{-Dx^{-m-1} + \cfrac{1}{Ex^{-1} + \cfrac{1}{-Fx^{-m-1} + \text{etc.}}}}}}$$

Without giving us any details, he tells us values for A, B, C, etc., starting with $A = \frac{1}{c}$ and $B = \frac{(m+3)c}{ac+b}$ and going to

$$F = \frac{(5m+11)c(ac-(m+2)b)(ac-(2m+4)b)}{(ac+b)(ac+(m+3)b)(ac+(2m+5)b)}.$$

From these values, he gets values for the products *AB*, *BC*, *CD*, *DE*, *EF*, *FG*, etc. And now, the work he did on continued fractions and interpolations pays off, for he knows

$$cxy = 1 + \cfrac{(ac+b)x^{m+2}}{-(m+3) + \cfrac{(ac-(m+2)b)x^{m+2}}{(2m+5) + \cfrac{(ac+(m+3)b)x^{m+2}}{-(3m+7) + \cfrac{(ac-(2m+4)b)x^{m+2}}{(4m+9) + \text{etc.}}}}}$$

This is a rather unexpected result. Who would have realized through all this that Euler was thinking about differential equations? He does similar continued fraction expansions for the solutions to two more differential equations, $ncx^m\, dx + cy^2\, dx + dy = 0$ and $x^{(\beta-2b)/b}\, dx = y^2\, dx + b\, dy$. Finally, we reach the end of this epic paper.

This has been an extremely long and tedious paper. We have skipped over large sections. The enthusiastic reader is again encouraged to try to tackle the Latin original. It is mostly formulas, and the formulas are much more challenging than the Latin.

References

[M] Miller, Jeff, Earliest Known Uses of Some of the Words of Mathematics, `http://members.aol.com/jeff570/mathword.html`, as revised June 24, 2002.

41

E-125: Consideratio progressionis cuiusdam ad circuli quadraturam inveniendam idoneae*

Consideration of some progressions appropriate for finding the quadrature of the circle

We are going to approximate π again. As before, we will learn techniques for estimating π, rather than learning π to some new and larger number of decimal places. For the first time, we will see Euler using special coefficients, the ones we now call Bernoulli numbers, to help him evaluate series. His main tools will be the relations between integrals and series that he developed in E-25 and E-47. At the end of the article, he adds some remarks on divergent series.

We begin with the integral for the arctangent. Euler describes it rather awkwardly as the arc length on a circle of radius 1 of the arc with tangent t given by the integral

$$\int \frac{dt}{1+tt}.$$

It does not bother Euler that he uses the same symbol t both in the integrand and as the upper bound of integration.

We approximate this integral with a finite sum by adding the values of the integrand at each of n points,

$$\frac{t}{n}, \frac{2t}{n}, \frac{3t}{n}, \ldots, \frac{nt}{n}$$

and multiplying by the appropriate interval length. This gives the sum

$$s = \frac{nt}{nn+tt} + \frac{nt}{nn+4tt} + \frac{nt}{nn+9tt} + \cdots + \frac{nt}{nn+nntt}.$$

Euler plans eventually to take n to be an infinite number.

Now, each of these terms can be rewritten as a geometric series.

$$\begin{aligned}
\frac{nt}{n^2+t^2} &= \frac{t}{n} - \frac{t^3}{n^3} + \frac{t^5}{n^5} - \frac{t^7}{n^7} + \text{etc.} \\
\frac{nt}{n^2+4t^2} &= \frac{t}{n} - \frac{2^2t^3}{n^3} + \frac{2^4t^5}{n^5} - \frac{2^6t^7}{n^7} + \text{etc.} \\
\frac{nt}{n^2+9t^2} &= \frac{t}{n} - \frac{3^2t^3}{n^3} + \frac{3^4t^5}{n^5} - \frac{3^6t^7}{n^7} + \text{etc.} \\
&\vdots \\
\frac{nt}{n^2+n^2t^2} &= \frac{t}{n} - \frac{n^2t^3}{n^3} + \frac{n^4t^5}{n^5} - \frac{n^6t^7}{n^7} + \text{etc.}
\end{aligned}$$

---

*_Comm. Acad. Sci. Imp. Petropol._ 11 (1739) 1750, 117–127; _Opera Omnia_ I.14, 350–363

Euler sums the columns here instead of the rows and gets

$$
s = \begin{cases}
+\frac{t}{n}(1^0 + 2^0 + 3^0 + \cdots + n^0) \\
-\frac{t^3}{n^3}(1^2 + 2^2 + 3^2 + \cdots + n^2) \\
+\frac{t^5}{n^5}(1^4 + 2^4 + 3^4 + \cdots + n^4) \\
-\frac{t^7}{n^7}(1^6 + 2^6 + 3^6 + \cdots + n^6) \\
+\frac{t^9}{n^9}(1^8 + 2^8 + 3^8 + \cdots + n^8) \\
-\frac{t^{11}}{n^{11}}(1^{10} + 2^{10} + 3^{10} + \cdots + n^{10}) \\
\text{etc. to infinity.}
\end{cases}
$$

People have long understood how to sum finite series of powers, like the ones that appear on the right-hand side here. Euler himself found new ways to get those formulas way back in E-25, then again in E-47. Euler reminds us of a few of those formulas,

$$1^0 + 2^0 + \cdots + n^0 = n,$$

$$1^2 + 2^2 + \cdots + n^2 = \frac{n^3}{3} + \frac{n^2}{2} + \frac{n}{6}$$

and so forth down to

$$1^8 + 2^8 + \cdots + n^8 = \frac{n^9}{9} + \frac{n^8}{2} + \frac{2n^7}{3} - \frac{7n^5}{15} + \frac{2n^3}{9} - \frac{n}{30}.$$

Euler substitutes these formulas for the sums in his value for s, grouped by powers of t.

$$
s = \begin{cases}
+t \\
-\frac{t^3}{3} - \frac{t^3}{2n} - \frac{t^3}{6n^2} \\
+\frac{t^5}{5} + \frac{t^5}{2n} + \frac{t^5}{3n^3} - \frac{t^5}{30n^4} \\
-\frac{t^7}{7} - \frac{t^7}{2n} - \frac{t^7}{2n^2} + \frac{t^7}{6n^4} - \frac{t^7}{42n^6} \\
+\frac{t^9}{9} + \frac{t^9}{2n} + \frac{2t^9}{3n^2} - \frac{7t^9}{15n^4} + \frac{2t^9}{9n^6} - \frac{t^9}{30n^8} \\
\text{etc.}
\end{cases}
$$

To rearrange these sums grouped by powers of n instead of by powers of t will use what we now call Bernoulli numbers,

$$\frac{1}{6}, \frac{1}{30}, \frac{1}{42}, \frac{1}{30}, \frac{5}{66}, \frac{691}{13 \cdot 210}, \frac{7}{6}, \frac{3617}{17 \cdot 30}, \frac{43867}{19 \cdot 42}, \frac{174611}{330}, \frac{854513}{6 \cdot 23}, \frac{236364091}{5 \cdot 546}.$$

Though these numbers have a variety of interesting properties, here Euler mentions only those related to these series. He tells us that s can be rearranged using these coefficients to be

$$\begin{aligned}
s = &+t - \frac{t^3}{3} + \frac{t^5}{5} - \frac{t^7}{7} + \frac{t^9}{9} - \frac{t^{11}}{11} + \text{etc.} \\
&- \frac{t^2}{2n}\left(t - t^3 + t^5 - t^7 + t^9 - t^{11} + \text{etc.}\right) \\
&- \frac{t^2}{6n^2}\left(t - 2t^3 + 3t^5 - 4t^7 + 5t^9 - 6t^{11} + \text{etc.}\right) \\
&- \frac{t^4}{30n^4}\left(t - 5t^3 + 14t^5 - 30t^7 + 55t^9 - 91t^{11} + \text{etc.}\right) \\
&- \frac{t^6}{42n^6}\left(t - \frac{28}{3}t^3 + 42t^5 - 132t^7 + \frac{1001}{2}t^9 - 728t^{11} + \text{etc.}\right) \\
&\text{etc.}
\end{aligned}$$

The series inside the parentheses are given by the rule

$$t - \frac{(m+1)(m+2)}{2\cdot 3}t^3 + \frac{(m+1)(m+2)(m+3)(m+4)}{2\cdot 3\cdot 4\cdot 5}t^5 - \text{etc.}$$

We now appeal to our knowledge of series expansions. If we take this last sum to be equal to v, and if we take the series to be an infinite one, then

$$\begin{aligned}
mv &= \frac{m}{1}t - \frac{m(m+1)(m+2)}{1\cdot 2\cdot 3}t^3 + \frac{m(m+1)(m+2)(m+3)(m+4)}{1\cdot 2\cdot 3\cdot 4\cdot 5}t^5 - \text{etc.} \\
&= \frac{(1-t\sqrt{-1})^{-m} - (1+t\sqrt{-1})^{-m}}{2\sqrt{-1}} \\
&= \frac{(1+t\sqrt{-1})^{m} - (1-t\sqrt{-1})^{m}}{2(1+tt)^m\sqrt{-1}}.
\end{aligned}$$

In this last expression, if we expand the binomials in the numerators, a great many terms cancel, leaving us with

$$mv = \frac{1}{(1+tt)^m}\left(\frac{mt}{1} - \frac{m(m-1)(m-2)}{1\cdot 2\cdot 3}t^3 + \frac{m(m-1)(m-2)(m-3)(m-4)}{1\cdot 2\cdot 3\cdot 4\cdot 5}t^5 - \text{etc.}\right).$$

We divide by m to get an expression for v, a different v for each value of m. We substitute these many series for the corresponding series in s to get a fantastic series of series,

$$\begin{aligned}
s = &\ t - \frac{t^3}{3} + \frac{t^5}{5} - \frac{t^7}{7} + \frac{t^9}{9} - \frac{t^{11}}{11} + \text{etc.} \\
&- \frac{t^3}{2n(1+tt)} \\
&- \frac{t^2}{2\cdot 6n^2(1+tt)^2}\cdot\frac{2t}{1} \\
&- \frac{t^4}{4\cdot 30n^4(1+tt)^4}\left(\frac{4t}{1} - \frac{4\cdot 3\cdot 2}{1\cdot 2\cdot 3}t^3\right) \\
&- \frac{t^6}{6\cdot 42n^6(1+tt)^6}\left(\frac{6t}{1} - \frac{6\cdot 5\cdot 4}{1\cdot 2\cdot 3}t^3 + \frac{6\cdot 5\cdot 4\cdot 3\cdot 2}{1\cdot 2\cdot 3\cdot 4\cdot 5}t^5\right)
\end{aligned}$$

$$
\begin{aligned}
&-\frac{t^8}{8\cdot 30n^8(1+tt)^8}\left(\frac{8t}{1}-\frac{8\cdot 7\cdot 6}{1\cdot 2\cdot 3}t^3+\frac{8\cdot 7\cdot 6\cdot 5\cdot 4}{1\cdot 2\cdot 3\cdot 4\cdot 5}t^5-\text{etc.}\right)\\
&-\frac{5t^{10}}{10\cdot 66n^{10}(1+tt)^{10}}\left(\frac{10t}{1}-\frac{10\cdot 9\cdot 8}{1\cdot 2\cdot 3}t^3+\text{etc.}\right)\\
&-\frac{69t^{12}}{12\cdot 13\cdot 210n^{12}(1+tt)^{12}}\left(\frac{12t}{1}-\frac{12\cdot 11\cdot 10}{1\cdot 2\cdot 3}t^3+\text{etc.}\right)\\
&-\frac{7t^{14}}{14\cdot 6n^{14}(1+tt)^{14}}\left(\frac{14t}{1}-\frac{14\cdot 13\cdot 12}{1\cdot 2\cdot 3}t^3+\text{etc.}\right)\\
&-\frac{3617t^{16}}{16\cdot 17\cdot 30n^{16}(1+tt)^{16}}\left(\frac{16t}{1}-\frac{16\cdot 15\cdot 14}{1\cdot 2\cdot 3}t^3+\text{etc.}\right)\\
&\text{etc.}
\end{aligned}
$$

It is easy to lose perspective amid these marvelous calculations. Perhaps we ought to pause to see where we are. We defined s as an approximation of the arctangent of t. At first we took s to be a finite sum, but now we also know s as an infinite sum. This last formula gives s as an infinite sum minus infinitely many finite sums.

But look closely. The infinite sum here is just the ordinary power series expansion for the arctangent of t. It is what s is supposed to be approximating. We can use this to get an exact representation of the arctangent of t. Euler lets the arctangent of t be z, and rewrites the sum above as

$$
\begin{aligned}
z=\arctan(t)=s&+\frac{t^3}{2n(1+tt)}\\
&+\frac{1}{6}\cdot\frac{tt}{2nn(1+tt)^2}\cdot 2t\\
&+\frac{1}{30}\cdot\frac{t^4}{4n^2(1+tt)^4}(4t-4t^3)\\
&+\frac{1}{42}\cdot\frac{t^6}{6n^6(1+tt)^6}(6t-20t^3+6t^5)\\
&+\text{etc.}
\end{aligned}
$$

Euler gives this series up to the term of degree 14.

As an example, Euler takes $t=1$, so that $z=\frac{\pi}{4}$. Then, for any positive integer n, he gets

$$
\begin{aligned}
\frac{\pi}{4}=&\frac{n}{n^2+1}+\frac{n}{n^2+4}+\frac{n}{n^2+9}+\frac{n}{n^2+16}+\cdots+\frac{n}{n^2+n^2}\\
&+\frac{1}{4n}+\frac{1}{6}\cdot\frac{1}{2\cdot 2n^2}-\frac{1}{42}\cdot\frac{1}{2^3\cdot 6n^6}+\frac{5}{66}\cdot\frac{1}{2^5\cdot 10n^{10}}-\frac{7}{6}\cdot\frac{1}{2^7\cdot 14n^{14}}\\
&+\frac{43867}{19\cdot 42}\cdot\frac{1}{2^9\cdot 18n^{18}}-\frac{854513}{6\cdot 23}\cdot\frac{1}{2^{11}\cdot 22n^{22}}+\text{etc.}
\end{aligned}
$$

This, he tells us, converges faster for large values of n. Note that the first line of this expression is a *finite* series, but the rest is an infinite series. If the first line were extended to an infinite series, then the series itself would diverge. Issues of convergence and divergence lurk just below the surface here, but Euler, with the skill that characterizes his calculations, avoids falling into these traps.

He demonstrates how fast this series converges by using it to approximate π for various values of n. We summarize his results in the following table. We mark the last correct digit by underlining it.

n	estimate of π
1	$3.\underline{1}646\ldots$
2	$3.14\underline{1}635\ldots$
3	$3.141159\underline{2}7216\ldots$
4	$3.14159265\underline{3}74\ldots$
5	$3.1415926535\,9\underline{0}0726\ldots$
6	$3.1415926535\,8979\underline{3}558$

Euler notes some issues of convergence. For example, if $t > 1$, then the series for z may not converge.

Euler asks to consider t and n to be infinite numbers with $n = pt$, and $z = \frac{\pi}{2}$. His formula for $z = \arctan(t)$ gives us $z = \frac{\pi}{2} = s + \frac{1}{2p}$, which makes

$$s = \frac{p}{p^2+1} + \frac{p}{p^2+4} + \frac{p}{p^2+9} + \frac{p}{p^2+16} + \frac{p}{p^2+25} + \text{etc.}$$

Here, because n is infinite, this is an infinite sum. But also, $s = \frac{\pi}{2} - \frac{1}{2p}$, so

$$\frac{\pi}{2p} - \frac{1}{pp} = \frac{1}{p^2+1} + \frac{1}{p^2+4} + \frac{1}{p^2+9} + \frac{1}{p^2+16} + \text{etc.}$$

This is not defined for $p = 0$, or it would be a new solution to the Basel Problem, but in case $p = 1$, it gives the sum of reciprocals of numbers one more than squares to be

$$\frac{1}{2} + \frac{1}{5} + \frac{1}{10} + \frac{1}{17} + \frac{1}{26} + \text{etc.} = \frac{\pi}{2} - \frac{1}{2}.$$

Now, Euler wants to see if he can say anything about divergent series. He denotes his series

$$\overset{1}{a} + \overset{2}{b} + \overset{3}{c} + \overset{4}{d} + \overset{5}{e} + \overset{6}{f} + \overset{7}{g} + \overset{8}{h} + \text{etc.}$$

where the terms, a, b, c, etc. depend on the index, x. Then he says we can write the term of index x as

$$a + \frac{x-1}{1}(b-a) + \frac{(x-1)(x-2)}{1\cdot 2}(c-2b+a) + \frac{(x-1)(x-2)(x-3)}{1\cdot 2\cdot 3}(d-3c+3b-a) + \text{etc.}$$

Though this particular calculation is mostly forgotten today, in Euler's time it was a well-known manipulation in the calculus of finite differences. It was used by Newton in his *Principia*, and, in some ways, it resembles Taylor's theorem for sequences. Note that for any finite integer value of x, this is a finite sum with x terms.

Euler demonstrates some of the versatility of this expression by using it to express the terms of index 0, -1, -2 and -3. He gets

$$\begin{aligned}
\text{term } 0 &= a + (a-b) + (a-2b+c) + (a-3b+3c-d) + \text{etc.}\\
\text{term } -1 &= a + 2(a-b) + 3(a-2b+c) + 4(a-3b+3c-d) + \text{etc.}
\end{aligned}$$

$$\text{term } -2 = a + 3(a-b) + \ \ 6(a-2b+c) + 10(a-3b+3c-d) + \text{etc.}$$
$$\text{term } -3 = a + 4(a-b) + 10(a-2b+c) + 20(a-3b+3c-d) + \text{etc.}$$

This time, he does not seem interested in taking fractional values of x. Instead, he sums the columns. He considers $a + a + \cdots$ as a geometric series, and gets its sum to be $\frac{a}{1-1}$. Similarly, he calculates that $1 + 2 + 3 + 4 + \cdots$ will be $\frac{1}{(1-1)^2}$, and that the sum of the triangular numbers $1 + 3 + 6 + 10 + \cdots$ will be $= \frac{1}{(1-1)^3}$. Of course, we would not be allowed to do this today.

He sums all these terms to get

$$\frac{a}{1-1} + \frac{a-b}{(1-1)^2} + \frac{a-2b+c}{(1-1)^3} + \frac{a-3b+3c-d}{(1-1)^4} + \text{etc.},$$

which rearranges as

$$\begin{aligned}
&+a\left(\frac{1}{1-1} + \frac{1}{(1-1)^2} + \frac{1}{(1-1)^3} + \frac{1}{(1-1)^4} + \text{etc.}\right)\\
&-b\left(\frac{1}{(1-1)^2} + \frac{2}{(1-1)^3} + \frac{3}{(1-1)^4} + \frac{4}{(1-1)^5} + \text{etc.}\right)\\
&+c\left(\frac{1}{(1-1)^3} + \frac{3}{(1-1)^4} + \frac{6}{(1-1)^5} + \frac{10}{(1-1)^6} + \text{etc.}\right)\\
&-d\left(\frac{1}{(1-1)^4} + \frac{4}{(1-1)^5} + \frac{10}{(1-1)^6} + \frac{20}{(1-1)^7} + \text{etc.}\right)\\
&+\text{etc.}
\end{aligned}$$

This, Euler thinks, is the same as the original sum, because he knows formulas to sum the quantities inside the parentheses. The usual geometric series formula, for example, tells us that the first quantity sums to

$$\frac{\frac{1}{1-1}}{1 - \frac{1}{(1-1)}} = \frac{1}{(1-1)-1} = -1.$$

He works it out so that the quantities inside the parentheses alternate between $+1$ and -1, so the series above becomes

$$-a - b - c - d - \text{etc.}$$

With this, Euler ends the article. Besides the intention announced in the title of pursuing results related to circles, we have seen another use of the Bernoulli numbers and also some of Euler's thoughts on divergent series.

42

E-128: Methodus facilis computandi angulorum sinus ac tangentes tam naturales quam artificiales*

An easy method for computing sines and tangents of angles both natural and artificial

Kurt Vonnegut once wrote a story in which the main character falls into a "chronosynclastic infundibulum." In the story, that means he comes un-stuck in time and space. This Euler paper seems to have fallen into its own chronosynclastic infundibulum. Euler presented it to the St. Petersburg Academy on December 15, 1739, and it was published in the *Commentarii* for that year. He had presented E-130 two months earlier, on October 22, but it did not appear in the *Commentarii* until the volume for 1740. This does not stop Euler from using results from E-130 here in E-128.

Both volumes of the *Commentarii* suffer long publication delays, and will not actually see print until 1750. By then, Euler is working in Berlin, though he is the primary editor of the *Commentarii*. In 1748, Euler publishes one of his best-known and most influential books, the *Introductio in analysin infinitorum*, or *Introduction to the analysis of the infinites*. Many consider it to be the world's first precalculus textbook.

This concerns us here because, in his capacity as editor of the *Commentarii*, Euler slips a reference to the *Introductio* into the opening paragraphs of E-128. Thus, as if it had fallen into Vonnegut's chronosynclastic infindibulum, E-128 appears in the volume for 1739, yet it cites a paper in the volume for 1740, as well as a book that will not appear until 1748. It has come un-stuck in time and space.

Euler opens by warning us that he will be using results from his recent book about the properties of series of the form

$$\frac{1}{1\pm p} \pm \frac{1}{4\pm p} + \frac{1}{9\pm p} \pm \frac{1}{16\pm p} + \frac{1}{25\pm p} \pm \text{etc.}$$

He means the *Introductio*, and he will cite the particular properties as he needs them. He will also use results from his "recent" paper. In this, he means E-130.

The rest of the article will be structured as a series of five problems with solutions and corollaries.

Problem 1. *To find a value for this expression given as the product of an infinite progression*

$$\frac{1+p}{1} \cdot \frac{4+p}{4} \cdot \frac{9+p}{9} \cdot \frac{16+p}{16} \cdot \frac{25+p}{25} \cdot \text{etc.}$$

Euler begins "Solution: Take the product to be $= s$, and then take the logarithms" of both sides. We get

$$\ln s = \ln(1+p) + \ln\left(1+\frac{p}{4}\right) + \ln\left(1+\frac{p}{9}\right) + \ln\left(1+\frac{p}{16}\right) + \ln\left(1+\frac{p}{25}\right) + \text{etc.}$$

Comm. Acad. Sci. Imp. Petropol. 11 (1739) 1750, 194–230; *Opera Omnia* I.14, 364–406

Each of these logarithms expands into a series to give

$$\begin{aligned}\ln s &= +\frac{p}{1} - \frac{p^2}{2} + \frac{p^3}{3} - \frac{p^4}{4} + \frac{p^5}{5} - \frac{p^6}{6} + \text{etc.}\\ &= +\frac{p}{4} - \frac{p^2}{2\cdot 4^2} + \frac{p^3}{3\cdot 4^3} - \frac{p^4}{4\cdot 4^4} + \frac{p^5}{5\cdot 4^5} - \frac{p^6}{6\cdot 4^6} + \text{etc.}\\ &\quad + \frac{p}{9} - \frac{p^2}{2\cdot 9^2} + \frac{p^3}{3\cdot 9^3} - \frac{p^4}{4\cdot 9^4} + \frac{p^5}{5\cdot 9^5} - \frac{p^6}{6\cdot 9^6} + \text{etc.}\\ &\quad + \frac{p}{16} - \frac{p^2}{2\cdot 16^2} + \frac{p^3}{3\cdot 16^3} - \frac{p^4}{4\cdot 16^4} + \frac{p^5}{5\cdot 16^5} - \frac{p^6}{6\cdot 16^6} + \text{etc.}\\ &\quad + \text{etc.}\end{aligned}$$

We differentiate this to get

$$\begin{aligned}\frac{ds}{s\,dp} &= 1 - p + p^2 - p^3 + p^4 - p^5 + \text{etc.}\\ &\quad + \frac{1}{4} - \frac{p}{4^2} + \frac{p^2}{4^3} - \frac{p^3}{4^4} + \frac{p^4}{4^5} - \frac{p^5}{4^6} + \text{etc.}\\ &\quad + \frac{1}{9} - \frac{p}{9^2} + \frac{p^2}{9^3} - \frac{p^3}{9^4} + \frac{p^4}{9^5} - \frac{p^5}{9^6} + \text{etc.}\\ &\quad + \frac{1}{16} - \frac{p}{16^2} + \frac{p^2}{16^3} - \frac{p^3}{16^4} + \frac{p^4}{16^5} - \frac{p^5}{16^6} + \text{etc.}\end{aligned}$$

Each line of this is a geometric series, so it simplifies to give

$$\frac{ds}{s\,dp} = \frac{1}{1+p} + \frac{1}{4+p} + \frac{1}{9+p} + \frac{1}{16+p} + \frac{1}{25+p} + \text{etc.}$$

Euler now reaches into the future to cite a result from E-130 and simplifies this to

$$\frac{ds}{s\,dp} = \frac{\pi\sqrt{p}-1}{2p} + \frac{\pi\sqrt{p}}{p(e^{2\pi\sqrt{p}-1})}.$$

Euler suggests we substitute $p = qq$, so that $dp = 2q\,dq$. Then he integrates to get

$$\ln s = \ln C - \pi q - \ln q + \ln\bigl(e^{2\pi q} - 1\bigr)$$

where C is a constant of integration. Thus

$$s = \frac{C(e^{2\pi q} - 1)}{e^{\pi q} q} = \frac{C(e^{2\pi\sqrt{p}} - 1)}{e^{\pi\sqrt{p}}\sqrt{p}}.$$

He knows that if $p = 0$ or $q = 0$, then $s = 1$, uses this to find that $C = \frac{1}{2\pi}$, and concludes that the product is given by

$$s = \frac{e^{2\pi\sqrt{p}} - 1}{2e^{\pi\sqrt{p}}\pi\sqrt{p}}.$$

Q. E. I.

This problem has six corollaries. Among them is

Corollary 1. *If in place of p is put 4p, then the expression is*

$$\frac{1+4p}{1}\cdot\frac{1+p}{1}\cdot\frac{9+4p}{9}\cdot\frac{4+p}{4}\cdot\frac{25+4p}{25}\cdot\frac{9+p}{9}\cdot\text{etc.}=\frac{e^{4\pi\sqrt{p}}-1}{4e^{2\pi\sqrt{p}}\pi\sqrt{p}}.$$

The other corollaries mostly exploit the observation that the original product is embedded as the even numbered factors in this expression.

Problem 2. *To find a value for this expression given as the product of an infinite progression*

$$\frac{1-p}{1}\cdot\frac{4-p}{4}\cdot\frac{9-p}{9}\cdot\frac{16-p}{16}\cdot\frac{25-p}{25}\cdot\frac{36-p}{p}\cdot\text{etc.}$$

The solution closely parallels the solution to problem 1 and leads to the conclusion that this product is

$$\frac{\sin\left(\pi\sqrt{p}\right)}{\pi\sqrt{p}}.$$

The corollaries are also similar. They include a product for the cosine that will be useful later:

$$\frac{1-4p}{1}\cdot\frac{9-4p}{9}\cdot\frac{25-4p}{25}\cdot\text{etc.}=\cos\pi\sqrt{p}.$$

Quotients of these two results lead, of course, to products for tangents, and we get our first concrete results that reflect the goals Euler stated in his title, to study sines and tangents.

We get a quick review of the sines, cosines and tangents of the basic angles π, $\frac{1}{2}\pi$, $\frac{1}{3}\pi$ and $\frac{1}{4}\pi$. We note that Euler calls these functions sin. A., cos. A. and tang. A., where the A.s emphasize that he is applying the functions to the arcs rather than to the angles.

In his product formula from Problem 2, Euler takes $p=\frac{m^2}{n^2}$ to get

$$\frac{n^2-m^2}{n^2}\cdot\frac{4n^2-m^2}{4n^2}\cdot\frac{9n^2-m^2}{9n^2}\cdot\frac{16n^2-m^2}{16n^2}\cdot\text{etc.}=\frac{n\sin\frac{m}{n}180^\circ}{m\pi}.$$

We solve this for π and take various values of m and n to get infinite products that give π. Euler's first example is $m=1$, $n=2$, which gives Wallis's formula

$$\begin{aligned}\pi&=2\cdot\frac{4}{3}\cdot\frac{16}{15}\cdot\frac{36}{35}\cdot\frac{64}{63}\cdot\frac{100}{99}\cdot\frac{144}{143}\cdot\text{etc.}\\&=2\cdot\frac{2\cdot2\cdot4\cdot4\cdot6\cdot6\cdot8\cdot8\cdot10\cdot10\cdot12\cdot12}{1\cdot3\cdot3\cdot5\cdot5\cdot7\cdot7\cdot9\cdot9\cdot11\cdot11\cdot13}\text{etc.}\end{aligned}$$

He gets other products by taking $m=1$ and $n=3$, $m=1$ and $n=4$ and finally $m=1$ and $n=6$.

These products all converge slowly, so they are not of much practical use. However, in the 18th century, people often used logarithms to aid in multiplication, so the logarithm of π is quite practical. Euler takes the log of both sides of his last product, the one for $m=1$ and $n=6$, to get a twelve decimal place value for $\ln\pi$.

It is almost time to calculate sines and tangents. To prepare for this, Euler recasts his results so that they give the trigonometric formulas directly. He takes $\pi=2q$, so that, as he says it, "q is the arc of 90 degrees." Then he gives products for sin. A., cosec. A., cos. A., sec. A., tang. A., and cot. A. The one for sin. A. is, for example,

$$\sin \frac{m}{n} q = \frac{m}{n} q \cdot \frac{4n^2 - m^2}{4n^2} \cdot \frac{16n^2 - m^2}{16n^2} \cdot \frac{36n^2 - m^2}{36n^2} \cdot \frac{64n^2 - m^2}{64n^2} \cdot \text{etc.}$$

Problem 3. *To find a general method for finding the sine and the cosine of whatever given angle.*

Solution. In a completely unexpected turn of events, Euler does not use his carefully derived product formulas for the trigonometric functions. Instead, he solves Problem 3 using ordinary Taylor series. He gives the formulas

$$\sin . A. \frac{m}{n} q = \frac{m}{n} q - \frac{m^3}{n^3} \cdot \frac{q^3}{1 \cdot 2 \cdot 3} + \frac{m^5}{n^5} \cdot \frac{q^5}{1 \cdot 2 \cdot 3 \cdot 4 \cdot 5} - \frac{m^7}{n^7} \cdot \frac{q^7}{1 \cdot 2 \cdot 3 \cdot 4 \cdot 5 \cdot 6 \cdot 7} + \text{etc.},$$

and similarly for cosine. Then he gives q to thirty decimal places and gives

$$\begin{aligned}
\sin \frac{m}{n} 90^\circ = \\
&+ \frac{m}{n} \cdot 1.57179\,63267\,94896\,61923\,13216\,916 - \frac{m^3}{n^3} \cdot 0.64596\,40975\,06246\,25365\,57565\,639 \\
&+ \frac{m^5}{n^5} \cdot 0.07969\,26262\,46167\,04512\,05055\,495 - \frac{m^7}{n^7} \cdot 0.00468\,17541\,35348\,68810\,06854\,639 \\
&+ \frac{m^9}{n^9} \cdot 0.00016\,04411\,84787\,35982\,18726\,609 - \frac{m^{11}}{n^{11}} \cdot 0.00000\,35988\,43235\,21208\,53404\,585 \\
&+ \frac{m^{13}}{n^{13}} \cdot 0.00000\,00569\,21729\,21967\,92681\,178 - \frac{m^{15}}{n^{15}} \cdot 0.00000\,00006\,68803\,51098\,11467\,232 \\
&+ \frac{m^{17}}{n^{17}} \cdot 0.00000\,00000\,06066\,93573\,11061\,957 - \frac{m^{19}}{n^{19}} \cdot 0.00000\,00000\,00043\,77065\,46731\,374 \\
&+ \frac{m^{21}}{n^{21}} \cdot 0.00000\,00000\,00000\,25714\,22892\,760 - \frac{m^{23}}{n^{23}} \cdot 0.00000\,00000\,00000\,00125\,38995\,405 \\
&+ \frac{m^{25}}{n^{25}} \cdot 0.00000\,00000\,00000\,00000\,51564\,552 - \frac{m^{27}}{n^{27}} \cdot 0.00000\,00000\,00000\,00000\,00181\,240 \\
&+ \frac{m^{29}}{n^{29}} \cdot 0.00000\,00000\,00000\,00000\,00000\,551 - \frac{m^{31}}{n^{31}} \cdot 0.00000\,00000\,00000\,00000\,00000\,001
\end{aligned}$$

This is only the first of almost thirty such exact calculations that will appear in this paper. After this, we will only describe them and not give them explicitly. The coefficients, given here to 28 decimal places, are just the value of q divided by the appropriate factorial number.

Euler gives us the corresponding expression for the cosine, then does a calculation to show that, for $\frac{m}{n} = 1$ the first formula gives the correct value for both the sine and the cosine to 27 decimal places. He demonstrates the method by taking $\frac{m}{n} = \frac{1}{10}$ to find the sine and the cosine of 9° to 28 decimal places.

Problem 4. *To find a general method for finding the tangent and the cotangent of any angle.*

Euler's solution contains no surprises. He again adapts a Taylor series expansion and gives a page-long formula, this time to only 13 decimal places, for the tangent and cotangent functions. He verifies that the expression gives the tangent of 45° as 1, this time exactly, to 13 decimal places.

Now we move to the last problem in the paper. We get to use all those product formulas we worked so hard to derive.

Problem 5. *To find the natural logarithm of the sine or the cosine of any given angle.*

Solution. Euler reminds us of the product formula for the sine that we developed earlier.

$$\sin \frac{m}{n} q = \frac{m}{n} q \cdot \frac{4n^2 - m^2}{4n^2} \cdot \frac{16n^2 - m^2}{16n^2} \cdot \frac{36n^2 - m^2}{36n^2} \cdot \text{etc.}$$

We take logarithms, then expand those into series to get

$$\begin{aligned} \ln \sin \frac{m}{n} q = \ln q + \ln \frac{m}{n} &- \frac{m^2}{4n^2} \left(1 + \frac{1}{4} + \frac{1}{9} + \frac{1}{16} + \text{etc.} \right) \\ &- \frac{m^4}{2 \cdot 4^2 n^4} \left(1 + \frac{1}{4^2} + \frac{1}{9^2} + \frac{1}{16^2} + \text{etc.} \right) \\ &- \frac{m^6}{3 \cdot 4^3 n^6} \left(1 + \frac{1}{4^3} + \frac{1}{9^3} + \frac{1}{16^3} + \text{etc.} \right) \\ &- \frac{m^8}{4 \cdot 4^4 n^8} \left(1 + \frac{1}{4^4} + \frac{1}{9^4} + \frac{1}{16^4} + \text{etc.} \right) \\ &- \text{etc.} \end{aligned}$$

Euler has estimated the series on the right in E-25, but since then he has calculated them far more exactly. He gives another formula for the $\ln \sin \frac{m}{n} 90°$, this time using terms all the way to degree 32, with coefficients to 22 decimal places. He checks this on a value he knows, the log of the sine of 45°.

He does the corresponding calculations for the cosine, using terms to degree 38 and giving 20 decimal places. In this formula the editors of the *Opera Omnia* have identified eight errors in the 19th and 20th decimal places. They note and correct a number of other errors elsewhere.

Euler does similar calculations to find the tangent, with terms to degree 28 and giving 14 decimal places. Euler tests this last formula and finds that log tan $45° = 1$ to all visible decimal places.

We see that in this paper Euler has delivered more than the title promised, giving cosines and logarithms of trigonometric functions in addition to the sines and tangents he promised. For the first time, Euler has shown the details of some of his very exact calculations. Except for the tedium of those calculations, this has been a pleasant paper, with his clearly explained manipulation of series. Also, it has been amusing to see how a paper Euler wrote in St. Petersburg, edited in Berlin, then published in St. Petersburg, dated 1739, can refer to another paper dated 1740 and a book dated 1748. Indeed, perhaps the paper did come unstuck in time and space, just as Vonnegut described.

43

E-129: Investigatio curvarum quae evolutae sui similes producunt*

Investigation of curves which produce evolutes that are similar to themselves

This author was at a mathematics conference recently when a graduate student asked him, "Grandpa, what's an evolute?" Evolutes have completely vanished from the calculus curriculum. In 1696, though, evolutes and their companion curves, involutes, were so important that L'Hôpital [L'H] devoted an entire chapter of his ten chapter *Analyse des infiniment petits* to the topic. Let's learn about evolutes and involutes in L'Hôpital's own words:

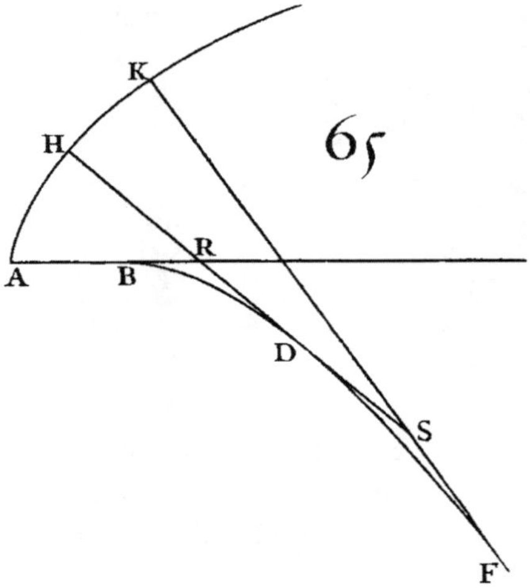

> If one imagines that an arbitrary curved line *BDF* (Figure 65), curving always in the same direction is enveloped or wrapped by a string *ABDF*, with one end of the string attached at *F* and the other end pulled along the tangent *BA*, and if one moves the end *A*, always holding the string taut along the curve *BDF*; then it is clear that the end *A* of the string will describe in its movement a curved line *AHK*. This done, the curve *BDF* will be called the *involute* (*dévelopée*) of the curve *AHK*. The straight parts of the string *AB*, *HD*, and *KF* are called the *radii of the involute*.

By Euler's time, the segment *AB* had been eliminated from the definition of the involute, and the endpoint of the string that traces the involute begins the process as a point on the curve itself rather than a point on a tangent line.

Involutes and evolutes are closely related. If *AHK* is the involute of the curve *BDF*, then we say that *BDF* is the evolute of the curve *AHK*. As we will see later, the construction of an evolute involves the fact that a radius of the involute is perpendicular to the involute *AHK* and tangent to the

**Comm. Acad. Sci. Imp. Petropol.* 12 (1740) 1750, 3–52; *Opera Omnia* I.27, 130–180

evolute *BDF* and that the length of the radius is related both to the arc length of the evolute and the radius of curvature of the involute.

Euler is interested in a problem about evolutes that G. W. Krafft had posed in 1727. Because Euler wrote this paper in 1740, but it only appeared in print in 1750, it is not surprising that the paper was mostly ignored when it finally did appear. Among over 3000 letters to and from Euler that survive today, not a single one of them mentions E-129, so small was its impact in its own time.

Krafft had asked which curves were their own evolutes and had shown that such curves include logarithmic spirals and cycloids. Euler, typically, extends the problem. Suppose A is a curve with evolute B. If the evolute of B is C, then he calls C the second evolute of A, and so forth. Krafft was interested in when a curve was similar to its first evolute. Euler is also interested in curves that are similar to their higher evolutes.

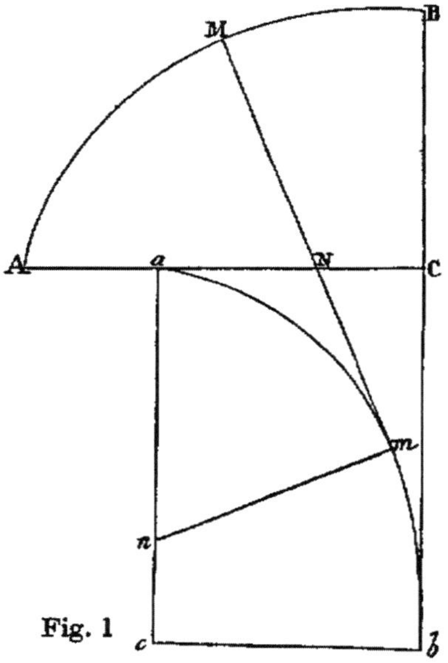

Fig. 1

We start with first evolutes. In Figure 1, the first curve is *AMB*, and its evolute is *amb*. There is some structure here that Euler expects us to know. The points A, M, and B on the original curve correspond to points a, m, and b respectively on the evolute in two separate ways. The first way is by the construction of the evolute. The line segment from a point on the first curve to its corresponding point on the evolute is perpendicular to the original curve and tangent to the evolute. This alone is enough to characterize evolutes. Evolutes have some important arc length properties. For example, the length of the arc *am* is equal to the difference between the lengths of the segments *Mm* and *Aa*. Also, the length *Mm* is the radius of curvature of the curve *AMB* at the point M. Equivalently, m is the center of the osculating circle of *AMB* at M.

The second way that points on *AMB* correspond to points on the evolute is by the similarity relation. In this relation, corresponding points are called *homologs*, and the correspondence is a *homology*. In Figure 1, these two correspondences are the same, but we will see in Figure 2 that this is not necessarily the case.

Euler calls the curve *AMB* the "curve being sought" and begins his analysis by extending *Aa* to *AC* and letting N be the point where *Mm* intersects *AC*. He constructs *ac* perpendicular to *AC* at a

and mn perpendicular to amb at m, with n being on ac. Then, for the curves *AMB* and *amb* to be similar, it is necessary that the two angles *ANM* and *anm* be equal.

We call this problem, finding a pair of curves for which the homology correspondence is the same as the evolute correspondence, the problem of *direct similarity*. Its complementary problem, finding a pair of curves for which the homology correspondence goes in the opposite direction as the evolute correspondence, is the problem of *inverse similarity*.

Euler discusses the inverse problem in Figure 2. Here, curve *bema* is the evolute of the curve, *AEMB*, and again the two are supposed to be similar. This time, though, Euler is thinking that one is a reflection of the other. This leaves Euler with a decision to make on his choice of notation in Figure 2. Either his homology relationships are all scrambled up, or corresponding points in the evolute relation will be scrambled. He finds a way, though, to scramble them both at once.

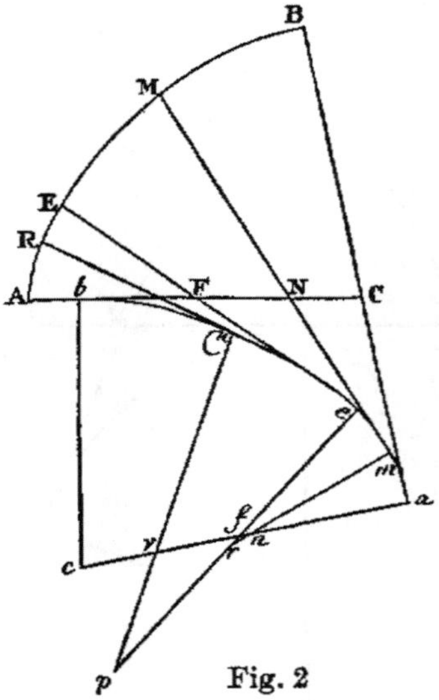

Fig. 2

The point C is where the normal to *AEMB* at B intersects the normal at A. The point c is similarly defined. Euler uses the lines *AC* and *ac* as axes. If the two curves are to be similar, then the two angles *bca* and *ACB* should be equal. Also segments should be proportional, so $AC : BC = ac : bc$.

Now we draw *Mm*, a radius of the osculating circle to *AEMB* at M. It will be tangent to *bema* at m and normal to *AEMB* at M. Draw the normal, *mn* to *bema* at m, intersecting *ca* at n. We have the angle formula $\angle ANM + \angle anm = ACB$.

On the curve *ab*, find the point μ (it looks a bit like a C in Euler's figure) and ν where the normal to *ab* at μ intersects *ac* such that $\angle a\nu\mu + \angle amn = \angle acb$. The point μ is a kind of complement to m. There will be some point for which these two will coincide, and we call that e. If f is the point where the normal to e intersects *ac*, then $\angle ace = 2\angle efa$. Let E be the point on *AEMB* that is homologous to the point e. Moreover, the points e and E will also be connected by the evolute relation; that is, the segment *eE* will be tangent to the curve *ab* at e and normal to the curve *AB* at E.

Extend *ef* as necessary and let p be the point where $\mu\nu$ intersects *ef*, and let r be the point where mn intersects ef. Note that in Euler's Figure 2, it is not entirely clear that the points f, n, and r are distinct, though they indeed are. The angles *mre* and μpe will be equal. There is a sense in which the arcs $e\mu$ and *em* are related as well, because their normals intersect in equal angles. In 1705, Johann Bernoulli, according to Euler, called this kind of relation *aeque ampli*, or "equally wide."

This property, Euler explains, is the one that makes the point homologous to m be the one connected to μ by the evolute relation, and vice versa.

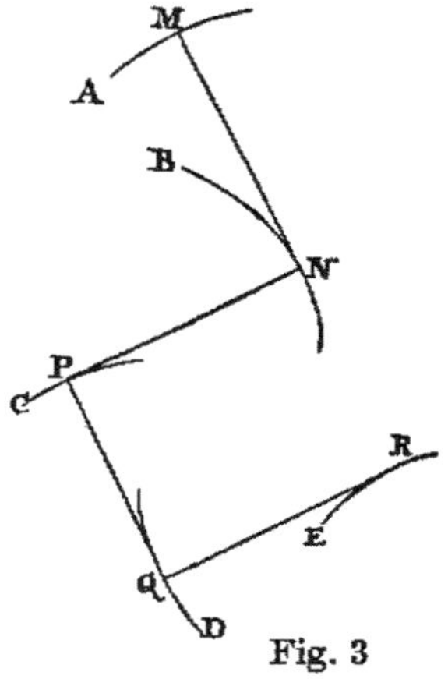

Fig. 3

Now we will look at higher order evolutes a bit. In Figure 3, *AM* is any curve. *BN* is the evolute of *AM*. *CP* is the evolute of *BN*, and so is the second evolute of *AM*, and so on to the third evolute *DQ* and the fourth evolute *ER*. Let the arc length of *AM* be s and the radius of curvature *MN* be $= r$. Then, for the first evolute, the arc $BN = a \pm r$. The sign depends on the orientation of the curve *BN*. In Figure 3, $BN = a + r$. Note that Euler does not tell us exactly what this new value a is, just as he does not tell us later when he introduces b, c, d, etc. in analogous contexts. In general, they are constants of integration, and for the given illustrations they usually turn out to be the radii of curvature at A, B, C, etc.

In Euler's day, people are quite familiar with the properties of evolutes and the radius of curvature, so Euler does not need to explain his claim that $NP = \frac{r\,dr}{ds}$. Moving on to the second evolute, these facts combine to give us that

$$\text{arc } CP = b - \frac{r\,dr}{ds}.$$

Going on to the third evolute, this gives us the radius of curvature

$$PQ = \frac{r}{ds} d \cdot \frac{r\,dr}{ds} \quad \text{and so} \quad DQ = c + \frac{r}{ds} d \cdot \frac{r\,dr}{ds}.$$

Then its radius of curvature is

$$QR = \frac{r}{ds} d \cdot \frac{r}{ds} d \cdot \frac{r\,dr}{ds}.$$

In case we still do not see the pattern, Euler tells us that the arc length

$$ER = e + \frac{r}{ds} d \cdot \frac{r}{ds} d \cdot \frac{r\,dr}{ds}$$

and its radius of curvature is

$$\frac{r}{ds} d \cdot \frac{r}{ds} d \cdot \frac{r}{ds} d \cdot \frac{r\,dr}{ds}.$$

Euler now has his tools lined up, so he poses his first problem. Referring to Figure 1, he asks "to find a curve AMB that is directly similar to its first evolute amb."

We follow the notation used in Figure 1, with AMB the curve being sought, amb its evolute, arc length $AM = s$, and the radius of curvature at $M = r$. Moreover, we can assume that the radius of curvature is less at A than it is at B, for, if it is not, we just start at B instead of at A. As he often does, Euler uses the symbol a in two ways when he asks us to let the length $Aa = a$. He also uses n again when he sets $am = ns$. This could be confusing. The value am is an arc length on the evolute, while s is an arc length on the original curve and n is the proportionality constant between the curve and its evolute to which it is assumed to be similar. Likewise, the radius of curvature at m is nr. By the elementary properties of evolutes, $a + ns = r$ and so $nr = \frac{r\,dr}{ds}$. For the curve AM being sought, $s = \frac{r-a}{n}$. Euler says that because the starting point of the curve is arbitrary, we can choose it with $a = 0$, so that $s = \frac{r}{n}$ and $r = ns$.

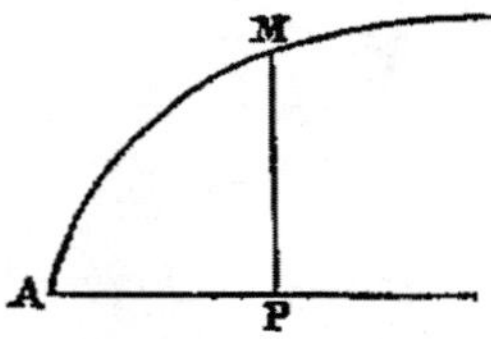

Fig. 4

We introduce a system of orthogonal coordinates (Figure 4). Let AMB be the curve being sought, $AP = x$, $PM = y$, arc length $AM = s$, so that $dx^2 + dy^2 = ds^2$. Let $dx = p\,ds$. Then $dy = ds\sqrt{1-pp}$ and the radius of curvature is $r = \frac{ds\sqrt{1-pp}}{dp}$. Be a little careful here, because we are used to having derivatives with respect to x, and here they are with respect to p and to s as well. Now, $r = ns$, so this last expression transforms to

$$\frac{n\,dp}{\sqrt{1-pp}} = \frac{ds}{s},$$

which integrates to give $n \arcsin p = \ln \frac{s}{a}$, where a is another constant of integration. Then

$$p = \sin\left(\frac{1}{n}\ln\frac{s}{a}\right) \quad \text{and} \quad \sqrt{1-pp} = \cos\left(\frac{1}{n}\ln\frac{s}{a}\right).$$

From this we get

$$dx = ds\sin\left(\frac{1}{n}\ln\frac{s}{a}\right) \quad \text{and} \quad dy = ds\cos\left(\frac{1}{n}\ln\frac{s}{a}\right).$$

Euler pauses to state a couple of lemmas about the integrals

$$\int ds\sin\int\frac{ds}{\sqrt{\alpha+\beta ss}} \quad \text{and} \quad \int ds\cos\int\frac{ds}{\sqrt{\alpha+\beta ss}}.$$

What the lemmas say will be clear when we use them.

In our problem,

$$\frac{1}{n}\ln\frac{s}{a} = \int\frac{ds}{ns}$$

and, taking $\alpha = 0$, $\beta = nn$, we see that Euler's lemmas give

$$x = \int ds \sin \int \frac{ds}{ns} = \frac{nns}{1+nn} \sin\left(\frac{1}{n}\ln\frac{s}{a}\right) - \frac{ns}{1+nn}\cos\left(\frac{1}{n}\ln\frac{s}{a}\right)$$

and

$$y = \int ds \cos\left(\frac{1}{n}\ln\frac{s}{a}\right) = \frac{nns}{1+nn}\cos\left(\frac{1}{n}\ln\frac{s}{a}\right) + \frac{ns}{1+nn}\sin\left(\frac{1}{n}\ln\frac{s}{a}\right).$$

Euler asks that we choose our coordinate system so that the constants of integration here will both be zero.

Through the magic of algebra, it turns out that

$$xx + yy = \frac{nnss}{1+nn},$$

where $xx + yy$ is the square of the chord AM subtended by the arc of length s.

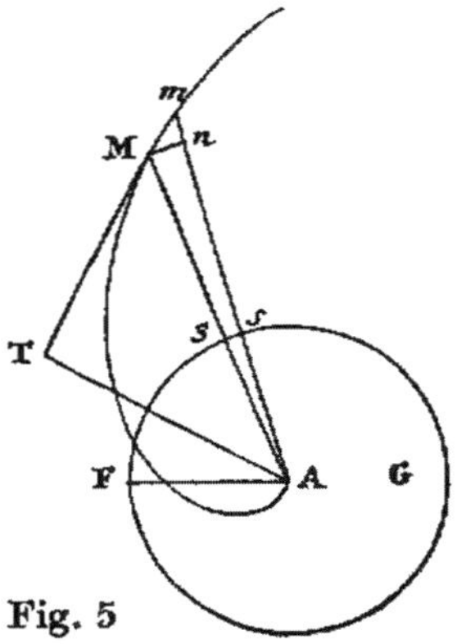

Fig. 5

A while back we found $dx = ds \sin\left(\frac{1}{n}\ln\frac{s}{a}\right)$ and $dy = ds\cos\left(\frac{1}{n}\ln\frac{s}{a}\right)$. These give $\frac{dx}{ds}$ and $\frac{dy}{ds}$ which, substituted into the long expressions for x and y and then simplified, yield

$$x\,ds = \frac{nns\,dx - ns\,dy}{1+nn} \quad \text{and} \quad y\,ds = \frac{nns\,dy + ns\,dx}{nn+1}.$$

Dividing these gives

$$\frac{x}{y} = \frac{n\,dx - dy}{n\,dy + dx}$$

or

$$nx\,dy + x\,dx = ny\,dx - y\,dy$$

or

$$n(y\,dx - x\,dy) = x\,dx + y\,dy.$$

Dividing by $xx + yy$ and integrating yields

$$n \arctan\frac{x}{y} = \ln\frac{\sqrt{x^2+y^2}}{a},$$

where a is yet another constant of integration.

Euler does not have polar coordinates, so he does a bit of work to show that this equation describes the logarithmic spiral. Those of us who do polar coordinates will recognize this easily.

Euler now poses his second problem, "to find a curve *AMB* (Figure 2) which is inversely similar to its evolute, *bma*."

We begin with that special point E on AB which is homologous to e, the opposite end of the radius of curvature at that point. Let the radius of curvature, $Ee = a$, yet another new value for a. For M a point on AB, let the arc length $EM = s$ and its radius of curvature $Mm = r$.

Then $em = r - a$. Let the "ratio of similitude" be n, as it was before. Then arc $e\mu = a - R$, $em = nS$ and $m = nR$.

The analysis parallels what we saw above. We find that $S = \frac{r-a}{n}$ and $R = a - ns$. We note that *EM* and *ER* are "equally wide." This leads, as before, to the differential equation

$$\frac{dS}{R} = \frac{dr}{na - nns} = \frac{ds}{r}.$$

This makes $r\,dr = na\,ds - nns\,ds$. Integrating this gives $rr = 2nas - nnss + aa$. Euler takes $s = 0$, which yields $r = a$, and then substitutes ns for $ns - a$. Then the curve becomes

$$rr = 2aa - nnss, \quad \text{or} \quad r = \sqrt{2aa - nnss}.$$

This solves the problem, but as before the answer is not in orthogonal coordinates so it is not easy to recognize what curve is being described here. To clarify this, we make the same substitutions as in Figure 4, namely $AP = x$, $PM = y$, $dx = p\,ds$, $dy = ds\sqrt{1-pp}$, and $r = \frac{ds\sqrt{1-pp}}{dp}$.

This time, our first differential equation will be

$$\frac{dp}{\sqrt{1-pp}} = \frac{ds}{\sqrt{2aa - nnss}},$$

which, integrated, gives

$$\arcsin p = \int \frac{ds}{\sqrt{2aa - nnss}} = \frac{1}{n}\arcsin\frac{ns}{a\sqrt{2}}.$$

This is difficult to solve unless n is an integer. In case $n = 1$, Euler finds that this gives the "common cycloid," a curve that turns out to be the solution to a great many of the problems Euler considers.

Otherwise, we work with

$$p = \sin\int\frac{ds}{\sqrt{2aa + nnss}} \quad \text{and} \quad \sqrt{1-pp} = \cos\int\frac{ds}{\sqrt{2aa - nnss}}.$$

Translating this into orthogonal coordinates makes

$$dx = ds\,\sin\int\frac{ds}{\sqrt{2aa - nnss}} \quad \text{and} \quad dy = ds\,\cos\int\frac{ds}{\sqrt{2aa - nnss}}.$$

Euler's lemmas about integrals apply here too, except, conveniently, in the case $n = 1$. Euler has dealt with that case separately, though. Here as before, he integrates to find x and y, then calculates $xx + yy$ to find that

$$xx + yy = \frac{n^4ss + 2aa - nnss}{(nn-1)^2}.$$

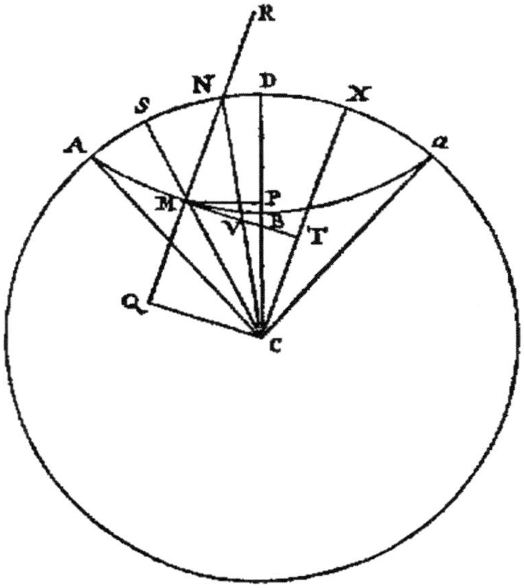

Fig. 6

Again, the left side of this equation suggests polar coordinates. Euler uses Figure 6 to try to untangle the right side. He takes a circle with center C and lets the curve be $AMBa$. He takes MT a tangent at M, with CT perpendicular to MT. To assign variables, let $CM = \sqrt{xx+yy} = z$, $CT = p$ and $MT = t = \sqrt{zz-pp}$. Then $ds = \frac{z\,dz}{t}$.

From this, after a good deal of calculation, Euler finds that the curve is a hypocycloid, if initial conditions put it inside the circle and an epicycloid if it is outside the circle. This solves Euler's second problem, so he moves on to the third.

We now seek curves that are similar to their second evolutes. Again, the similarity can be either direct or inverse, and he chooses first "to find curves which are directly similar to their second evolutes." We refer to Figure 7, where AM is the curve being sought, BN is its first evolute, and am is the second evolute. We note that from B to N, the radius of curvature of BN might increase, as shown in the figure, or it might decrease. Thus two different kinds of solutions arise. Euler first considers the case in which the radius of curvature is increasing.

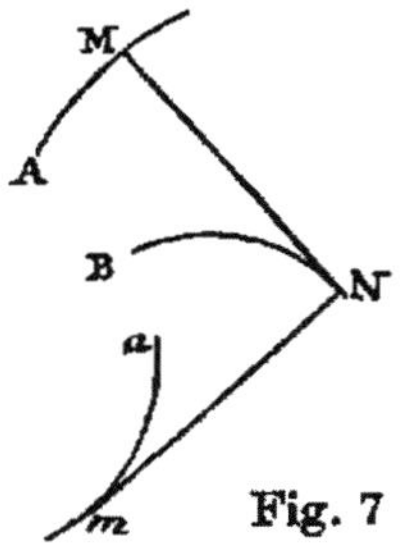

Fig. 7

Take arc length $AM = s$ and the radius of curvature $MN = r$. Again $BN = r - a$, and the radius of curvature of the first evolute will be $Nm = \frac{r\,dr}{ds}$. The arc length of the second evolute will be $am = \frac{r\,dr}{ds} - b$ and, as we saw back with Figure 3, at the point m, its radius of curvature will be $\frac{r}{ds} d \cdot \frac{r\,dr}{ds}$, where the "$d\cdot$" in the middle of the expression means we are to take the differential of the term that follows. Let the ratio of similarity be n, so that $am = ns$ and at m the radius of curvature will be nr. Thus, we have arc length and radius of curvature two ways each, so we get

$$\frac{r\,dr}{ds} - b = ns \quad \text{and} \quad nr = \frac{r}{ds}d \cdot \frac{r\,dr}{ds}.$$

Because the position of A on the curve is arbitrary, we can locate A so that $b = 0$ and get $\frac{r\,dr}{ds} = ns$. Integrating gives $rr = nss + aa$, where a is the radius of curvature at the starting point A.

If $a = 0$, Euler tells us that this all turns out to be another logarithmic spiral. This should not surprise us, for if a curve is similar to its evolute, it should be similar to its second evolute as well.

So, taking $a \neq 0$, we define our variables as we did in Figure 4—$AP = x$, $PM = y$, etc.—and find that

$$\frac{dp}{\sqrt{1-pp}} = \frac{ds}{\sqrt{ns+aa}}.$$

Here, Euler notes that aa might be positive or negative, so he replaces aa with ab. This leads to

$$p = \sin \int \frac{ds}{\sqrt{nss+ab}} \quad \text{and} \quad \sqrt{1-pp} = \cos \int \frac{ds}{\sqrt{nss+ab}}.$$

The forms of these equations are familiar now, but a close look will reveal that they are not quite what we have seen before. Knowing p and $\sqrt{1-pp}$ gives us dx and dy. Euler integrates these, again using his integration lemmas, to get expressions for x and y. From that he finds

$$xx + yy = \frac{ab}{(1+n)^2} + \frac{nss}{1+n}.$$

Solving this for s gives

$$s = \sqrt{\frac{(1+n)(xx+yy)}{n} - \frac{ab}{n(1+n)}}.$$

Without showing any of the calculations, Euler gives us a list of formulas for the arc length, s, for different levels of evolutes:

For the first evolute $\quad s = \sqrt{\frac{(1+n)(xx+yy)}{n} + \frac{ab}{1+n}}.$

For the second evolute $\quad s = \sqrt{\frac{(1+n)(xx+yy)}{n} - \frac{nab}{1+n}}.$

For the third evolute $\quad s = \sqrt{\frac{(1+n)(xx+yy)}{n} - \frac{n^2ab}{1+n}}.$

Euler takes us back to Figure 5 to try to untangle the formula describing the curve itself. There, he takes straight line $AM = \sqrt{xx+yy} = z$ (*not* the length of the arc *AM*), the perpendicular to the tangent at M as $AT = p$, the tangent itself as $MT = \sqrt{zz-pp} = t$ and the curve element $ds = \frac{z\,dz}{t}$. From this, he finds that $r = (1+n)p$ and $s = \frac{1+n}{n}t$ and claims that these properties are enough to define the curve.

Euler now distinguishes two cases, depending on whether ab is positive or negative. In case ab is positive, he sets

$$\frac{ab}{(1+n)^2} = cc$$

so that

$$t = \sqrt{\frac{n}{1+n}(zz - cc)} \quad \text{and} \quad p = \sqrt{\frac{1}{1+n}(zz + ncc)}$$

and, eventually, as in Figure 8,

$$\tan CMT = \frac{p}{t} = \sqrt{\frac{zz + ncc}{nzz - ncc}}.$$

Again, this is almost in polar coordinates, though with the principal axis in an unfamiliar position and angles measured clockwise instead of counterclockwise. Euler does not give this curve a name but says that it "is confused with the logarithmic spiral."

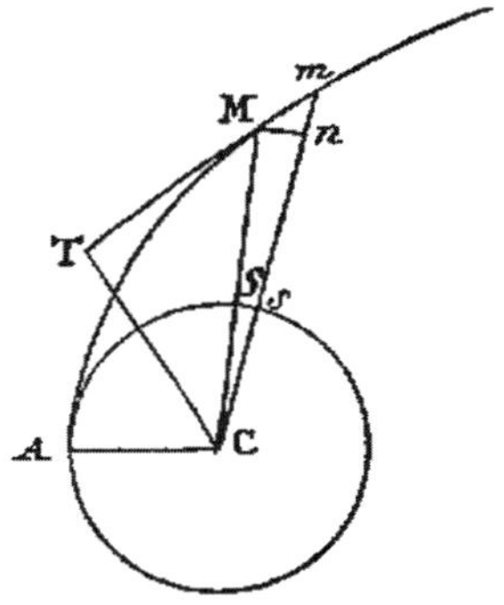

Fig. 8

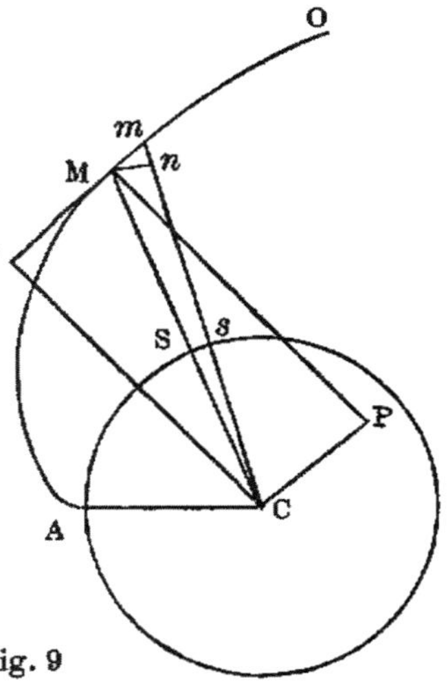

Fig. 9

On the other hand, in case ab is negative, we set $-ab/(1+n)^2 = cc$ and make the necessary changes to t, p, and $\tan CMT$. He gets another curve, illustrated in Figure 9, that he also describes as being "confused with the logarithmic spiral."

Recall that with Figure 7 we considered the case in which the radius of curvature in the first evolute grows as we move from B to N. Now, in Figure 10, we consider the case where the radius of curvature decreases from B to N. It does not take long for Euler to return to the equations for the epicycloids and the hypocycloids.

This completes Euler's analysis of the problem of a curve directly similar to its second evolute. We note that there were four cases to be considered here, depending on whether the radius of curvature increases or decreases from B to N and whether ab is positive or negative.

Euler turns to his next problem, "to find curves (Figure 11) that are inversely similar to their second evolutes."

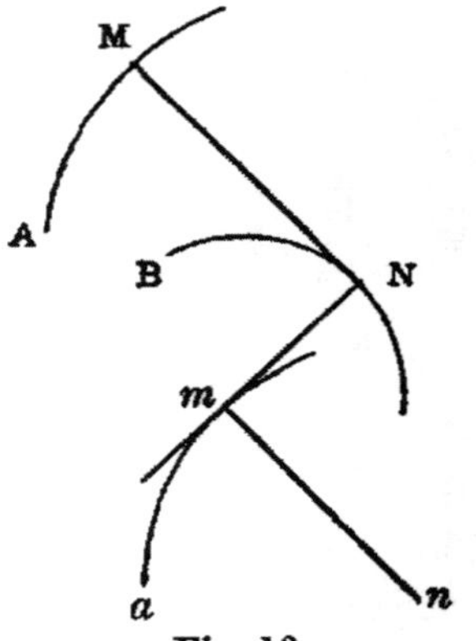

Fig. 10

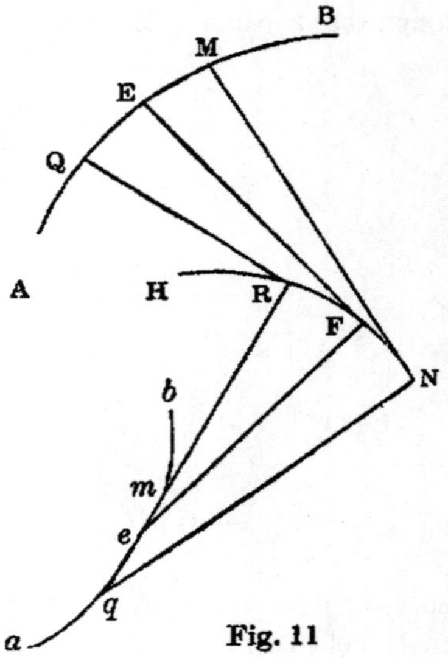

Fig. 11

We let *AEB* be the curve we are seeking, *HN* its first evolute and *aeb* the second evolute that is supposed to be inversely similar to *AEB*. Moreover, E is to be the point on *AEB* that relates to *e* two ways, both by homology and by the two radii of curvature.

As before, Euler distinguishes two cases, depending on whether the radius of curvature increases or decreases as we go from H to N. We consider the increasing case first.

Around E, we take two points, Q and M so that *EQ* and *EM* are equally wide. Then *EQ* will be similar to *em* and *EM* will be similar to eq.

Fearlessly reusing variable names, Euler lets the radii of curvature $EF = a$ and $Fe = b$. Also let arc lengths $EM = s$ and $EQ = S$ and radii of curvature $MN = r$ and $QR = R$. Then, because the arcs are equally wide, $\frac{ds}{r} = \frac{dS}{R}$.

Again, n is the ratio of similarity, so $em = ns$, $eq = nS$, and the radius of curvature at m is nr, and at q it is nR. The properties of evolutes give $FN = r - a$, $FR = a - R$,

$$Nq = \frac{r\,dr}{ds} \quad \text{and} \quad Rm = -\frac{R\,dr}{ds} = -\frac{r\,dR}{ds}.$$

In the second evolute, we have

$$eq = \frac{r\,dr}{ds} - b \quad \text{and} \quad em = b + \frac{r\,dR}{ds},$$

and, as we noted earlier, the radius of curvature at q is

$$\frac{r}{ds} d \cdot \frac{r\,dr}{ds},$$

and at m it is

$$\frac{r}{ds} d \cdot \frac{r\,dR}{ds}.$$

These, combined with the ratio of similarity and the change in orientation between the curve and its second evolute give

$$ns = b + \frac{rdR}{ds},\ nS = \frac{r\,dr}{ds} - b,\ nr = \frac{r}{ds}d\cdot\frac{r\,dR}{ds},\quad \text{and}\quad nR = \frac{r}{ds}d\cdot\frac{r\,dr}{ds}.$$

From this analysis we get three simultaneous equations

I. $r\,dS = R\,ds$
II. $ns\,ds = r\,dR + b\,ds$
III. $nS\,ds = r\,dr - b\,ds.$

These give us

$$n^2 s\,ds^3 = r^3\,d^3r + 4r^2\,dr\,ddr + r\,dr^3.$$

Holding ds constant and taking C to be a constant of integration, Euler integrates to get

$$n^2s^2\,ds^2 + C\,ds^2 = 2r^3\,ddr + r^3\,dr^2,$$

but this is difficult to solve. It takes Euler several pages of calculations, in which he makes some mistakes in substitutions that are noted by the editors of the *Opera Omnia*, before he finally gets a parametric expression for the original curve, namely

$$x = \int ds \sin\int\frac{ds}{r} \quad\text{and}\quad y = \int ds\cos\int\frac{ds}{r}.$$

This is not a very satisfying solution.

Euler turns next to the case where the radius of curvature of the first evolute is decreasing. The analysis is similar, with some signs reversed, and the result is even more complicated. It is not a familiar curve.

At the end of this analysis, Euler notes that if a curve is inversely similar to its second evolute, then it will be directly similar to its fourth evolute. He did not make the corresponding remark about curves inversely similar to their first evolutes being directly similar to their second involutes.

Euler turns his attention to third evolutes. He does not distinguish the directly similar case from the inversely similar case. If a curve is similar to its third evolute, and if n is the ratio of similarity, then we have

$$\pm ns = \frac{r}{ds}d\cdot\frac{r\,dr}{ds} = \frac{r^2\,ddr + r\,dr^2}{ds^2}$$

or

$$\pm ns\,ds^2 = r^2\,ddr + r\,dr^2.$$

He uses the techniques he has been developing since E-10 to reduce this second order differential equation to a first order one by holding ds constant, then substituting $s = e^{\int(u\,au)/y}$ and $r = e^{\int(u\,dy)/y}u$. This gives

$$\pm n\,du = y\,dy + 3uy\,du = u^3\,du,$$

which has a general solution

$$y = z - u^2 - u\sqrt[3]{\pm n} - \sqrt[3]{u^2}.$$

Euler takes $\pm n = m^3$ and eventually writes his solution as

$$r\,dr = mr\,ds + m^2 s\,ds = ce^{\int (m\,ds)/r}\,ds.$$

This curve does not seem to have a name, either, but at least this solution is given in terms of properties intrinsic to the curve itself, its arc length and its radius of curvature. This kind of solution might be useful in mechanics, where the value of r is related to the centripetal force acting on an object traversing the curve.

Now, suddenly, two thirds of the way through this fifty-page article, Euler seems to have another idea. All of his analyses have involved a relation between r and s. Recall that his problem to find a curve similar to its first evolute started with the equation $\frac{r\,dr}{ds} = \pm ns$. Each of his other problems has begun with a similar equation. Instead of looking for a relation between r and s, he thinks it might be a good idea to look for a relation between s and $\frac{ds}{r}$. He takes $\frac{ds}{r} = dv$, and offers a table of equations with which to start the analysis to find curves similar to various evolutes:

To find a curve similar to its own	use the equation
first evolute	$\pm ns\,dv = ds$
second evolute	$\pm ns\,dv^2 = dds$
third evolute	$\pm ns\,dv^3 = d^3s$
fourth evolute	$\pm ns\,dv^4 = d^4s$
fifth evolute	$\pm ns\,dv^5 = d^5s$
sixth evolute	$\pm ns\,dv^6 = d^6s$

Euler tells us, but gives no details, that these will lead to solutions in orthogonal coordinates as $x = \int ds \sin v$ and $y = \int ds \cos v$ and that the resulting differential equations will be solved with the following substitutions:

I.	$\dfrac{ds}{dv} = e^{gv}g$	$\dfrac{ds}{s\,dv} = g = \pm n$
II.	$\dfrac{dds}{dv^2} = e^{gv}g^2$	$\dfrac{dds}{s\,dv^2} = g^2 = \pm n$
III.	$\dfrac{d^3s}{dv^3} = e^{gv}g^3$	$\dfrac{d^3s}{s\,dv^3} = g^3 = \pm n$
IV.	$\dfrac{d^4s}{dv^4} = e^{gv}g^4$	$\dfrac{d^4s}{s\,dv^4} = g^4 = \pm n$

Euler works with these equations for some time, then summarizes his results in terms of s and v, giving two cases to each problem. He puts these in a table.

To find a curve similar to its:	solve	and get the solution
first evolute	$ds = +fs\,dv$	$s = Ce^{fv}$
or	$ds = -fs\,dv$	$s = Ce^{-fv}$

second evolute	$d^2s = +f^2s\,dv^2$	$s = Ce^{fv} + De^{-fv}$
or	$d^2s = -f^2s\,dv^2$	$s = C\sin(fv + \gamma)$
third evolute	$d^3s = +f^3s\,dv^3$	$s = Ce^{fv} + De^{-(/12)fv}\sin\left(\frac{fv\sqrt{3}}{2} + \delta\right)$
or	$d^3s = -f^3s\,dv^3$	$s = Ce^{fv/2}\sin\left(\frac{fv\sqrt{3}}{2} + \gamma\right) + De^{-fv}$
fourth evolute	$d^4s = +f^4s\,dv^4$	$s = Ce^{fv} + D\sin(fv + \delta) + Ee^{-fv}$
or	$d^4s = -f^4s\,dv^4$	$s = Ce^{fv/\sqrt{2}}\sin\left(\frac{fv}{\sqrt{2}} + \gamma\right) + De^{-\frac{fv}{\sqrt{2}}}\sin\left(\frac{fv}{\sqrt{2}} + \delta\right)$
fifth evolute	$d^5s = +f^5s\,dv^5$	$s = Ce^{fv} + De^{fv(\sqrt{5}-1)/4}\sin\left(\frac{fv(\sqrt{10}+2\sqrt{5})}{4} + \delta\right)$ $+ Ee^{-\frac{fv(\sqrt{5}-1)}{4}}\sin\left(\frac{fv(\sqrt{10}-2\sqrt{5})}{4} + \varepsilon\right)$
or	$d^5s = -f^5s\,dv^5$	etc.
sixth evolute	$d^6s = +f^6s\,dv^6$	
or	$d^6s = -f^6s\,dv^6$	

Finally, Euler notes that the pattern here involves the trigonometric functions of $\frac{\pi}{n}$, and that because he cannot give algebraic equations for trigonometric functions of $\frac{\pi}{7}$, he cannot give solutions of this kind for curves similar to their seventh evolutes.

And so we come to the end of a long and difficult article. We have had to skip a good deal. Perhaps the main impression of this paper is that Euler regards a differential equation as an acceptable solution to problems like this. He does not usually find it necessary to give an explicit solution. Second, there is this surprise ending, that high order evolutes involve trigonometric functions of $\frac{\pi}{n}$. Subsequent discoveries have showed that such values depend on whether or not an n-gon is constructible. One wonders, if Euler had lived to know Gauss's results, would he have tried to construct a curve similar to its 17th evolute?

References

[L'H] L'Hôpital, Le Marquis de, *Analyse des infiniment petits, pour l'intelligence des lignes courbes, nouvelle édition*, Jombert, Paris, 1781. Reprint of the first edition of 1696.

44

E-130: De seriebus quibusdam considerationes*

Considerations about certain series

Ever since his solution of the Basel problem made him famous, Euler had an interest in series given as the sum of reciprocals of powers. He had a continuing suspicion that most such series sum to a value that somehow involved π, a question that is still unresolved today. In this paper, he is motivated by series of the form

$$1 \pm \frac{1}{3^n} + \frac{1}{5^n} \pm \frac{1}{7^n} + \frac{1}{9^n} \pm \frac{1}{11^n} + \text{etc.}$$

where, for even values of n, he means to resolve the ambiguous sign as addition, and for odd values to resolve it as subtraction.

If we are looking for π, then a good place to start is with the series for trigonometric functions. Euler's notation for π is still not universally accepted, so he has to remind his readers yet again that π denotes the length of a 180-degree arc on a circle of radius 1. Then he denotes an arc of the circle by s, the sine of that arc by y, the cosine by x and the tangent by t. He reminds us of the power series for the trigonometric functions

$$y = s - \frac{s^3}{1 \cdot 2 \cdot 3} + \frac{s^5}{1 \cdot 2 \cdot 3 \cdot 4 \cdot 5} - \frac{s^7}{1 \cdot 2 \cdot \cdots \cdot 7} + \text{etc.},$$

$$x = 1 - \frac{s^2}{1 \cdot 2} + \frac{s^4}{1 \cdot 2 \cdot 3 \cdot 4} - \frac{s^6}{1 \cdot 2 \cdot \cdots \cdot 6} + \text{etc.}$$

Then, since $t = \frac{y}{x}$ so $0 = xt - y$,

$$0 = t - s - \frac{s^2 t}{1 \cdot 2} + \frac{s^3}{1 \cdot 2 \cdot 3} + \frac{s^4 t}{1 \cdot 2 \cdot 3 \cdot 4} - \frac{s^5}{1 \cdot 2 \cdot \cdots \cdot 5} - \frac{s^6}{1 \cdot 2 \cdot \cdots \cdot 6} + \text{etc.}$$

or

$$0 = 1 - \frac{s}{t} + \frac{s^2}{1 \cdot 2} + \frac{s^3}{1 \cdot 2 \cdot 3t} + \frac{s^4}{1 \cdot 2 \cdot 3 \cdot 4} - \frac{s^5}{1 \cdot 2 \cdot \cdots \cdot 5t} - \frac{s^6}{1 \cdot 2 \cdot \cdots \cdot 6} + \text{etc.}$$

As he continues to gather his tools by recounting familiar results, Euler gives a list of about a dozen arcs that have the same sine, y, as a given arc $\frac{m}{n}\pi$. Knowing this, he can use the same factoring idea that worked so spectacularly in E-41 when he solved the Basel problem. If we rewrite our series for sine by dividing by y, then subtracting 1, we can factor the resulting series to get

$$0 = 1 - \frac{s}{1y} + \frac{s^3}{1 \cdot 2 \cdot 3y} - \frac{s^5}{1 \cdot 2 \cdot 3 \cdot 4 \cdot 5y} + \frac{s^7}{1 \cdot 2 \cdot \cdots \cdot 7y} - \text{etc.}$$

$$= \left(1 - \frac{ns}{m\pi}\right)\left(1 + \frac{ns}{(n+m)\pi}\right)\left(1 - \frac{ns}{(n-m)\pi}\right)\left(1 + \frac{ns}{(2n-m)\pi}\right)\left(1 - \frac{ns}{(2n+m)\pi}\right) \text{etc.}$$

**Comm. Acad. Sci. Imp. Petropol.* 12 (1740) 1750, 53–96; *Opera Omnia* I.14, 407–462

Now we match terms in the expansion of the product with terms in the sum. The terms where s appears to the first power give us

$$\frac{1}{y} = \frac{n}{m\pi} + \frac{n}{(n-m)\pi} - \frac{n}{(n+m)\pi} - \frac{n}{(2n-m)\pi} + \frac{n}{(2n+m)\pi} + \frac{n}{(3n-m)\pi} - \text{etc.}$$

We note that there are issues of convergence and divergence here that Euler is completely ignoring.

Taking the sum of the products of these terms taken two at a time gives zero, the coefficient of degree 2 in the series. In fact, taking the sum of products of any even number of terms will give zero as well. Euler meticulously tabulates the sums of the products of terms taken up to eight at a time.

We will be doing this kind of calculation again, so Euler generalizes a bit, handicapped again because subscripts still are not widely used. He starts with a series $a + b + c + d + e +$ etc. and denotes the sum of these terms by α. The sum of the products taken two at a time will be β, three at a time will be γ, and so on for δ, ε, ξ and so forth.

He further takes the sum of the series to be A, the sum of the squares to be B, the cubes C, etc. Euler has done all this before in E-41, but his notation has changed a bit, so he reminds us of Newton's relations among these values.

$$\begin{aligned} A &= \alpha \\ B &= \alpha A - 2\beta \\ C &= \alpha B - \beta A + 3\gamma \\ D &= \alpha C - \beta B + \gamma A + 4\delta \\ &\text{etc.} \end{aligned}$$

Quite soon, in E-158, Euler will invent generating functions. His next step demonstrates that he has the command of the combinatorial aspects of series manipulations that he will need to make that invention. He notes that the quotient of two series involving the Greek letters gives a series involving the upper case letters. He tells us

$$\frac{\alpha - 2\beta z + 3\gamma z^2 - 4\delta z^3 + 5\varepsilon z^4 - 6\zeta z^5 + 7\eta z^6 - \text{etc.}}{1 - \alpha z + \beta z^2 - \gamma z^3 + \delta z^4 - \varepsilon z^5 + \zeta z^6 - \text{etc.}} = A + Bz + Cz^2 + Dz^3 + Ez^4 + Fz^5 + \text{etc.}$$

He sets the denominator here Z and rewrites this sum as $\frac{-dZ}{Z\,ds}$.

We apply this notation to the series at hand, which Euler rewrites as

$$\frac{n}{\pi}\left(\frac{1}{m} + \frac{1}{n-m} - \frac{1}{n+m} - \frac{1}{2n-m} + \frac{1}{2n+m} + \frac{1}{3n-m} - \frac{1}{3n+m} - \text{etc.}\right).$$

Taking the sums of products one, two, three, and so forth at a time, we get

$$\begin{aligned} A &= \frac{1}{1y} \\ B &= \frac{A}{1y} \\ C &= \frac{B}{1y} - \frac{1}{1\cdot 2y} \end{aligned}$$

$$D = \frac{C}{1y} - \frac{A}{1\cdot 2\cdot 3y}$$
$$E = \frac{D}{1y} - \frac{B}{1\cdot 2\cdot 3y} + \frac{1}{1\cdot 2\cdot 3\cdot 4y}$$
etc.

Now, applying his formulas involving the Greek letters, he gets some results about π.

$$\frac{1}{m} + \frac{1}{n-m} - \frac{1}{n+m} - \frac{1}{2n-m} + \frac{1}{2n+m} + \text{etc.} = \frac{A\pi}{n}$$
$$\frac{1}{m^2} + \frac{1}{(n-m)^2} - \frac{1}{(n+m)^2} - \frac{1}{(2n-m)^2} + \frac{1}{(2n+m)^2} + \text{etc.} = \frac{B\pi^2}{n^2}$$
$$\frac{1}{m^3} + \frac{1}{(n-m)^3} - \frac{1}{(n+m)^3} - \frac{1}{(2n-m)^3} + \frac{1}{(2n+m)^3} + \text{etc.} = \frac{C\pi^3}{n^3}$$
etc.

Let us now fix y and take

$$Z = 1 - \frac{z}{1y} + \frac{z^3}{1\cdot 2\cdot 3y} - \frac{z^5}{1\cdot 2\cdot \cdots \cdot 5y} + \frac{z^7}{1\cdot 2\cdot \cdots \cdot 7y} - \text{etc.} = 1 - \frac{1}{y}\sin z.$$

This makes

$$dZ = \frac{-dz\cos z}{y} = A + Bz + Cz^2 + Dz^3 + Ez^4 + \text{etc.}$$

By previous work we get

$$Az + Bz^2 + Cz^3 + Dz^4 + Ez^5 + \text{etc.} = \frac{-dZ}{Z} = \frac{z\cos z}{y - \sin z}.$$

Taking $z = \frac{p\pi}{n}$, we express this series as a sum of series

$$+\frac{p}{m} + \frac{p}{n-m} - \frac{p}{n+m} - \frac{p}{2n-m} + \frac{p}{2n+m} + \text{etc.}$$
$$+\frac{p^2}{m^2} + \frac{p^2}{(n-m)^2} + \frac{p^2}{(n+m)^2} + \frac{p^2}{(2n-m)^2} + \frac{p^2}{(2n+m)^2} + \text{etc.}$$
$$+\frac{p^3}{m^3} + \frac{p^3}{(n-m)^3} - \frac{p^3}{(n+m)^3} - \frac{p^3}{(2n-m)^3} + \frac{p^3}{(2n+m)^3} + \text{etc.}$$
etc.

The columns here are geometric series, and they sum to give

$$\frac{p}{m-p} + \frac{p}{n-m-p} - \frac{p}{n+m+p} - \frac{p}{2n-m+p} + \frac{p}{2n+m-p} + \text{etc.}$$

This series, in turn, sums to

$$\frac{p\pi\cos\frac{p\pi}{n}}{ny - n\sin\frac{p\pi}{n}} = \frac{p\pi\cos\frac{p\pi}{n}}{n\sin\frac{m\pi}{n} - n\sin\frac{p\pi}{n}},$$

because $y = \sin\frac{m\pi}{n}$.

If we let $m - p = a$ and $m + p = b$ and divide by p, then we get the series

$$\frac{1}{a} + \frac{1}{n-b} - \frac{1}{n+b} - \frac{1}{2n-a} + \frac{1}{2n+a} + \frac{1}{3n-b} - \frac{1}{3n+b} - \text{etc.}$$

Associating pairs of terms, and making appropriate substitutions in the formula for the sum, gives

$$\begin{aligned}\frac{1}{a} &+ \frac{2b}{n^2-b^2} - \frac{2a}{4n^2-a^2} + \frac{2b}{9n^2-b^2} - \frac{2a}{16n^2-a^2} + \frac{2b}{25n^2-b^2} - \text{etc.}\\ &= \frac{\pi\cos\frac{(b-a)\pi}{2n}}{n\sin\frac{(b+a)\pi}{2n} - n\sin\frac{(b-a)\pi}{2n}}.\end{aligned}$$

We will use this formula a little later.

Now we turn to a particular example. Take $y = 1$, $m = 1$ and $n = 2$. We get the series

$$\begin{aligned}&\frac{1}{1} + \frac{1}{1} - \frac{1}{3} - \frac{1}{3} + \frac{1}{5} + \frac{1}{5} - \frac{1}{7} - \text{etc.} = \frac{A\pi}{2},\\ &\frac{1}{1^2} + \frac{1}{1^2} + \frac{1}{3^2} + \frac{1}{3^2} + \frac{1}{5^2} + \frac{1}{5^2} + \frac{1}{7^2} + \text{etc.} = \frac{B\pi^2}{2^2}\\ &\text{etc.}\end{aligned}$$

This gives values for A, B, C, etc.

$$\begin{aligned}A &= 1\\ B &= \frac{A}{1} = 1\\ C &= \frac{B}{1} - \frac{1}{1\cdot 2} = \frac{1}{1\cdot 2}\\ D &= \frac{C}{1} - \frac{A}{1\cdot 2\cdot 3} = \frac{2}{1\cdot 2\cdot 3}\\ E &= \frac{D}{1} - \frac{B}{1\cdot 2\cdot 3} + -\frac{1}{1\cdot 2\cdot 3\cdot 4} = \frac{5}{1\cdot 2\cdot 3\cdot 4}\\ &\text{etc.}\end{aligned}$$

Euler finds coefficients down to N and O, which correspond to sums of 13th and 14th powers, respectively. Note that the value of B leads to a solution of the Basel problem. For $B = 1$, we get

$$1\cdot\frac{\pi^2}{2^3} = 1 + \frac{1}{3^2} + \frac{1}{5^2} + \frac{1}{7^2} + \text{etc.}$$

As we saw in Euler's first solution to the Basel problem in E-41, we can find the sum of the reciprocals of all squares from the reciprocals of the odd squares by multiplying by $\frac{4}{3}$. This gives a slightly different solution to the Basel problem from the one given before.

The same series arguments we have just seen work also for the series for arctangent. Euler takes

$$s = \tan\left(\frac{\pi}{4} + \frac{z}{2}\right)$$

so that

$$\frac{\pi}{4} + \frac{z}{2} = \arctan s = \int \frac{ds}{1+ss}.$$

Euler supposes that $s = A + Bz + Cz^2 + Dz^3 + Ez^4 + \text{etc.}$ and finds that

$$\begin{aligned} A &= 1 \\ B &= \frac{A^2+1}{2} \\ C &= \frac{2AB}{4} \\ D &= \frac{2AC+B^2}{6} \\ E &= \frac{2AD+2BC}{8} \\ &\text{etc.} \end{aligned}$$

He further finds values for the Greek letters,

$$\begin{aligned} \alpha &= 1 \\ \beta &= \frac{\alpha^2+1}{2} \\ \gamma &= \alpha\beta \\ \delta &= \alpha\gamma + \beta^2 \\ \varepsilon &= \alpha\delta + 3\beta\gamma \\ &\text{etc.} \end{aligned}$$

This leads fairly easily to a new series,

$$\frac{\pi \cos \frac{p\pi}{2}}{4 - 4\sin \frac{p\pi}{2}} = \frac{1}{1-p} - \frac{1}{3+p} + \frac{1}{5-p} - \frac{1}{7+p} + \frac{1}{9-p} - \text{etc.}$$

We return now to a sum we developed earlier, the one with differences of squares in the denominators. We take $a = b = m$ and get

$$\frac{2m}{n^2-m^2} - \frac{2m}{4n^2-m^2} + \frac{2m}{9n^2-m^2} - \frac{2m}{16n^2-m^2} + \frac{2m}{25n^2-m^2} - \text{etc.} = \frac{\pi}{n\sin\frac{m\pi}{n}} - \frac{1}{m}.$$

On the other hand, if we take $a = -m$ and $b = +m$, we have the same terms, but without the alternating signs, and they sum to

$$\frac{\pi\cos\frac{m\pi}{n}}{n\sin\frac{m\pi}{n}} + \frac{1}{m}.$$

If we let $m = 2i + 1$ and $n = 2$, so that the cosine in the numerator will vanish, we have

$$\frac{1}{2(2i+1)^2} = \frac{1}{4-(2i+1)^2} + \frac{1}{16-(2i+1)^2} + \frac{1}{36-(2i+1)^2} + \frac{1}{64-(2i+1)^2} + \text{etc.}$$

Euler likes this particular series, for some reason. By taking $n = 1$ and $mm = p$ he is able to sum

$$\frac{1}{1-p} \pm \frac{1}{4-p} + \frac{1}{9-p} \pm \frac{1}{16-p} + \text{etc.}$$

He extends this to a number of related series.

Euler does the same kind of analysis starting with the series $1 - \frac{1}{3} + \frac{1}{5} - \frac{1}{7} + \frac{1}{9} - \text{etc.}$ and uses it to derive again the sums of the reciprocals of the even powers, what we now call $\zeta(2n)$, up to 24th powers. These values all have the form

$$\zeta(2n) = \frac{2^{2n-1}}{(2n+1)!} \cdot \frac{p}{q} \cdot \pi^{2n}.$$

Euler, of course, uses neither ζ nor the factorial notation. He calls $\frac{p}{q}$ the "middle fraction" and notes that it follows an irregular pattern, $\frac{1}{2}, \frac{1}{6}, \frac{1}{6}, \frac{3}{10}, \frac{5}{6}, \frac{691}{210}, \frac{35}{2}$, etc. He does a little bit to explain the pattern. He further estimates $\zeta(n)$ to nine decimal places for values of n from 2 to 13, including odd values of n.

In passing, Euler gets an expression that allows him to assign meaning to some divergent series. He observes that these series all sum to zero:

$$\begin{aligned} 1 - 3 + 5 - 7 + 9 - \text{etc.} &= 0 \\ 1 - 3^3 + 5^3 - 7^3 + 9^3 - \text{etc.} &= 0 \\ 1 - 3^5 + 5^5 - 7^5 + 9^5 - \text{etc.} &= 0 \\ 1 - 3^7 + 5^7 - 7^7 + 9^7 - \text{etc.} &= 0 \end{aligned}$$

Euler does the same kind of analysis on yet another series, $1 - \frac{1}{2} + \frac{1}{3} - \frac{1}{4} + \frac{1}{5} - \text{etc.}$, and exploits the fact that this series sums to $\ln 2$. Along the way, he considers divergent series. His first one is familiar, $1 - 1 + 1 - 1 + \text{etc.} = \frac{1}{2}$. The others all sum to zero:

$$\begin{aligned} &1 - 2^2 + 3^2 - 4^2 + \text{etc.} \\ &1 - 2^4 + 3^4 - 4^4 + \text{etc.} \\ &1 - 2^6 + 3^6 - 4^6 + \text{etc.} \end{aligned}$$

It is a little bit disturbing that he also finds the alternating sums of odd powers. which he has previously found to sum to zero. This time the sums are

$$\begin{aligned} 1 - 3 + 5 - 7 + 9 - \text{etc.} &= \frac{1}{4} \\ 1 - 3^3 + 5^3 - 7^3 + 9^3 - \text{etc.} &= \frac{-1}{8} \\ 1 - 3^5 + 5^5 - 7^5 + 9^5 - \text{etc.} &= \frac{1}{4} \\ 1 - 3^7 + 5^7 - 7^7 + 9^7 - \text{etc.} &= \frac{-17}{16}. \end{aligned}$$

We move to another series.

$$\frac{1}{2 \cdot 1} + \frac{1 \cdot 3}{2 \cdot 4 \cdot 2} + \frac{1 \cdot 3 \cdot 5}{2 \cdot 4 \cdot 6 \cdot 3} + \frac{1 \cdot 3 \cdot 5 \cdot 7}{2 \cdot 4 \cdot 6 \cdot 8 \cdot 4} + \text{etc.}$$

He takes

$$s = \frac{x}{2\cdot 1} + \frac{1\cdot 3x^2}{2\cdot 4\cdot 2} + \frac{1\cdot 3\cdot 5x^3}{2\cdot 4\cdot 6\cdot 3} + \frac{1\cdot 3\cdot 5\cdot 7x^4}{2\cdot 4\cdot 6\cdot 8\cdot 4} + \text{etc.}$$
$$= \int \frac{dx}{x\sqrt{1-x}} - \ln x\,.$$

This time, he uses more integration techniques and evaluates

$$\int \frac{dx}{x\sqrt{1-x}} = c - \ln\left(1+\sqrt{1-x}\right) + \ln\left(1-\sqrt{1-x}\right).$$

He evaluates his constant of integration to be $c = 2\ln 2$, then sets $x = 1$. This makes $s = 2\ln 2$, and gives the series

$$\frac{1}{2\cdot 1} + \frac{1\cdot 3}{2\cdot 4\cdot 2} + \frac{1\cdot 3\cdot 5}{2\cdot 4\cdot 6\cdot 3} + \frac{1\cdot 3\cdot 5\cdot 7}{2\cdot 4\cdot 6\cdot 8\cdot 4} + \text{etc.} = 2\ln 2.$$

Euler ends this article, "But since we transmute these series to simple series, the sums of which it is evident cannot be reduced, we break off this activity having found several expressions which are equal to the proposed series

$$1 - \frac{1}{2^3} + \frac{1}{3^3} - \frac{1}{4^3} + \frac{1}{5^3} - \text{etc."}$$

This concludes one of the longer articles from Euler's early work. It has been a technical sequel and extension of E-41, his solution to the Basel problem of six years earlier. It would have been nice to dwell a bit more on certain details, like his passing remarks on divergent series, but those are couched in some otherwise tedious calculations. As with other such papers laden with technical details, we encourage the interested reader to brave the original Latin. The language is easier than the mathematics.

Interlude: 1740

World events

Euler's protégé Anders Johann Lexell (1740–1784) was born in Sweden. When Euler died in 1783, Lexell took Euler's position at the Academy in St. Petersburg, but he became ill and died just a year later.

Also in 1740 Frederick II ascended the throne in Prussia and began trying to hire Euler to join the Academy of Sciences there. Euler initially declined, but changed his mind a year later. That same year Maria Theresa inherited the Hapsburg Empire, now Austria, Belgium and Hungary. Frederick contested Maria Theresa's ascension and on December 16 he invaded Silesia, precipitating the War of the Austrian Succession.

On October 28, Empress Anna Ivanova of Russia died, leaving a power vacuum in Russia and setting off a chain of events that would result in Euler leaving Russia for Berlin the next year.

In Euler's life

In 1740 Euler was asked to cast a horoscope for the young Prince Ivan. Euler declined, and the Academy's astronomer had to do it instead. Ivan was soon arrested by his political enemies and imprisoned for 20 years.

Euler won a one-fourth share of the Paris Prize, on the ebb and flow of the tides, sharing the prize with Daniel Bernoulli, Colin Maclaurin, and Jesuit mathematician Antoine Cavaleri.

Euler's fifth child Karl, was born on July 26. Karl grew up to be an officer in the Russian navy.

Euler's other work

Euler wrote three papers on astronomy in 1740, E-38, Orbitae solaris determinatio, "Determination of the solar orbit," E-39, Solutio problematum quorundam astronomicorum, "Solution of certain problems of astronomy," and E-131, Emendatio tabularum astronomicarum per loca planetarum geocentrica, "Revised astronomical tables for the geocentric location of planets."

Euler's mathematics

Euler wrote the second volume of his arithmetic text, the *Rechen-Kunst*, E-35. He gave a solution to the Chinese Remainder Theorem in E-36, Solutio problematis arithmetici de inveniendo numero, qui per datos numeros divisus, relinquat data residua, "Solution of problems in arithmetic of finding a number, which, when divided by given numbers leaves given remainders." There was also a paper on the theory of equations, E-157, De extractione radicum ex quantitatibus irrationalibus, "On the extraction of roots of irrational quantities."

45

E-36: Solutio problematis arithmetici de inveniendo numero, qui per datos numeros divisus, relinquat data residua*

Solution of problems in arithmetic of finding a number, which, when divided by given numbers leaves given remainders

In this paper, Euler addresses another well-known problem, the one we now know as the Chinese Remainder Theorem. In this case, the theorem is well named, for the Chinese completely solved it almost 2000 years ago. The theorem certainly should not be named after Euler. It seems peculiar that Euler would attack a problem that had already been solved and whose solution was well known, but Euler adds some generality to the solution, often attributed to Gauss about 65 years later. [D] This is Euler's only paper on the subject.

Euler begins with a common "problem in ordinary arithmetic books," to find a number which, when divided by 2, 3, 4, 5 or 6, leaves a remainder 1 but which leaves no remainder if divided by 7. Exactly this problem had been posed, and all its infinitely many solutions accounted for, by Ibn al-Haitam about the year 1000. [D]

Euler also mentions that the problem has applications to magic squares and attributes those applications to La Hire and to Sauveur. Euler himself will do some work on magic squares in the 1750s while he is in Berlin.

Euler begins with the case of just one remainder. If the problem is to find a number z that leaves a remainder a when divided by p, then "the solution is very easy; it will be $z = ma + p$, denoting by m any whole number." Euler notes that this gives all possible solutions.

Euler moves to "the following case, in which two divisors with their remainders are proposed." He asks that we "find a number z, which divided by a leaves p, and divided by b leaves q." So there are integer values m and n such that $z = ma + p$ and $z = nb + q$. This makes $ma + p = nb + q$, so

$$n = \frac{ma + p - q}{b}.$$

Setting $p - q = v$, this is

$$n = \frac{ma + v}{b}.$$

This implies that b divides $ma + v$. If $a > b$, then put $a = \alpha b + c$, (here, and elsewhere, we are to assume that α is an integer and that the remainder, c, is the smallest possible positive remainder) so that

$$n = m\alpha + \frac{mc + v}{b}.$$

This means that b must divide $mc + v$.

*Comm. Acad. Sci. Imp. Petropol. 7 (1734/5) 1740, 46–66; Opera Omnia I.2, 18–32

Now Euler is going to develop a recursive algorithm, but he will do it without subscripts. Though Euler occasionally used subscripts, (we saw an example in E-31) it took over a hundred years before they became common mathematical notation. So, instead of calling $\frac{mc+v}{b}$ something like q_1, Euler calls it A. This makes $m = \frac{Ab-v}{c}$.

Two things can happen here. Either c divides v, or it does not.

If c divides v, then we are close to a solution to the original problem, for we may take $A = 0$ and $m = -\frac{v}{c}$, where m will be a whole number. Then $z = -\frac{av}{c} + p$. Euler leaves it to the reader to check that this value of z satisfies both equations, and that other solutions may be found by adding or subtracting the least common multiple of a and b.

The other thing that might happen is that c does not divide v. Then divide b by c and call the quotient β and the remainder d, so that $b = \beta c + d$. Then

$$\frac{Ab - v}{c} = A\beta + \frac{Ad - v}{c} = m,$$

and it is required that $\frac{Ad-v}{c}$ be a whole number. We can call it B, so that $A = \frac{Bc+v}{d}$. Now, again, two things can happen. If d divides v, then take $B = 0$ and get a solution because $A = \frac{v}{d}$ so $m = \frac{\beta v}{d}$ and $z = a\frac{bv}{d} + p$. On the other hand, if d does not divide v, then it is necessary to iterate again. Introduce C, e and γ in place of B, d and β, which, in turn, had been introduced in place of A, c and α and repeat as necessary.

Euler is sensitive to the fact that he may be asking us to undertake a task that never ends. He has learned from his readings of Fermat how to deal with such things. Euler reminds us that $a > b$. Further, $b > c$, $c > d$, etc., and all of them are greater than zero. The process cannot go on forever. It must stop sometime, and the step at which it stops will produce a solution. This is, of course, a form of what we now call mathematical induction. This particular variation of mathematical induction is called Fermat's Method of Infinite Descent.

Euler understands that perhaps the method is lost in the explanation, so he shows us a way to perform the calculation without getting too confused. He organizes the method into a tableau, which he regards as mostly self-explanatory. (See the tableau at the top of page 353.)

This process terminates when k divides v. The number k is a multiple of the greatest common divisor of a and b, and we can see the usual greatest common divisor algorithm working its way down the far right column. Euler notes a necessary condition for his two-remainder problem to have a solution is that the greatest common divisor of a and b must divide the v, the difference between the given remainders p and a.

Once the tableau has been completed, we must perform back-substitution to get an answer. A value for m will suffice. For those of us who don't want to work the back-substitution out for ourselves, Euler gives us explicit formulas, depending on how many steps were necessary in the tableau above. If we get to stop after just one iteration, then

$$m = \frac{Ab - v}{c}.$$

If it takes two iterations, then

$$m = \frac{Bb - \beta v}{d}.$$

If it takes six, then

$$m = \frac{Fb + v(\delta\varepsilon\zeta + \beta\gamma\delta\varepsilon\zeta + \beta\varepsilon\zeta + \beta\gamma\zeta + \zeta + \delta + \beta\gamma\delta + \beta)}{h}.$$

$n = \dfrac{ma+v}{b}$	$b \mid a \mid \alpha$	$a = \alpha b + c$
$m = \dfrac{Ab-v}{c}$	$c \mid b \mid \beta$	$b = \beta c + d$
$A = \dfrac{Bc+v}{d}$	$d \mid c \mid \gamma$	$c = \gamma d + e$
$B = \dfrac{Cd-v}{e}$	$e \mid d \mid \delta$	$d = \delta e + f$
$C = \dfrac{De+v}{f}$	$f \mid e \mid \varepsilon$	$e = \varepsilon f + g$
$D = \dfrac{Ef-v}{g}$	$g \mid f \mid \zeta$	$f = \zeta g + h$
$E = \dfrac{Fg+v}{h}$	$h \mid g \mid \eta$	$g = \eta h + i$
$F = \dfrac{Gh-v}{i}$	$i \mid h \mid \theta$	$h = \theta i + k$
$G = \dfrac{Hi+v}{k}$	k	

Euler stops there.

Euler gives another table for m, depending on the first time that $\frac{v}{b}$, $\frac{v}{c}$ $\frac{v}{d}$, etc. is a whole number and then another giving z directly. These rules come up later, so we will list them all here.

If this number is an integer	Then z is given by
$\dfrac{v}{b}$	$z = q + \dfrac{bv}{b}1$
$\dfrac{v}{c}$	$z = q - \dfrac{bv}{c}1$
$\dfrac{v}{d}$	$z = q + \dfrac{bv}{d}(\alpha\beta + 1)$
$\dfrac{v}{e}$	$z = q - \dfrac{bv}{e}(\alpha\beta\gamma + \alpha + \gamma)$
$\dfrac{v}{f}$	$z = q + \dfrac{bv}{f}(\alpha\beta\gamma\delta + \alpha\beta + \alpha\delta + \gamma\delta + 1)$
$\dfrac{v}{g}$	$z = q - \dfrac{bv}{g}(\alpha\beta\gamma\delta\varepsilon + \alpha\beta\gamma + \alpha\beta\varepsilon + \alpha\delta\varepsilon + \gamma\delta\varepsilon + \alpha + \gamma + \varepsilon)$

Note that the signs following q alternate. The pattern among the Greek symbols is less obvious. A few minutes of study reveals that each term is formed from the first by removing 0, 1 or 2 pairs of consecutive factors. Once we notice this, the pattern becomes clear.

Euler is ready to put all this in just a few words.

> To find a number z which divided by a leaves a remainder p and divided by b leaves q, I put $p - q = v$ and we will have the following rule:

> Perform the operation for finding the greatest common divisor between a and b, up until the remainder that is found is a divisor of that number v. The quotient from the division of this v by that divisor is kept and called Q, at which point this part of the calculation is broken off. Then the quotients that arise from this division, α, β, γ etc., are written in a series and a new series is constructed from them:
>
> $$1,\ \alpha,\ \alpha\beta + 1,\ \alpha\beta\gamma + \alpha + \gamma \text{ etc.}$$
>
> ... Under this new series are written alternating signs $+ - + -$ etc. and the last term with its sign is multiplied by Q and then by the smaller of the given divisors b; and the result is added to q the remainder left by the divisor b. What results from this sum will be the number that is sought.

This gives one solution. Euler adds that if z is one solution, then so also are the numbers $ab + z$, $2ab + z$, and $mab + z$. Then he improves upon this remark by letting the least common multiple of a and b be M, and noting that all numbers $mM + z$ are also solutions and that, by adding or subtracting multiples of M, the smallest solution may be found.

It is finally time for an example. We are to find a number which divided by 103 leaves 87, and divided by 57 leaves 25. This makes $a = 103$, $b = 57$, $p = 87$, $q = 25$, and so $v = 62$. Euler does a calculation that looks like this:

$$\begin{array}{r|r|r|r|l}
57 & 103 & 1 & & \\
 & 57 & & & \\
\hline
 & 46 & 57 & 1 & \\
 & & 46 & & \\
\hline
 & & 11 & 46 & 4 \\
 & & & 44 & \\
\hline
 & & & 2 &
\end{array}
\qquad \frac{62}{2} = 31 = Q$$

The Euclidean algorithm stops when the remainder, 2, divides 62. The quotients are $\alpha = 1$, $\beta = 1$, $\gamma = 4$, and the sequence $1, \alpha, \alpha\beta + 1, \alpha\beta\gamma + \alpha + \gamma$ etc. that we are to calculate goes

$$1,\ 1,\ 2,\ 9.$$

The sign attached to the 9, the term we want, is negative. Finally, the number we seek is $q - 9Qb = 25 - 9 \cdot 31 \cdot 57 = -15878$. That is negative, so we can add multiples of ab or $103 \cdot 57$ until we get the smallest positive answer, 1735. Because $103 \cdot 57$ is the least common multiple of a and b all possible solutions have the form $m \cdot 103 \cdot 57 + 1735$.

His next example is to find a number that leaves a remainder 10 when divided by 41 and remainder 28 when divided by 29. Euler throws a small twist into his solution by taking $a = 41$, $b = 29$ and $p = 10$, as expected, but taking $q = -1$. This choice of q works out a little more easily, with smaller numbers and gives a positive solution, 666, the first time.

Euler's next example has $a = 24$, $b = 15$, $p = 13$ and $q = 9$, so that $v = 4$. It shows how the algorithm fails because 3, the greatest common divisor of a and b, does not divide $v = 4$.

We now get another form of the rule, based on the values a, b, c etc. generated in the Euclidean algorithm. Without any explanation, Euler tells us that

$$z = q + abv\left(\frac{1}{ab} - \frac{1}{bc} + \frac{1}{cd} - \frac{1}{de} + \frac{1}{ef} - \text{etc.}\right).$$

He demonstrates that this rule works for the example $a = 16$, $b = 9$, $p = 1$, $q = 7$ and taking $v = -6$. He finds the smallest positive solution to be 97, using both this formula and the algorithm he had explained above.

We move on to special cases. Our first special case is when $b = a - 1$, and remainders p and q are required. Because a and b are relatively prime, a solution will always exist. The Euclidean algorithm stops after just one step, with $c = 1$, and gives the solution

$$z = p - av = p - ap + aq.$$

Euler attributes this observation to Stifel. Euler also gives us an example, to find a number that can be divided by 100 to leave remainder 75 and by 101 to leave remainder 37. Here, $a = 101$, $b = 100$, $p = 37$, $q = 75$ and $v = -38$, so $z = 3875$.

The next special case is when $p = q$, so that $v = 0$. Then $z = p$ is the smallest possible solution. If M is the least common multiple of a and b, the set of all solutions is given by $mM + p$, where m is any integer.

This special case leads to a quick solution to the problem that first motivated this paper, the "common problem from ordinary arithmetic books," to find a number which divided by 2, 3, 4, 5 or 6 leaves remainder 1, but divided by 7 leaves no remainder. The first part of the problem is done easily: because 60 is the least common multiple of 2, 3, 4, 5 and 6, a solution must be of the form $60m + 1$. From there, the usual method, with $a = 60$, $b = 7$, $p = 1$, $q = 0$ and $v = 1$, leads to the smallest positive solution 301 and the general solution $z = 420m + 301$.

This method also easily solves a common twist on this same problem, to divide by 2, 3, 4, 5 and 6 and leave remainders of 1, 2, 3, 4 and 5, respectively, but to divide by 7 and leave no remainder. The trick is to notice that all of the remainders are equal if we see they are all congruent to -1. Then the solution, 119, follows easily.

Euler uses these two examples to suggest general problems involving more than two divisors and remainders. When none of these special cases apply, he tells us to solve the problems a pair at a time and demonstrates with another example, to divide by 7 and leave remainder 6, by 9 and leave remainder 7, by 11 and leave remainder 8 and by 17 to leave remainder 1. The first step is to divide by 7 and leave 6 and by 9 to leave 8. The usual method gives the general solution $z = 63m + 34$. The second step is to divide by 63 and leave 34, and by 11 and leave 8, and the final step is to use that general solution and also to divide by 17 and leave 1. Euler finds that the general solution to the whole problem is $z = 11781m + 1735$.

As an application, Euler proposes to find the year in which Christ was born. He says that the solar years go in 28-year cycles and lunar years go in 19-year cycles, and the *Indictio Romana* or census occurred every 15 years. So, if we knew p, the year in the solar cycle, q, the year in the lunar cycle, and r, the year in the cycle of the census, and knowing that the least common multiple of 15, 19 and 28 is 7980, the possible solutions are $4845p + 4200q + 6916r + 7980m$. Because in Euler's time, the world was thought to be only a few thousand years old, this leads to a single solution.

This solution is unsatisfactory on two counts. First, some of the information, p and q, required for the solution is unavailable. Perhaps Euler thinks that theologians will eventually come upon the data. Second, he is using a technique he has not yet explained. In the last section of this article he explains where those coefficients 4845, 4200 and 6916 came from.

Euler claims that if the divisors, a, b, c, d, e etc. are relatively prime, and if remainders p, q, r, s, t etc. are required, then the solutions are given by

$$Ap + Bq + Cr + Ds + Et + mabcde,$$

where A is a number divisible by the product $bcde$, but when divided by a leaves remainder 1. Similarly, B is divisible by $acde$, and divided by b leaves 1, etc.

It is clear why this works. The first term, Ap, for example, divided by a, leaves p, whereas all other terms are multiples of a. Likewise, each of the divisors, b, c, d, e, divide all but one of the terms, and for that exceptional term, they leave the required remainder. Readers familiar with Hermite polynomials in numerical analysis will recognize this as a similar idea. [B+F, p. 134]

This paper seems to have been written for fun. Euler was motivated by a problem in an elementary book. In a couple of years we will see another paper, E-53, the Königsburg Bridge Problem, also written just for fun. Later, in 1759, Euler will write E-309, on the Knight's Tour, which begins, "I found myself one day at a party where, over a game of chess, someone proposed this question: . . . " These three papers give us a glimpse of the joy that Euler found in mathematics, and that he did mathematics in part simply because he loved it.

References

[B+F] Burden, Richard L. and J. Douglas Faires, *Numerical Analysis*, 7th ed., Brooks/Cole, 2001.

[D] Dickson, Leonard Eugene, *Theory of Numbers*, v. 2, p. 59–62, Chelsea, NY, 1952.

46

E-157: De extractione radicum ex quantitatibus irrationalibus*

On the extraction of roots of irrational quantities

Euler opens this article with, "In the old days, Analysts devoted a great deal of study to the doctrine of irrational quantities, or surds." It is an old topic that filled thousands of pages in 16th century algebra books and, as Euler notes, occupied an important place in Isaac Newton's *Arithmetica universalis*.

We begin with topics that were probably familiar to Euler's contemporaries but have been largely forgotten today. We want to find square roots of binomials, $A \pm B$, where A and B are either rational or square roots.

If $AA - BB = CC$, then the square root of $A + B$ is

$$= \sqrt{\frac{A+C}{2}} + \sqrt{\frac{A-C}{2}}$$

and the square root of $A - B$ is

$$\sqrt{\frac{A+C}{2}} - \sqrt{\frac{A-C}{2}}.$$

He takes a moment to explain how we can treat both of these simultaneously by using what he calls the "ambiguous sign," $\pm$. We get no reasons this is true, though it is easy to verify, nor do we get any clues about how it might be discovered. We do get an example, though. To find the square root of $54 \pm \sqrt{980}$, we set $A = 54$ and $B = \sqrt{980}$. Then $AA - BB = 1936$, which is the square of 44. Then $\frac{A+C}{2} = 49$ and $\frac{A-C}{2} = 5$, so the two roots are $7 \pm \sqrt{5}$.

In case A and B are both square roots, that is the binomial has the form $\sqrt{a} \pm \sqrt{b}$, then Euler hopes that $a(a-b) = cc$ is a square, and he takes $a - b = d$. Then the square root of the binomial is given by

$$\frac{\sqrt{\frac{c+d}{2}} \pm \sqrt{\frac{c-d}{2}}}{\sqrt[4]{d}}.$$

Again, there is no derivation, but we do get another couple of examples.

In the first one, we take the square root of the binomial $\sqrt{12} \pm 3$. Note that the radical part, $\sqrt{12}$ is larger than the rational part, 3. Here, $a = 12$, $b = 9$, and so $a - b = 3 = d$ and $a(a-b) = 36 = cc$, so $c = 6$. This makes the square root of the binomial

$$\frac{\sqrt{\frac{9}{2}} \pm \sqrt{\frac{3}{2}}}{\sqrt[4]{3}} = \frac{3 \pm \sqrt{3}}{\sqrt[4]{12}}.$$

Comm. Acad. Sci. Imp. Petropol. 13 (1741/3) 1751, 16–60; *Opera Omnia* I.6, 31–77

He continues with a similar example, $4\sqrt{3}+3\sqrt{5}$, finding the root to be $\frac{\sqrt{15}+3}{\sqrt[4]{12}}$.

We move on to more serious topics. "It is a much more difficult task to extract a cube root or any higher power of such binomials." Euler describes in detail a method of Newton from *Arithmetica universalis*. If $A \pm B$ is the given binomial and we seek the cth root, then we begin by finding the smallest number n whose cth power is divisible by $AA-BB$. That is to say, Euler explains, $AA-BB$ divides n^c. Let Q be the quotient of n^c by $AA-BB$. Now, find r, a "convenient integer" number near $\sqrt[c]{(A+B)\sqrt{Q}}$. Third, divide the expression $A\sqrt{Q}$ by its maximum rational integer divisor. We shall identify that divisor when we do the examples. This will leave an irreducible irrational quotient that is to be set equal to s. Fourth, let t be an integer which is near the expression $\frac{rr+n}{2rs}$. With this preparation, Newton's method tells that the desired root of the given binomial $A \pm B$ will be

$$\frac{ts \pm \sqrt{ttss+n}}{\sqrt[2c]{Q}}.$$

Euler agrees that it would be a good idea to work an example, for this is a complicated rule. We will try to find the "radix supersolida" of the binomial $5\sqrt{5}+11$. This term "supersolida" is long obsolete, having enjoyed the last of its popularity about 1600. It heralds from a time when each power or root had its own geometric name. A fourth root was described as a square root of a square root. A fifth root is "beyond the solid root," that is, the first root beyond cube roots that is not reducible to simpler roots. The only remnant of this old vocabulary is that we still call the second and third powers and roots "squares" and "cubes." This is the only time we have seen Euler use this old vocabulary. This reflects the fact that this topic, which he regards as arithmetic, is very old, as he observed in his opening sentence.

Euler has chosen his example carefully, and already knows that the root he seeks is $\frac{\sqrt{5}+1}{\sqrt[5]{16}}$.

We begin by setting $A = 5\sqrt{5}$ and $B = 11$. Because we want a fifth root, $c = 5$. Then $AA - BB = 4$, and the smallest fifth power divisible by four is 32. This makes $n = 2$ and

$$Q = \frac{n^c}{AA-BB} = \frac{32}{4} = 8.$$

Thus we have $\sqrt[c]{(A+B)\sqrt{Q}} = \sqrt[5]{(5\sqrt{5}+11)\sqrt{8}} = \sqrt[5]{5\sqrt{40}+11\sqrt{8}}$.

Euler calculates. $5\sqrt{40} = 31.622$ and $11\sqrt{8} = 31.112$. (Both times here, as well as elsewhere in this paper, he rounds down when he probably should have rounded up.) Consequently $5\sqrt{40} + 11\sqrt{8} = 62.734$, where the fifth root is 2.288. A "convenient integer" near this is 2, so take $r = 2$.

For the third step, $A\sqrt{Q} = 5\sqrt{40} = 10\sqrt{10}$. The largest rational factor of this is 10, and the quotient is $\sqrt{10}$, so we take $s = \sqrt{10}$.

Fourth, $\frac{rr+n}{2rs} = \frac{6}{4\sqrt{10}} = \frac{3}{2\sqrt{10}}$. This evaluates to 0.474. Still, Euler tells us that the integer that comes near this value is $t = 1$. It would seem that zero might be closer, and it is not clear why Euler chooses $t = 1$ instead.

In any case, we have $ts = \sqrt{10}$ and $\sqrt{ttss+n} = \sqrt{12}$.

Finally, the tools are gathered and Euler can apply Newton's formula. The fifth root we seek ought to be

$$\frac{ts \pm \sqrt{ttss+n}}{\sqrt[2c]{Q}} = \frac{\sqrt{10}+\sqrt{12}}{\sqrt[10]{8}}.$$

Euler notes that this answer is clearly incorrect, for it contains a factor of $\sqrt{3}$, and the correct answer, $\frac{\sqrt{5}+1}{\sqrt[5]{16}}$, contains no such factor.

So, Euler concludes his analysis of Newton's method with a surprise: not only is it rather difficult, but also it is incorrect! Euler says that this is not the only case for which it is wrong, but that it almost always gives an incorrect answer. Euler sets out to find a correct method.

He takes $A \pm B$ to be his binomial and n to be the degree of the root he seeks. He assumes that both AA and BB are integers with $AA > BB$, and he guesses that the form of the nth root will be $\frac{x \pm y}{\sqrt[2n]{p}}$, where x and y are not necessarily integers. A bit of thought about signs leads to the observation that

$$\sqrt[n]{A+B} = \frac{x+y}{\sqrt[2n]{p}}$$

means

$$\sqrt[n]{A-B} = \frac{x-y}{\sqrt[2n]{p}}.$$

If we multiply these two equations, we get

$$\sqrt[n]{AA-BB} = \frac{xx-yy}{\sqrt[n]{p}} \quad \text{or} \quad xx-yy = \sqrt[n]{(AA-BB)p}.$$

Here, $(AA-BB)p$ must be an nth power, say r^n, so that

$$p = \frac{r^n}{AA-BB}$$

and $xx-yy=r$.

Now, squaring and adding the same formulas that we multiplied before gives us

$$\sqrt[n]{(A+B)^2} + \sqrt[n]{(A-B)^2} = \frac{(x+y)^2+(x-y)^2}{\sqrt[n]{p}}$$

so that

$$xx+yy = \frac{1}{2}\sqrt[n]{(A+B)^2p} + \frac{1}{2}\sqrt[n]{(A-B)^2p}.$$

This makes the integer $xx+yy$ the sum of two irrational quantities, so there is a sense in which they have the same irrational part. Euler makes this idea clear by taking s and t to be the integers closest to $\sqrt[n]{(A+B)^2p}$ and $\sqrt[n]{(A-B)^2p}$ respectively. One integer will be larger than its radical, and the other will be smaller. That is

$$\sqrt[n]{(A+B)^2p} = s \pm \text{fraction} \quad \text{and} \quad \sqrt[n]{(A-B)^2p} = t \mp \text{fraction},$$

and the two fractions, both positive and less than $\frac{1}{2}$, will be the same.

It follows that $xx+yy = \frac{s+t}{2}$. Knowing that $xx-yy=r$, we get $xx = \frac{2r+s+t}{4}$ and $yy = \frac{s+t-2r}{4}$. So the root we wanted is

$$\frac{\sqrt{2r+s+t} \pm \sqrt{s+t-2r}}{2\sqrt[2n]{p}}.$$

Euler makes a few remarks about where Newton went wrong, then tests his method by finding the (correct) fifth root of $5\sqrt{5}+11$. Here $A = 5\sqrt{5}$, $B = 11$ and $n = 5$. Euler's calculations give $AA - BB = 4$, $r = 2$ and $p = 8$. Then

$$(A+B)^2 = 246 + 110\sqrt{5} \quad \text{and} \quad (A-B)^2 = 246 - 110\sqrt{5}.$$

As decimals he gets

$$(A+B)^2 p = 3935.7333 \quad \text{and} \quad (A-B)^2 p = 0.2666.$$

His calculations are a little off in the third decimal place, but it does not affect his answers. Euler finds the fifth roots of these decimals to be 5.23 and 0.76. The correct fifth roots are 5.24 and 0.76, which, to two decimal places, sum to 6. Anyway, $s+t=6$. Euler does not actually note that $s=5$ and $t=1$, though according to his method they would. However, his answer does not depend on s and t separately, but only on their sum.

Now, we have s, t, r and p. Substituting into Euler's formula gives $\frac{\sqrt{10}+\sqrt{2}}{2\sqrt[10]{8}}$, which is easily seen to equal what we know to be the correct answer, $\frac{\sqrt{5}+1}{\sqrt[5]{16}}$.

Euler does another example, the 7th root of $139\sqrt{3}+19\sqrt{7}$, which he finds to be $\frac{\sqrt{7}+\sqrt{3}}{\sqrt[7]{64}}$.

According to Euler, this method seems to work in every example he has tried. He does not seem to have enough confidence in his derivation to assert that it always works, though. Perhaps somebody should check. He does admit a critical defect, that it "does not accommodate binomials in which there are imaginary quantities." Until now, we have not seen Euler show this kind of commitment to imaginary quantities, or, as we call them now, complex numbers. He has introduced them when they have been useful, but it has never bothered him before when a technique did not apply to imaginary quantities.

We now change gears a bit. We will find roots of a binomial $A \pm B$ by introducing a polynomial in z and finding roots of that polynomial. Euler takes $z = \sqrt[n]{(A+B)^2 p} + \sqrt[n]{(A-B)^2 p}$, where one or the other of the quantities A and B may be imaginary. He continues to assume that his binomial has an nth root of the form $\frac{x \pm y}{\sqrt[2n]{p}}$. To start this new analysis, Euler reminds us of a formula of de Moivre, that if

$$\alpha = z^n - nz^{n-2}\sqrt[n]{\beta} + \frac{n(n-3)}{1\cdot 2}z^{n-4}\sqrt[n]{\beta^2} - \frac{n(n-3)(n-5)}{1\cdot 2\cdot 3}z^{n-6}\sqrt[n]{\beta^3}$$
$$+ \frac{n(n-3)(n-5)(n-7)}{1\cdot 2\cdot 3\cdot 4}z^{n-8}\sqrt[n]{\beta^4} - \text{etc.},$$

then its root is given by

$$z = \sqrt[n]{\frac{\alpha + \sqrt{\alpha^2 - 4\beta}}{2}} + \sqrt[n]{\frac{\alpha - \sqrt{\alpha^2 - 4\beta}}{2}}.$$

To make this apply to our problem, Euler must separate real and imaginary quantities, so he rewrites z to get

$$z = \sqrt[n]{(A^2+B^2)p + 2pAB} + \sqrt[n]{(A^2+B^2)p - 2pAB}.$$

Matching terms, he sees that $\alpha = 2p(AA+BB)$, which is an integer, and $\sqrt{\alpha\alpha - 4\beta} = 4pAB$, which could be an imaginary integer. Also, $\beta = pp(AA-BB)^2 = r^{2n}$, which is an integer.

De Moivre's formula gives us

$$z^n - nr^2z^{n-2} + \frac{n(n-3)}{1\cdot 2}r^4z^{n-4} - \frac{n(n-4)(n-5)}{1\cdot 2\cdot 3}r^6z^{n-6} + \text{etc.} = 2p(AA+BB).$$

Because $xx+yy = \frac{1}{2}z$ and $xx-yy=r$, we see that $x = \frac{\sqrt{z+2r}}{2}$ and $y = \frac{\sqrt{z-2r}}{2}$. This makes the nth root of the binomial $A \pm B$ equal

$$\frac{\sqrt{z+2r} \pm \sqrt{z-2r}}{2\sqrt[2n]{p}}.$$

Euler returns to his favorite binomial to demonstrate this technique. He seeks the fifth root of $A+B = 5\sqrt{5}+11$. Note that the technique works if one of the terms of the binomial is imaginary, but it does not require that one be imaginary. Here $n=5$, $A=5\sqrt{5}$ and $B=11$, as before. He gets the polynomial $z^5 - 20z^3 + 80z = 16\cdot 246$, and finds that it has a real root, $z=6$. This leads easily to the root we expect for this binomial.

Euler notes that to find the root of the polynomial in z, it is often easier to take $z=ru$ and to solve instead a polynomial in u. It is still an nth degree polynomial, but the coefficients are smaller.

Euler leads us through a few more examples. He does the seventh root of $139\sqrt{3}+91\sqrt{7}$ again, and then, to demonstrate how the method works with imaginary terms, he finds the fifth root of $+1-\sqrt{-3}$ and the seventh root of $13\sqrt{3}+\sqrt{-5}$.

Euler will want to use roots of binomials to dissect angles. To prepare for this, he announces two lemmas.

I. If $u = \cos\frac{1}{n}\arcsin a$, then

$$u = \frac{\sqrt[n]{\sqrt{1-aa}+a\sqrt{-1}} + \sqrt[n]{\sqrt{1-aa}-a\sqrt{-1}}}{2}.$$

II. If $v = \sin\frac{1}{n}\arcsin a$, then

$$v = \frac{\sqrt[n]{\sqrt{1-aa}+a\sqrt{-1}} - \sqrt[n]{\sqrt{1-aa}-a\sqrt{-1}}}{2\sqrt{-1}}.$$

Now, if

$$a\sqrt{-1} = \frac{2AB}{AA-BB}$$

so that

$$\sqrt{1-aa} = \frac{AA+BB}{AA-BB},$$

then the lemmas produce the equations:

$$\text{I. } \cos\frac{1}{n}\arcsin\frac{2AB}{(AA-BB)\sqrt{-1}} = \frac{\sqrt[n]{(A+B)^2}+\sqrt[n]{(A-B)^2}}{2\sqrt[n]{AA-BB}}.$$

$$\text{II. } \sin\frac{1}{n}\arcsin\frac{2AB}{(AA-BB)\sqrt{-1}} = \frac{\sqrt[n]{(A+B)^2}-\sqrt[n]{(A-B)^2}}{2\sqrt{-1}\sqrt[n]{AA-BB}}.$$

Euler combines these lemmas with his previous results and takes

$$\alpha = \arcsin \frac{2AB}{(AA - BB)\sqrt{-1}} = \arccos \frac{AA + BB}{AA - BB}.$$

He gets

$$x + y = \sqrt{r}\left(\cos\frac{\alpha}{n} + \sqrt{-1}\sin\frac{\alpha}{n}\right) \quad \text{and} \quad x - y = \sqrt{r}\left(\cos\frac{\alpha}{n} - \sqrt{-1}\sin\frac{\alpha}{n}\right).$$

Combining these formulas reduces the root of our binomial $A \pm B$ to

$$\frac{\sqrt{1 + \cos\frac{\alpha}{n}} + \sqrt{-1 + \cos\frac{\alpha}{n}}}{\sqrt[2n]{\frac{2^n}{AA-BB}}}.$$

We would like to take roots of expressions that are not binomials. What we know now about solvability makes us sure that Euler will enjoy only limited success in this effort, but he is encouraged that he can find the square root of

$$-\frac{7}{3} + \frac{1 - \sqrt{-3}}{6}\sqrt[3]{\frac{7}{2}\left(-1 - 3\sqrt{3}\right)} + \frac{1 + \sqrt{-3}}{6}\sqrt[3]{\frac{7}{2}\left(-1 + 3\sqrt{3}\right)}.$$

Euler begins his next analysis by supposing we want to find the nth root of some irrational quantity x. Call the root y so that $x = y^n$. Then Euler says it is easy to find an algebraic equation that has x as a root, and he assumes that the equation has the form

$$x^m + ax^{m-1} + bx^{m-2} + cx^{m-3} + dx^{m-4} + \text{etc.} = 0.$$

Then y will satisfy

$$y^{mn} + ay^{mn-n} + by^{mn-2n} + cy^{mn-3n} + \text{etc.} = 0.$$

Now, in a key step that Euler does not fully explain, he tells us that this last form can be factored into n equations, each of degree m, one of which will be

$$y^m + Ay^{m-1} + By^{m-2} + Cy^{m-3} + \text{etc.} = 0.$$

Thus the problem of finding an nth root is transformed into a problem of solving an mth degree polynomial. It may be simpler, or it may not. Moreover, the mth degree polynomial will have m roots, only one of which will be the nth root of x being sought.

As a kind of example, Euler considers the case where x is the root of a quadratic, $xx + ax + b = 0$ and we seek the nth root of x. By the quadratic formula, we know the value of x. Then y must satisfy $y^{2n} + ay^n + b = 0$. This must factor into n quadratic factors, each of which will have the form $yy + uy\sqrt[2n]{b} + \sqrt[n]{b} = 0$. Without the benefit of subscripts or explanation, Euler tells us that these n different values of u are the roots of the equation

$$u^n - nu^{n-2} + \frac{n(n-3)}{1 \cdot 2}u^{n-4} - \frac{n(n-4)(n-5)}{1 \cdot 2 \cdot 3}u^{n-6} + \text{etc.} \pm \frac{a}{\sqrt{b}} = 0.$$

We can use the quadratic formula to give y in terms of u.

Euler offers an example: to find the seventh root of $45\sqrt{5} - 7\sqrt{-7}$. Then $a = -82\sqrt{5}$ and $b = 4 \cdot 3^7$, so $-\frac{a}{\sqrt{b}} = \frac{41\sqrt{5}}{\sqrt{3^7}}$ and the polynomial in u is $u^7 - 7u^5 + 14u^3 - 7u + \frac{41\sqrt{5}}{\sqrt{3^7}} = 0$. Standard techniques of the time, or your favorite computer algebra system today, can check that $u = -\frac{\sqrt{5}}{\sqrt{3}}$ is a root of this expression. Then the quadratic formula gives the seventh root we seek: $y = \frac{\sqrt{5}+\sqrt{-7}}{\sqrt[7]{64}}$.

Suppose that x is the root of a cubic. Euler starts by looking for a square root of x, given that x satisfies $x^3 + ax^2 + bx + c = 0$. Its square root, y, will then satisfy $y^6 + ay^4 + by^2 + c = 0$, and we learned before that this polynomial must factor into two cubics. The key part of this step is to show that the factors are of the forms $y^3 + py^2 + qy + r = 0$ and $y^3 - py^2 + qy - r = 0$. With this, the rest of the case is straightforward. He tests his analysis on a root of the equation $x^3 + 3xx + 6x - 25 = 0$. He chose this example because the English mathematician Colson had examined it in 1707.

Euler's next case is to find cube roots of roots of cubics. Again, the trick is to factor the resulting 9th degree polynomial, and it is not too difficult, though it does involve the complex cube roots of unity.

By this time, Euler hopes that his readers understand the analysis and realize that the tricky step is to factor those polynomials of degree mn. To make this easier, he gives formulas for the coefficients of the factors for cubics in case $n = 1$ through 6 and for quartics in case $n = 1$ to 4. He assures us that higher roots can be handled in the same way, but it might be significant to note that he makes no such guarantees in case x is the root of a quintic or higher degree equation.

Euler has noticed, and perhaps the reader has noticed as well, that taking nth roots seems to involve nth roots of unity. So, Euler turns to examining the properties of the roots of the equation $x^n - 1 = 0$. He notes that if α is one such root, then α^2, α^3, α^4, α^5, etc. are also roots. He denotes the roots with Greek letters, α, β, γ, etc. and gives complex values for each of these roots up to tenth roots.

He gives special treatment to fifth roots of unity. He finds

$$\alpha = \frac{-1 - \sqrt{5} + \sqrt{-10 + 2\sqrt{5}}}{4}$$

and also gives explicit complex values for β, γ and δ. He does this for the other roots of unity as well, but for the fifth roots he notes that $\alpha^2 = \delta$, $\alpha^3 = \gamma$, $\alpha^4 = \beta$, $\alpha^5 = 1$, and similarly for β, γ and δ.

Seventh roots also get special consideration. Euler plows through six and a half pages of dense calculations to get explicit values for the seventh roots of unity. Two of those roots are given by the astonishing expression

$$\frac{-1 \pm \sqrt{-7}}{6} - \frac{1}{6\sqrt{-7}}\left(\sqrt{-7} \pm (-2 + \sqrt{-3})\right)\sqrt[3]{7\frac{-1 + 3\sqrt{-3}}{2}}$$
$$- \frac{1}{6\sqrt{-7}}\left(\sqrt{-7} \pm (-2 - \sqrt{-3})\right)\sqrt[3]{7\frac{-1 - 3\sqrt{-3}}{2}}.$$

A footnote in the *Opera Omnia* reports that in 1771 Vandermonde found explicit representations for the eleventh roots of unity. Among 12th to 16th roots, only the 13th root presents any difficulty, because the rest are all composite numbers. That leaves 17 as the next interesting case. In 1801,

Gauss made one of his first contributions to mathematics when he showed that the 17th roots of unity are constructible with ruler and compass.

So Euler's excursion into the old problem of extracting roots of binomials has brought us to the "modern" topic of cyclotomic polynomials, though at this point, Euler has only an inkling of the role they will eventually play.

Interlude: 1741

World events

The War of the Austrian Succession continued through 1741. Elizabeth I became Empress of Russia, though it took several years for the unrest there to calm down.

Vitus Bering died in December on an expedition to Alaska and Siberia. His expedition showed that Asia and North America were not attached and also that California was not an island. The news of Bering's discoveries was announced to the world in a letter to the Royal Society of London. That letter was written by Leonhard Euler, and was one of the only times Euler ever mentioned America.

Also, Anders Celsius developed the temperature scale used today in most of the world.

In Euler's life

The political instability in Russia caused Euler to change his mind about accepting the offer from Frederick II, so Euler left St. Petersburg to spend 25 years in Berlin. His initial salary in Berlin was 1600 thalers, which Clifford Truesdell estimated to be worth about $80,000 in 1982.

In this political turmoil and with the efforts of moving, Euler did not produce much in 1741, and what he did write is difficult to place accurately. It is hard to know if he wrote in St. Petersburg, in Berlin, or perhaps even along the way. We will include all the works here and discuss our doubts about particular ones later.

Euler's other work

Euler wrote only one scientific article in 1741, E-132, Methodus determinandi gradus meridiani pariter ac paralleli telluris, secundum mensuram a celeb. de Maupertuis cum sociis institutam, "Method of determining the degree of an equal or parallel meridian on the earth, as found in the second measurement of Maupertuis with his expedition."

Euler's mathematics

There were three mathematical articles in 1741. E-63, Démonstration de la somme de cette suite $1 + \frac{1}{4} + \frac{1}{9} + \frac{1}{16} + \frac{1}{25} +$ etc., "Proof of the sum of this series $1 + \frac{1}{4} + \frac{1}{9} + \frac{1}{16} + \frac{1}{25} +$ etc." is a completely different solution to the Basel problem. It was probably written in Berlin. In E-158, Observationes analyticae variae de combinationibus, "Several analytic observations on combinations," Euler invents generating functions and solves a problem of Philip Naudé. Finally, E-790, Commentatio de matheseos sublimioris utilitate, "On the utility of higher mathematics," was written in 1741 but not published for over a hundred years. It is a fitting way to end this book, and we provide a new English translation of the essay.

47

E-63: Démonstration de la somme de cette suite $1 + \frac{1}{4} + \frac{1}{9} + \frac{1}{16} + \frac{1}{25} + \frac{1}{36} + \text{etc.}$*

Proof of the sum of this series $1 + \frac{1}{4} + \frac{1}{9} + \frac{1}{16} + \frac{1}{25} + \frac{1}{36} + \text{etc.}$

Euler's greatest mathematical triumph in his first St. Petersburg years was his solution to the Basel Problem, the sum of the reciprocals of the square integers. He gave three solutions in E-41, but not everyone was satisfied. The first two solutions depended on delicate operations on infinite series, and his critics were right to be suspicious of these.

People were even more skeptical of Euler's third solution, the one that is best remembered today. It depends on expressing the function $\frac{\sin x}{x}$ both as its Taylor series

$$1 - \frac{x^2}{3!} + \frac{x^4}{5!} - \frac{x^6}{7!} + \frac{x^8}{9!} - \text{etc.}$$

and as an infinite product

$$\left(1 - \frac{x}{\pi}\right)\left(1 + \frac{x}{\pi}\right)\left(1 - \frac{x}{2\pi}\right)\left(1 + \frac{x}{2\pi}\right)\left(1 - \frac{x}{3\pi}\right)\left(1 + \frac{x}{3\pi}\right)\text{etc.}$$

Euler gives us only one reason to believe that the product is actually equal to the Taylor series: that it has the same roots as $\frac{\sin x}{x}$. This certainly is a *necessary* condition, but it is not *sufficient*. As we have seen before, particularly in Euler's solution to the Königsburg Bridge problem and in his work on the calculus of variations, he sometimes treats necessary conditions as if they were sufficient conditions. In this case, the function $e^x \frac{\sin x}{x}$ has the same zeros as $\frac{\sin x}{x}$, yet they can't both be given by the same infinite product. Euler is correct; the infinite product does indeed give $\frac{\sin x}{x}$. But his critics are right too; Euler's reasons are not strong enough.

Euler was sensitive to these criticisms. Here in E-63 he answers them with an entirely different solution to the Basel problem, one that is both complete and correct. It uses tools no more controversial than integration by parts, the binomial theorem, and simple recursive relationships. This solution is mostly forgotten today. People prefer the elegance of the third solution from E-41.

Even in 1907 Paul Stäkel, one of the editors of the *Opera Omnia*, wrote an article "Eine vergessene abhandlung Leonhard Eulers über die Summe der reziproken Quadrate de natürlichen Zahlen" (A forgotten article of Leonhard Euler about the sum of the reciprocals of the squares of the natural numbers).

Euler wrote this article in French in 1741, the year he left St. Petersburg for Berlin. It is his first article in French, which suggests that he may have composed it after he arrived in Berlin, where Frederick the Great expected his courtiers to speak and write in French. We include it here in part because it *may* have been written in St. Petersburg and in part to give some closure to the story of the Basel Problem.

Journal littéraire d'Allemagne, de Suisse et du Nord* (La Haye) **2:1, 1743, 115–127; *Opera Omnia* I.14, 177–186

The article appeared in the *Journal littéraire d'Allemagne, de Suisse et du Nord* in 1743, a trendy publication popular among the courtiers, the *salonistes* and the intellectual elite of the time. Today we might call it a "place to see and to be seen." That Euler chose to publish here is another clue that he may have written it in Berlin, not St. Petersburg.

Euler begins by asking us to consider a circle of radius 1. He takes s to be arc length, and takes $x = \sin s$, or, equivalently, $s = \arcsin x$. Then, working with differentials as he always does,

$$ds = \frac{dx}{\sqrt{1-xx}} \quad \text{and} \quad s = \int \frac{dx}{\sqrt{1-xx}}.$$

Now, he multiplies these together to get

$$s\,ds = \frac{dx}{\sqrt{1-xx}} \int \frac{dx}{\sqrt{1-xx}}.$$

He integrates both sides from $x = 0$ to $x = 1$. On the left, the antiderivative is $\frac{ss}{2}$. As x goes from 0 to 1, s goes from 0 to $\frac{\pi}{2}$, so he gets on the left $\frac{\pi\pi}{8}$.

On the right, Euler dives fearlessly into an intricate series calculation. He writes

$$\frac{1}{\sqrt{1-xx}} = (1-xx)^{-1/2}$$

and applies the generalized binomial theorem to expand the radical as an infinite series. He gets

$$(1-xx)^{-1/2} = 1 + \frac{1}{2}x^2 + \frac{1\cdot 3}{2\cdot 4}x^4 + \frac{1\cdot 3\cdot 5}{2\cdot 4\cdot 6}x^6 + \frac{1\cdot 3\cdot 5\cdot 7}{2\cdot 4\cdot 6\cdot 8}x^8 + \text{etc.}$$

This is a bit of a tricky step, but it really is the familiar binomial theorem, that

$$(1+a)^n = 1 + \frac{n}{1}a + \frac{n(n-1)}{1\cdot 2}a^2 + \frac{n(n-1)(n-2)}{1\cdot 2\cdot 3}a^3 + \cdots$$

in the case $a = xx$ and $n = \frac{-1}{2}$. If n is a positive integer, then the numerators eventually become zero, and we get a finite sum, but the theorem is still true if n is a fraction. Euler's series converges whenever $|x| < 1$. He integrates and multiplies to get

$$s\,ds = \frac{x\,dx}{\sqrt{1-xx}} + \frac{1}{2\cdot 3}\frac{x^3\,dx}{\sqrt{1-xx}} + \frac{1\cdot 3}{2\cdot 4\cdot 5}\frac{x^5\,dx}{\sqrt{1-xx}} + \frac{1\cdot 3\cdot 5}{2\cdot 4\cdot 7}\frac{x^7\,dx}{\sqrt{1-xx}} + \cdots.$$

He knows that if he integrates both sides from $x = 0$ to $x = 1$, he will have $\frac{\pi\pi}{8}$. Integrating an individual term on the right, using integration by parts, produces

$$\int \frac{x^{n+2}\,dx}{\sqrt{1-xx}} = \frac{n+1}{n+2}\int \frac{x^n\,dx}{\sqrt{1-xx}} - \frac{x^{n+1}}{n+2}\sqrt{1-xx}.$$

Because the second term is zero at both endpoints, he can ignore it, and he discovers a nice reduction formula. He summarizes the integral result with a list:

$$\int_0^1 \frac{x\,dx}{\sqrt{1-xx}} = 1$$

$$\int_0^1 \frac{x^3\,dx}{\sqrt{1-xx}} = \frac{2}{3}\int_0^1 \frac{x\,dx}{\sqrt{1-xx}} = \frac{2}{3}$$

$$\int_0^1 \frac{x^5\,dx}{\sqrt{1-xx}} = \frac{4}{5}\int_0^1 \frac{x^3\,dx}{\sqrt{1-xx}} = \frac{2\cdot 4}{3\cdot 5}$$
$$\int_0^1 \frac{x^7\,dx}{\sqrt{1-xx}} = \frac{6}{7}\int_0^1 \frac{x^5\,dx}{\sqrt{1-xx}} = \frac{2\cdot 4\cdot 6}{3\cdot 5\cdot 7}.$$

So, the integral of the expression above (the one that begins $s\,ds$) is

$$\frac{\pi\pi}{8} = \int_0^1 \frac{x\,dx}{\sqrt{1-xx}} + \frac{1}{2\cdot 3}\int_0^1 \frac{x^3\,dx}{\sqrt{1-xx}} + \frac{1\cdot 3}{2\cdot 4\cdot 5}\int_0^1 \frac{x^5\,dx}{\sqrt{1-xx}} + \frac{1\cdot 3\cdot 5}{2\cdot 4\cdot 6\cdot 7}\int_0^1 \frac{x^7\,dx}{\sqrt{1-xx}} + \text{etc.}$$

Substituting the values for the integrals gives

$$\frac{\pi\pi}{8} = 1 + \frac{1}{3\cdot 3} + \frac{1}{5\cdot 5} + \frac{1}{7\cdot 7} + \frac{1}{9\cdot 9} + \text{etc.}$$

The series on the right is the sum of the reciprocal of the odd squares. We saw this before in E-41, and Euler uses the same trick to conclude that

$$\frac{\pi\pi}{6} = 1 + \frac{1}{4} + \frac{1}{9} + \frac{1}{16} + \frac{1}{25} + \frac{1}{36} + \cdots.$$

This is a different solution to the Basel problem that does not depend on infinite products. In fact, all it requires to meet modern standards of rigor is that we fill in some routine steps and notice that certain series are absolutely convergent, so that we can do things like multiply two different series together.

By most objective standards, this is the "best" of Euler's four solutions to the Basel Problem. Why, then, is it not the one remembered today? Could it be that this was not the first solution? In such matters we usually remember the first solution, not the fourth. This can't be the reason because the solution we do remember is the third, not the first.

Could it be because it was in French and published in the *Journal littéraire*? This is plausible, for at the time most "serious" mathematics was published in Latin and in scientific journals, not literary ones.

It is more likely, though, that we remember the third solution because we find it more "elegant." It has the brilliant (though flawed) interaction between the infinite sum and the infinite product. It can be explained on a blackboard in just a few lines. It is the proof from *The Book*.[1] In contrast, this fourth proof is sterile and mechanical, although it suggests that Euler had an intuitive concern about the rigor of the infinite product. The use of the binomial theorem for the exponent $\frac{1}{2}$ is clever, and the recursive relation on the integrals is nice, but they pale before the beauty of Euler's infinite product.

Why do we tell the story of the Basel Problem as we do? Because the third proof is the most beautiful one, and beautiful mathematics makes better stories.

[1] Twentieth century mathematician Paul Erdős (1913–1996) reportedly said that God has a book "that contains the best proofs of all mathematical theorems, proofs that are elegant and perfect." He also said that as a mathematician "You don't have to believe in God, but you should believe in the *Book*."

References

[H] Hoffman, Paul, *The Man Who Loved Only Numbers: The Story of Paul Erdős and the Search for Mathematical Truth*, Hyperion, New York, 1998.

48

E-158: Observationes analyticae variae de combinationibus*

Several analytic observations on combinations

We finally reach the last of the papers from Euler's first St. Petersburg period. Euler presented it to the Academy on April 6, 1741, about ten weeks before he left for Berlin on June 19 [J]. In this paper, Euler invents what we now call generating functions. We start in familiar territory, the various series formed by the products of the terms of a given series, although Euler does introduce a new flavor to this menu. Then, we will come to generating functions and use them to find the partition numbers.

We begin with a given sequence of numbers,

$$a,\ b,\ c,\ d,\ e,\ f,\ g,\ h,\ \text{etc.}$$

The sequence may be finite or infinite, and the terms may or may not be distinct. As we have done before, we denote the sum of these numbers by A, the sum of their squares by B, their cubes by C, fourth powers by D, and so forth.

The sum is also denoted by α. The sum of pairs of different terms is denoted by β, of triples by γ, of quadruples by δ, and so forth.

The series we have not seen with Euler before are the series of products taken with repetition. For these, Euler uses Fraktur. $\mathfrak{A}$ denotes the series itself. $\mathfrak{B} = a^2 + ab + b^2 + ac + bc + c^2 + \text{etc.}$ denotes the series of products of two terms, with repetition of factors allowed. Then $\mathfrak{C}$ is the series of products of triples, $\mathfrak{D}$ the series of products of quadruples, and so forth.

We note that $A = \alpha = \mathfrak{A}$. Euler reminds us of the relations between the Greek letters and the upper case letters, beginning with

$$\begin{aligned} \alpha &= A, \\ \beta &= \frac{\alpha A - B}{2}, \\ \gamma &= \frac{\beta A - \alpha B + C}{3}, \\ \delta &= \frac{\gamma A - \beta B + \alpha C - D}{4}, \\ &\text{etc.} \end{aligned}$$

He gives similar relations between upper case and Fraktur and between Greek letters and Fraktur.

$$\mathfrak{A} = A, \qquad\qquad \mathfrak{A} = \alpha,$$

$$\mathfrak{B} = \frac{\mathfrak{A}A + B}{2}, \qquad\qquad \mathfrak{B} = a\mathfrak{A} - \beta,$$

Comm. Acad. Sci. Imp. Petropol. 13 (1741/3) 1750, 64–93; *Opera Omnia* I.2, 163–193

$$\mathfrak{C} = \frac{\mathfrak{B}A + \mathfrak{A}B + C}{3}, \qquad \mathfrak{C} = \alpha\mathfrak{B} - \beta\mathfrak{A} + \gamma,$$

$$\mathfrak{D} = \frac{\mathfrak{C}A + \mathfrak{B}B + \mathfrak{A}C + D}{4}, \qquad \mathfrak{D} = \alpha\mathfrak{C} - \beta\mathfrak{C} + \gamma\mathfrak{A} - \delta,$$

$$\mathfrak{E} = \frac{\mathfrak{D}A + \mathfrak{C}B + \mathfrak{B}C + \mathfrak{A}D + E}{5}. \qquad \mathfrak{E} = \alpha\mathfrak{D} - \beta\mathfrak{C} + \gamma\mathfrak{B} - \delta\mathfrak{A} + \varepsilon.$$

Euler now asks us to keep these relationships in mind as we contemplate the expression

$$P = \frac{az}{1-az} + \frac{bz}{1-bz} + \frac{cz}{1-cz} + \frac{dz}{1-dz} + \frac{ez}{1-ez} + \text{etc.}$$

Each term here can be expanded into a geometric series, then the series rearranged to get

$$\begin{aligned} P &= +z(a+b+c+d+e+\text{etc.}) \\ &\quad + z^2(a^2+b^2+c^2+d^2+e^2+\text{etc.}) \\ &\quad + z^3(a^3+b^3+c^3+d^3+e^3+\text{etc.}) \\ &\quad + z^4(a^4+b^4+c^4+d^4+e^4+\text{etc.}) \\ &\quad + \text{etc.} \\ &= Az + Bz^2 + Cz^3 + Dz^4 + Ez^5 + \text{etc.} \end{aligned}$$

Similarly, if

$$Q = \frac{az}{1+az} + \frac{bz}{1+bz} + \frac{cz}{1+cz} + \frac{dz}{1+dz} + \frac{ez}{1+ez} + \text{etc.}$$

then

$$Q = Az - Bz^2 + Cz^3 - Dz^4 + Ez^5 - \text{etc.}$$

Now we consider the expression

$$R = (1+az)(1+bz)(1+cz)(1+dz)(1+ez)\ \text{etc.}$$

Evaluating the coefficients of the powers of z, we see that the definitions of the values of the Greek letters imply

$$R = 1 + \alpha z + \beta z^2 + \gamma z^3 + \delta z^4 + \varepsilon z^5 + \text{etc.}$$

Likewise he presents S both as a product and as a sum:

$$\begin{aligned} S &= (1-az)(1-bz)(1-cz)(1-dz)(1-ez)\ \text{etc.} \\ &= 1 - \alpha z + \beta z^2 - \gamma z^3 + \delta z^4 - \varepsilon z^5 + \text{etc.} \end{aligned}$$

We can take logarithms of R and S and it follows easily that

$$Q = \frac{z\,dR}{R\,dz} \quad \text{and} \quad P = \frac{-z\,dS}{S\,dz}.$$

From this, and using the relations between capital letters and Greek letters, Euler leads us through calculations showing

$$Q = \frac{\alpha z + 2\beta z^2 + 3\gamma z^3 + 4\delta z^4 + 5\varepsilon z^5 + \text{etc.}}{1 + \alpha z + \beta z^2 + \gamma z^3 + \delta z^4 + \varepsilon z^5 + \text{etc.}}$$

and

$$P = \frac{\alpha z - 2\beta z^2 + 3\gamma z^3 - 4\delta z^4 + 5\varepsilon z^5 - \text{etc.}}{1 - \alpha z + \beta z^2 - \gamma z^3 + \delta z^4 - \varepsilon z^5 + \text{etc.}}.$$

Euler continues to manipulate these series to find relations among them. He recalls that $Q = \frac{z\,dR}{R\,dz}$, so $\int \frac{Q\,dz}{z} = \ln R$. There is a similar relation between S and P. Then he takes k to be the base of the natural logarithms (perhaps because he is already using e as a term of the series). Writing these as series we get

$$1 + \alpha z + \beta s^2 + \gamma z^3 + \delta z^4 + \text{etc.} = k^{Az - \frac{1}{2}Bz^2 + \frac{1}{3}Cz^3 - \frac{1}{4}Dz^4 + \text{etc.}}$$
$$1 - \alpha z + \beta s^2 - \gamma z^3 + \delta z^4 - \text{etc.} = k^{-Az - \frac{1}{2}Bz^2 - \frac{1}{3}Cz^3 - \frac{1}{4}Dz^4 - \text{etc.}}$$

Now, we look at

$$\frac{1}{S} = \frac{1}{(1 - az)(1 - bz)(1 - cz)(1 - dz)\,\text{etc.}}$$

and similarly $\frac{1}{R}$. He expands the factors of $\frac{1}{S}$ as geometric series, then multiplies all those geometric series and collects terms to get

$$\frac{1}{S} = 1 + \mathfrak{A}z + \mathfrak{B}z^2 + \mathfrak{C}z^3 + \mathfrak{D}z^4 + \mathfrak{E}z + \text{etc.}$$

Almost the identical derivation gives $\frac{1}{R}$ as the alternating version of this series. Multiplying both sides of the series for $\frac{1}{S}$ by S gives

$$1 = (1 - \alpha z + \beta z^2 - \gamma z^3 + \delta z^4 - \text{etc.})(1 + \mathfrak{A}z + \mathfrak{B}z^2 + \mathfrak{C}z^3 + \mathfrak{D}z^4 + \mathfrak{E}z^5 + \text{etc.}).$$

By multiplying these series and matching coefficients we can recover the relations between Fraktur and Greek that we saw earlier. The relations can also be recovered from $\frac{1}{R}$.

If we take $\frac{1}{R} = T$ and $\frac{1}{S} = V$, then $\frac{dR}{R} = -\frac{dT}{T}$ and $\frac{dS}{S} = -\frac{dV}{V}$ so $P = \frac{z\,dV}{V\,dz}$ and $Q = -\frac{z\,dT}{T\,dz}$. A bit of integration allows Euler to recover the relations between Fraktur and capital letters as well.

Euler sustains his rapid pace with results like this. He takes the logarithm of the equation

$$1 = (1 - \alpha z + \beta z^2 - \gamma z^3 + \delta z^4 - \text{etc.})(1 + \mathfrak{A}z + \mathfrak{B}z^2 + \mathfrak{C}z^3 + \mathfrak{D}z^4 + \mathfrak{E}z^5 + \text{etc.}).$$

Then he adds the series for R and S, adds the series for $\frac{1}{R}$ and $\frac{1}{S}$, and after a couple of pages finds that he can represent R in five different ways:

$$\begin{aligned} R &= 1 + \alpha z + \beta z^2 + \gamma z^3 + \delta z^4 + \text{etc.} \\ &= \frac{1}{1 - \mathfrak{A}z + \mathfrak{B}z^2 - \mathfrak{C}z^3 + \mathfrak{D}z^4 - \text{etc.}} \\ &= \frac{\alpha z + \beta z^2 + \gamma z^3 + \delta z^4 + \text{etc.}}{\mathfrak{A}z - \mathfrak{B}z^2 + \mathfrak{C}z^3 - \mathfrak{D}z^4 + \text{etc.}} \end{aligned}$$

$$= \frac{\alpha z + 2\beta z^2 + 3\gamma z^3 + 4\delta z^4 + \text{etc.}}{Az - Bz^2 + Cz^3 - Dz^4 + \text{etc.}}$$
$$= \frac{Az - Bz^2 + Cz^3 - Dz^4 + \text{etc.}}{\mathfrak{A}z - 2\mathfrak{B}z^2 + 3\mathfrak{C}z^3 - 4\mathfrak{D}z^4 + \text{etc.}}$$

Euler tells us that there are five more relations for S that we get by substituting $-z$ for z. Euler believes that such relations apply to any sequence a, b, c, d, etc., regardless of issues of convergence. He is interested in applying them to a particular sequence, the geometric sequence $n, n^2, n^3, n^4, n^5, n^6$ etc. This makes

$$A = \frac{n}{1-n},$$
$$B = \frac{nn}{1-nn},$$
$$C = \frac{n^3}{1-n^3},$$
$$D = \frac{n^4}{1-n^4}.$$

He gives us series and infinite products for P, Q, R and S as well.

Things start to get really interesting with the Greek letters, though. Euler analyzes the first four of these.

I. α is the sum of the individual terms, so

$$\alpha = n + n^2 + n^3 + n^4 + n^5 + n^6 + n^7 + \text{etc.}$$

II. β is the sum of products of two distinct terms, so

$$\beta = n^3 + n^4 + 2n^5 + 2n^6 + 3n^7 + 3n^8 + 4n^9 + 4n^{10} + \text{etc.}$$

Euler explains that the coefficient of the term of degree k is the number of ways that k can be written as the sum of two unequal parts. As an example he notes that the coefficient of n^{10} is 4. This means that there are four such ways of writing 10:

$$10 = 1 + 9, \qquad 10 = 3 + 7,$$
$$10 = 2 + 8, \qquad 10 = 4 + 6.$$

III. γ is the sum of products of three unequal terms of the series α.

$$\gamma = n^6 + n^7 + 2n^8 + 3n^9 + 4n^{10} + 5n^{11} + 7n^{12} + 8n^{13} + \text{etc.}$$

The coefficient on the term of degree k is the number of ways that k can be written as the sum of three distinct, or as Euler said, unequal parts. For example, the coefficient 7 of n^{12} arises since

$$12 = 1 + 2 + 9, \qquad 12 = 1 + 5 + 6,$$
$$12 = 1 + 3 + 8, \qquad 12 = 2 + 3 + 7,$$
$$12 = 1 + 4 + 7, \qquad 12 = 2 + 4 + 6,$$
$$12 = 3 + 4 + 5.$$

IV. δ is the sum of the product of four different factors of the series α.

$$\delta = n^{10} + n^{11} + 2n^{12} + 3n^{13} + 5n^{14} + 6n^{15} + 9n^{16} + \text{etc.}$$

As before, the coefficient of the term of degree k is the number of ways that k can be written as the sum of four unequal parts. As his example, Euler gives the nine ways that 16 can be partitioned into four unequal parts. He also gives the series for ε, ξ and η and tells us that by now he expects us to be able to interpret the coefficients without any more hints from him. From the series for η, he exhibits the 11 ways that 34 can be written as the sum of seven distinct parts, because it is directly related to the question that Philip Naudé asked Euler in a letter written in September of 1740. Naudé asked "to find in how many ways a given number can be given by the addition of integer numbers that are themselves unequal to each other." Naudé's question included the fact that there are 522 ways to divide the number 50 into such unequal parts, so the coefficient on the term of η of degree 50 will be 522.

Euler's representation of η stops with the term of degree 34, not enough to show that Naudé's analysis was correct, but Euler has done all that work on formulas relating Greek, Fraktur and capital letters, and he knows formulas for all the capital letters for his geometric sequence n, n^2, n^3, n^4, n^5, n^6, etc. Just a little more work gives him the apparently simple equations that describe this sequence:

$$\begin{aligned} \alpha &= A, \\ \beta &= AB, \\ \gamma &= ABC, \\ \delta &= ABCD, \\ \varepsilon &= ABCDE. \end{aligned}$$

These are not as useful in answering Naudé's question as they might seem. Naudé's example depends not on knowing the series for η but on knowing the 50th term of that series. For this, Euler sets out to find some relations on the coefficients.

Because we have explicit formulas for A, B, C, etc., it is easy to see that

$$\begin{aligned} \alpha &= \frac{n}{1-n}, \\ \beta &= \frac{n^3}{(1-n)(1-nn)}, \\ \gamma &= \frac{n^6}{(1-n)(1-nn)(1-n^3)}, \\ \delta &= \frac{n^{10}}{(1-n)(1-nn)(1-n^3)(1-n^4)}, \end{aligned}$$

etc.

These can be expanded to give

$$\begin{aligned} \alpha &= \frac{n}{1-n}, \\ \beta &= \frac{n^2}{1-n-n^2+n^3}, \end{aligned}$$

$$\gamma = \frac{n^6}{1-n-n^2+n^4+n^5-n^6},$$
$$\delta = \frac{n^{10}}{1-n-n^2+2n^5-n^8-n^9+n^{10}},$$
etc.

Euler does not yet know how to convert such forms to recursive relations, so he has to do it the hard way. To find the coefficients of β, he supposes that

$$\beta = \mathfrak{a}n^3 + \mathfrak{b}n^4 + \mathfrak{c}n^5 + \mathfrak{d}n^6 + \mathfrak{e}n^7 + \mathfrak{f}n^8 + \mathfrak{g}n^9 + \text{etc.}$$

He already knows that $\beta = \frac{\alpha n^2}{1-nn}$ so that $\beta = \beta nn + \alpha nn$. From β it is easy to write βnn, and we know that $\alpha n^2 = n^3 + n^4 + n^5 + n^6 + n^7 + n^8 + n^9 + \text{etc.}$ Adding these and matching coefficients, we get recursive relationships that give values

$$\begin{aligned} \mathfrak{a} &= 1, & \mathfrak{e} &= 3,\\ \mathfrak{b} &= 1, & \mathfrak{f} &= 3,\\ \mathfrak{c} &= 2, & \mathfrak{g} &= 4,\\ \mathfrak{d} &= 2, & \mathfrak{h} &= 4,\end{aligned}$$
etc.

Euler does not work through the details, but he assures us that, using $\gamma = \gamma n^3 + \beta n^3$ and $\delta = \delta n^4 + \gamma n^4$, we can get recursive relations for the values of the coefficients for γ, δ, etc.

This offers hope that he could calculate Naudé's number with some reasonable amount of effort, but Euler is not yet satisfied. He introduces a "function-like" notation. He has not used the $f(x)$ notation in seven years, since E-45, and with this he takes a step back in that direction. He denotes by $m^{(\mu)i}$ the number of ways a number m can be produced as a sum of μ parts that are unequal. The "i" stands for *inaequales* or unequal. Then $m^{(\mu)i}$ will be the coefficient of n^m in whichever series $\alpha, \beta, \gamma, \delta$, etc. it is that corresponds to μ.

Although he cannot give the name, as a Greek letter, of the series that corresponds to μ, he can give the function explicitly as

$$\frac{n^{\mu(\mu+1)/1\cdot 2}}{(1-n)(1-n^2)(1-n^3)(1-n^4)\cdots(1-n^\mu)}.$$

In this the term of degree m is $m^{(\mu)i}n^m$. If we divide the formula by n^μ, we get

$$\frac{n^{\mu(\mu-1)/(1\cdot 2)}}{(1-n)(1-n^2)(1-n^3)(1-n^4)\cdots(1-n^\mu)}.$$

Here, the term of degree m is $(m+\mu)^{(\mu)i}n^m$. Now, if we subtract the second formula from the first, we find

$$\frac{n^{\mu(\mu-1)/(1\cdot 2)}}{(1-n)(1-n^2)(1-n^3)(1-n^4)\cdots(1-n^{\mu-1})}.$$

Note that the denominator changed a bit. Here, the term of degree m is $m^{(\mu-1)i}n^m$. Matching terms, we get the relation

$$m^{(\mu-1)i} = (m+\mu)^{(\mu)i} - m^{(\mu)i},$$

which transforms into a recursive relation

$$(m+\mu)^{(\mu)i} = m^{(\mu)i} + m^{(\mu-1)i}.$$

This reduces problems of partitioning $m+\mu$ into simpler ones of partitioning just m. Euler notes a few shortcuts for sufficiently small values of m. If $m < \frac{\mu\mu+\mu}{2}$, then $m^{(\mu)i} = 0$, and if $m = \frac{\mu\mu+\mu}{2}$, then $m^{(\mu)i} = 1$.

There are other easy cases, like $\mu = 1$ or 2, that Euler ignores. Instead, he looks at the function

$$\frac{1}{(1-n)(1-n^2)(1-n^3)(1-n^4)\cdots(1-n^\mu)}$$

and suggests we rewrite it as a series

$$1 + pn + qn^2 + rn^3 + sn^4 + tn^5 + \text{etc.}$$

Euler states, "and from the generation of this are seen the coefficients of those powers of n." This is the only time in this article Euler uses a word like "generating function," but it seems to be where these functions got their name.

He notes that the coefficient $m^{(\mu)i}$ appears in this series in the term with exponent $m - \frac{\mu(\mu+1)}{2}$. The coefficients of this series, Euler tells us, give the number of ways that a given index, m, can be written as a sum of the numbers $1, 2, 3, 4, 5, 6, \ldots, \mu$. This leads Euler to make another comment on Naudé's example. The claims that the number of ways that $m = 50$ can be written as the sum of $\mu = 7$ unequal parts is the same as the number of ways that $50 - 28$ or 22 can be written as the sum of the numbers 1, 2, 3, 4, 5, 6 and 7. Thus, says Euler, two different problems, sums of μ unequal parts and sums of the numbers 1 to μ, have the same solutions.

Now we look at the question of writing a number m as the sum of integers, where we are allowed to repeat numbers. Recall that, for the geometric sequence $n, n^2, n^3, n^4, n^5, n^6$, etc. we have

$$R = (1+nz)(1+n^2z)(1+n^3z)(1+n^4z)\text{etc.},$$
$$S = (1-nz)(1-n^2z)(1-n^3z)(1-n^4z)\text{etc.}$$

and

$$\frac{1}{R} = 1 - \mathfrak{A}z + \mathfrak{B}z^2 - \mathfrak{C}z^3 + \mathfrak{D}z^4 - \mathfrak{E}z^5 + \text{etc.},$$
$$\frac{1}{S} = 1 + \mathfrak{A}z + \mathfrak{B}z^2 + \mathfrak{C}z^3 + \mathfrak{D}z^4 + \mathfrak{E}z^5 + \text{etc.}$$

Euler evaluates some of these coefficients.

I. $\mathfrak{A} = n + n^2 + n^3 + n^4 + n^5 + \text{etc.}$

The first case is often very easy.

II. $\mathfrak{B} = n^2 + n^3 + 2n^4 + 2n^5 + 3n^6 + 3n^7 + 4n^8 + 4n^9 + \text{etc.}$

Here, the coefficients are the number of ways their corresponding exponents can be written as the sum of two integers, allowing repetition. Euler gives the four such ways to write 8:

$$8 = 1+7, \quad 8 = 2+6, \quad 8 = 3+5, \quad 8 = 4+4.$$

III. $\mathfrak{C} = n^3 + n^4 + 2n^5 + 3n^6 + 4n^7 + 5n^8 + 7n^9 + \text{etc.}$

The coefficients give the ways to write numbers as the sum of three integers. Euler gives us the seven such ways to write 9.

IV. $\mathfrak{D} = n^4 + n^5 + 2n^6 + 3n^7 + 5n^8 + 6n^9 + 9n^{10} + \text{etc.}$

Euler gives us no discussion on this series but tells us that the other series, $\mathfrak{E}$, $\mathfrak{F}$, $\mathfrak{G}$, etc. can be found in the same way.

These ideas, Euler tells us, help to solve another problem posed by Naudé, namely to find how many ways a given number m can be partitioned into μ parts, whether equal or unequal. Euler denotes this number by $m^{(\mu)}$, which omits the superscripted i, and so does not require that the parts be unequal. Then $m^{(\mu)}$ will be the coefficient of the term of degree m in whichever Fraktur series corresponds to μ.

To describe these coefficients, Euler uses his two representations for $\frac{1}{S}$, one as an infinite quotient and the other as the Fraktur series. The usual manipulations, multiplying by $1 - nz$, and matching terms, lead to

$$\mathfrak{A} = \frac{n}{1-n},$$
$$\mathfrak{B} = \frac{\mathfrak{A}n}{1-n^2} = \frac{n^2}{(1-n)(1-n^2)},$$
$$\mathfrak{C} = \frac{\mathfrak{B}n}{1-n^3} = \frac{n^3}{(1-n)(1-n^2)(1-n^3)},$$
$$\mathfrak{D} = \frac{\mathfrak{C}n}{1-n^4} = \frac{n^4}{(1-n)(1-n^2)(1-n^3)(1-n^4)},$$
$$\text{etc.}$$

This means that

$$\alpha = \mathfrak{A},$$
$$\beta = n\mathfrak{B},$$
$$\gamma = n^3\mathfrak{C},$$
$$\delta = n^6\mathfrak{D},$$
$$\varepsilon = n^{10}\mathfrak{E},$$
$$\text{etc.}$$

Now, we can match terms with our previous results to compare the values of $m^{(\mu)}$ with those of $m^{(\mu)i}$. We find that

$$m^{(1)} = m^{(1)i},$$
$$m^{(2)} = (m+1)^{(2)i},$$
$$m^{(3)} = (m+3)^{(3)i},$$
$$m^{(4)} = (m+6)^{(4)i},$$

and, in general,

$$m^{(\mu)} = \left(m + \frac{\mu(\mu-1)}{2}\right)^{(\mu)i}.$$

This can be turned around to give

$$m^{(\mu)i} = \left(m - \frac{\mu(\mu-1)}{2}\right)^{(\mu)}.$$

Then, the recursive relation for $m^{(\mu)i}$ translates into $m^{(\mu)} = (m-\mu)^{(\mu)} + (m-1)^{(\mu-1)}$.

Euler says that this can be used to answer Naudé's questions about partitions of 50. The number 50 can be written as the sum of seven parts, allowing parts to be equal or unequal, in 8946 ways. That is the same as the number of ways $50 + 21 = 71$ can be written as the sum of seven unequal parts and also the same as the number of ways that $71 - 28 = 43$ can be written as the sum of 1, 2, 3, 4, 5, 6 and 7s.

Euler is finished with Naudé's questions for now, but he has a couple of closing observations. He writes "I have observed that this product of infinitely many factors

$$(1-n)(1-n^2)(1-n^3)(1-n^4)(1-n^5)\text{ etc.}$$

if it is multiplied together and written out, it produces this series

$$1 - n - n^2 + n^5 + n^7 - n^{12} - n^{15} + n^{22} + n^{26} - n^{35} - n^{40} + n^{51} + \text{etc.}$$

where the powers of n that occur are those contained in the form $\frac{3xx \pm x}{2}$." He also explains that the sign on the corresponding term is negative when x is odd and positive when x is even.

Finally, the reciprocal of this series,

$$\frac{1}{(1-n)(1-n^2)(1-n^3)(1-n^4)(1-n^5)\text{ etc.}}$$

expands to produce the series

$$1 + 1n + 2n^2 + 3n^3 + 5n^4 + 7n^5 + 11n^6 + 15n^7 + 22n^8 + \text{etc.}$$

He notes that the coefficients of these terms give the number of ways a particular exponent can be partitioned into any number of parts. As an example, he notes that the coefficient of n^5 is 7, so 5 can be partitioned in seven ways. He gives them as

$$5 = 5, \qquad 5 = 3 + 2, \qquad 5 = 2 + 2 + 1,$$

$$5 = 4 + 1, \qquad 5 = 3 + 1 + 1, \qquad 5 = 2 + 1 + 1 + 1,$$

$$5 = 1 + 1 + 1 + 1 + 1.$$

With this, the paper and the first St. Petersburg era draw to a close. We would have liked Euler to discover here how the reciprocal of a series gives the recursive relation among the coefficients of the series. He came so very close, but that discovery will wait another ten years, until E-191, written in 1750/51 and published in 1753. It has been a fascinating and important paper. This answers some important questions, going considerably beyond what Naudé had asked, and contains the seeds of better answers yet to come. It is a fitting way to end Euler's First St. Petersburg years.

References

[J] Juskevic, A. P. and E. Winter, *Leonhard Euler und Christian Goldbach, Briefwechsel 1729–1764*, Academie-Verlag, Berlin, 1965. Letter 38, 21 July/1 August 1741.

49

E-790: Commentatio de matheseos sublimioris utilitate*

On the utility of higher mathematics

> Nobody doubts the usefulness of mathematics; for they are indispensable to many arts and sciences, of which we have daily need.

This is how Euler begins his essay, "On the utility of higher mathematics," a defense of the use of calculus, rather than just algebra, to study science. Euler wrote the essay in 1741, the year he moved from St. Petersburg to Berlin, but it was not published until 1847. Except for letters, probably never intended for publication, this essay was the first article Euler wrote that was not published during his lifetime. Though about 300 of Euler's works were published posthumously, most of those were written in the last 20 years of his life.

The year 1741was a dangerous one to speak German in St. Petersburg. The death of Empress Anna Ivanova in 1740 left Russia in political chaos. Her successor, Elizabeth, assumed the throne in 1741, but it took a number of years of intrigue and unrest before she was able to assert real control. Some factions were vigorously, even violently, opposed to the large and prosperous German community in St. Petersburg and to the Academy of Sciences. An arsonist even set fire to a part of the Academy library. A suspect was arrested and convicted, but popular anti-German sentiment was so strong that the authorities were afraid to punish him and he was released.

In 1740, Euler had been offered a position in Berlin, working in the academy of Frederick II of Prussia, later to be known as Frederick the Great. Euler had turned down the offer, but feeling that he and his family were unsafe in Russia, he asked and received permission to accept the offer and left soon thereafter. He would spend 25 years in Berlin before returning to a much-changed St. Petersburg in 1766.

In all this turmoil, it is not surprising that Euler did little work in 1741. He wrote just three scientific articles and this essay. It is hard to know exactly where and when he did this. He might have written it in St. Petersburg, or in Berlin, or even while traveling between the two.

That it is in Latin suggests that he wrote it in St. Petersburg. In Berlin, such a non-technical essay would more likely have been in French. However, the essay is intended for a popular audience. He probably would not have been writing such an essay in Russia under the circumstances. We might guess he would compose it to try to appease the Russian mobs, but then he would have written it in Russian, and he would have seen that it was published.

If he wrote this essay in Berlin, it would be a nice way to introduce himself to his new employers and serve as a gentle description of his ideas of how mathematics and science fit together and the importance of his work. Still, popular works were usually in French, the language of the Court. Until more information surfaces, we can only guess where and when Euler actually wrote E-790.

The essay is only about eight pages long. Euler tries to convince the reader that, though many useful problems can be solved using just the "inferior parts of mathematics," arithmetic and algebra,

---

* *Journal für die reine und angewandte Mathematik* **35**, 1847, 106–116; *Opera Omnia* III.2, 392–415

some important problems require higher mathematics, especially calculus. He gives examples from mechanics, hydraulics, astronomy, artillery and navigation, without going into technical details. It is a simple and straightforward explanation of what Euler thought he was accomplishing with the way he did science and why he thought it was important. It is both reflective and forward-looking.

The essay remained unpublished until 1847, when it appeared in *Journal für die reine und angewandte Mathematik*, also known as "Crelle's Journal." Something about the essay captured the popular imagination of the time, and translations into French, Spanish and German quickly followed. To close this book, we add to these an English translation.

E-790A: Sur l'utilité de mathématiques supérieures*

On the utility of higher mathematics[1]

Nobody doubts the usefulness of mathematics; for they are indispensable to many arts and sciences, of which we have daily need. Nevertheless one generally believes that this character of utility belongs only to the inferior parts and to elementary mathematics. When considering the part that is justly called "higher," we fail no less to find useful applications. It is like a single strand of a spider's web, without any use because it is so fine. However, mathematics in general has unknown quantities as its object of research. To this end, it shows us methods, so to speak, paths that lead us to truth. It digs out the most hidden truths and brings them to light. Therefore, on the one hand, they give more vigor to the spirit and on the other hand they extend the field of our understanding. Can one possibly pay too much attention to such a result? The truth is by itself a great prize; despite all the truths that hold amongst them, there is not a one that is bereft of utility, even when it seems at first that this truth may be without use. One may object that higher mathematics may penetrate too deeply into the search for the truth. This is more of a eulogy than a criticism.

But we will not dwell on these abstract merits. We can basically prove that higher analysis has rules that are no less incontestable than the elementary mathematics that goes by the name "useful science." It has an even wider usage and mathematics, far from being too advanced, leaves on the contrary much to be desired in the interest of these same sciences, for which the most basic rudiments would seem to suffice. I wish therefore to show in this memoir that if elementary mathematics are useful, then the higher mathematics are no less so, and even that the degree of utility is always increasing, as one betters oneself in the study of that science; and that this science, finally, is too little advanced in its most common applications. To make my point, I will undertake a review of the most basic sciences, whose necessity is beyond doubt, such as mechanics, hydrostatics, astronomy, artillery, physics and physiology. I will prove, with evidence, that the more useful these sciences are, the greater need they have of higher analysis; that if sometimes the fruit we seek is beyond our hopes, it is almost always because the more subtle mathematics are not sufficiently advanced.

I begin with mechanics, and by this I do not mean the part that analyzes the most complicated movements and deduces them from the first principles of motion. There is no doubt that the subtlest

---

*Translated from Latin to French by Ed. Lévy. *Nouvelles Annales de Mathématiques* **12**, 1853, 5–21; *Opera Omnia* III.2, 400–407.

[1]Translated from French to English by Robert E. Bradley and C. Edward Sandifer.

analysis is nothing short of indispensable. But although this part of mechanics is extremely useful, it normally incurs the reproach that I wish to wash out of higher mathematics. Thus I wish to speak here about elementary mechanics necessary for the sciences that make machines of all types for our everyday use and which enjoys a reputation of great utility. In this part of mechanics we consider machines from a point of view of equilibrium. We only determine force or power equal to the weight that must be supported with the aid of the machine. But in practice we must consider the movement of the weights, and by and large it is completely ignored. Authors who have treated this part of mechanics teach us what force is necessary for each machine to support the weight in a state of equilibrium; but when the weight has to be moved, they are content to tell us only that we need a greater force. So even though in reality the weight must be put in motion, they do not tell us if this motion would be retarded or accelerated; they take no regard at all to the circumstances that produce this motion. Also the practitioners, for want of a better word, know quite well that the machine rarely responds as they wish. But still, these deceptions are made to count against the theory, and the machines that are invented inspire little confidence because they have not received the sanction of practice. This elementary theory of machines is thus imperfect. We recognize the need for a more certain theory that is in better accord with practice. But it is not the common mechanics that must provide such a service; it only treats the principles of statics. It has no other object than equilibrium. It is unable to give an explanation of motion. If you wish to perfect a theory of machines, study the movement that follows from a broken equilibrium; determine the force that causes the motion, and above all the external causes that resist the motion, such as friction and air resistance. It is therefore to higher mechanics that we must turn to analyze the most complicated motions. It is here that one has need of infinitesimal calculus and of higher analysis, and even analysis barely serves to explain the motions of the simplest machines. I have demonstrated all of this with the latest evidence, in a Memoir which I published in St. Petersburg [E-96] on simple and compound machines; I determined with higher analysis, the motions and their effects in all possible cases, and as a great number, perhaps even an infinity of similar machines might serve the same end, I gave the method for discovering which among these produces its effect with the smallest loss of time or of force, a problem whose solution is of continuous application in life, and this solution rests on the most profound theories of analysis and infinitesimal calculus. Mechanics provides us with a host of other arguments to demonstrate or prove that higher mathematics provides us with a great many applications in everyday life; but the few lines which precede appear to me to suffice admirably to demonstrate, as I had already proposed, that higher mathematics are indispensable to mechanics, and even to elementary mechanics, as everyone would agree, could not sustain itself, nor even take a step, without their application.

I pass then to hydrostatics, in which I also include hydraulics, a science which daily renders such services to mankind that no one can contest. Let us focus our attention particularly on the part to which we attribute these services, on ordinary hydrostatics, called "elementary." It is above all here that the practitioners complain that the results correspond so rarely to the theory. [Ironic note: Euler's own hydraulic theory ten years later is going to mismatch experience rather dramatically as well.] These complaints are far from being devoid of foundation; for the theory of flowing waters which they explain in the schools is almost entirely erroneous, and one should be astonished that they are not in greater disagreement with experience. It will be then in the general interest to substitute an exact theory for this false theory, but elementary mathematics will be much too weak for this task. Only the assistance of higher analysis permits us to approach an undertaking. We may easily convince ourselves of this by reading the excellent book that the celebrated Daniel

Bernoulli has published on hydrodynamics [1738]. He leads us to discover the natural laws that regulate moving fluids and facilitate their applications. Afterwards his father, with the ingenious spirit for which he was already famous, demonstrated the same laws from other principles and believed himself to corroborate the true theory of fluids in motion. In these two treatises, the infinitesimal calculus is encountered at every step. It is therefore to our ignorance of higher analysis that we must address ourselves if we are destined finally to approach a true theory of hydraulics. It is therefore by the progress of analysis that this theory may be elevated to its highest point of perfection and consequently to its greatest utility.

That astronomy is one of the most useful parts of mathematics everyone would easily agree. Now this utility is based on the exactness of the theory, to the agreement of this theory with the celestial phenomena, thus it is evident that this utility increases with the refinement of the science. As long as the true system of celestial bodies and their movements was unknown, arithmetic and the elements of geometry and optics sufficed for astronomy. However, in discovering the actual laws of the motion of celestial bodies, Kepler [1619] himself felt above all that elementary mathematics was no longer at the level of astronomy. Newton [1687] later miraculously perfected the work of Kepler; and for this, what tools of calculation from the arsenal of advanced mathematics did he not borrow? None, one cannot doubt, after having gone through his incomparable work. There we learn that the planets trace out ellipses around the sun and that the areas described by their radius vectors are proportional in time. Therefore, in order to construct tables of planetary motion, one must know the quadrature of the ellipse, which is certainly beyond elementary mathematics. Other problems, most useful and most necessary, serve to determine the very orbits of the planets according to their observations, and they demand even more imperiously the help of higher analysis. We can even less escape their use in research about the trajectories of comets. (see my *Mélanges de Berlin*, Volume VII) [1742]. On the other hand, the theory of the moon, which was extended and improved by the proofs of Newton, as rigorous as they are pleasing, nevertheless could not be brought to a good end. It is only through the completion of this theory requiring the solution of several difficult problems of mechanics that infinitesimal analysis, however advanced it might appear, does not yet suffice. Finally, we know that observations require corrections because of refraction. Now a table of refractions cannot be constructed by the aid of experience alone. The theory must determine, for a given height, the effects of refraction, and the theory must borrow the delicate calculations from higher analysis. The celebrated Bougeur [1729] clearly showed this to us in his memoir published in Paris on this subject. From all of the above, one may first conclude that astronomy has great need of infinitesimal analysis and also that analysis itself is far from being advanced enough for its applications to astronomy.

Artillery is ordinarily numbered as one of the branches of mathematics and it is under this title that it renders the greatest service to the art of war. Other than several well known problems of geometry, which have as their goal to deduce from the diameter the weight of the cannon ball, and vice versa, one principally considers the path of the cannon ball, and from there we derive the rules which one should follow when aiming the cannon so that the ball strikes the given location. Now, we suppose in this research that the projectile describes a parabola, as Galileo has shown [1638]. But this does not conform to the truth because the motion does not take place in a vacuum. We are thus led into a great error by the rules and tables based on this hypothesis, as their authors themselves will admit. They ignore the error in the calculation of their theory, and they imagine that they have no value in practice worth correcting. Now, the air seems to us to be too subtle a fluid to produce a tangible resistance. And yet in very fast motion such as those of cannonballs and bombs, the air resistance is big enough that the projectile describes a curve very different from a parabola. To correct this significant error, and to compensate for the improper use of the parabola, it is necessary to obtain the true curve along which the projectile moves through the air. Newton [1687] appears to have made a great effort to discover it, and despite his great ability in higher analysis, he was not able to resolve this problem. He left the honor of this discovery to the celebrated Jean Bernoulli [1713]. We see how someone who wants to resolve questions of artillery ought to be versed in higher mathematics. On other accounts, artillery is not worthy, even to this day, of the name of science, so great is its ignorance of its governing principles. Other than the motion of projectiles, it has not yet sufficiently studied the force and the action of gunpowder, and it is on this that the science pivots. Only in our time has an able Englishman Robins found, by a series of profound results, the true theory of the force of gunpowder. He first calculates the force that the ignition of the powder causes and the speed that it imparts to the ball. Then he determines the actual motion of the projectile. Experiments have made some contribution, without a doubt, to his results, but if he had not had higher analysis at his disposal, it would have been impossible for him to conceive of these experiments, nor to conclude anything from them.

A few words will suffice for navigation; for no one, I imagine, would dare to question the utility of higher mathematics here. If we consider the journey of a boat on the ocean, we will think first of the loxodromic curve, the invention of which assuredly may not be attributed to elementary mathematics. This curve is used to solve most of the problems that present themselves to anyone who wants to study the art of setting the course of a ship. The complete theory of navigation, a theory that lays down the principles of the construction and operation of ships is so arduous, demanding such a deep knowledge of hydrostatics and mechanics that the help of higher analysis is of prime necessity. The determination of the position that a ship will occupy in the ocean requires considerable calculation. If one wishes to deduce the shape that the ship should have, to measure the cargo that it ought to carry so that its equilibrium is stable, so that the ship withstands the pull of the sails and avoids capsizing, then one must perform the deepest calculations. Finally, if one wishes to discover the art of arranging the sails and steering the ship to its goal despite a contrary wind, one would never succeed without the aid of higher analysis. We find the last word on all this by reading the excellent work of Bernoulli [Johann Bernoulli, 1714]. I have treated the same material more fully in two books that I have written on the science of navigation. [E 110 and 111, *Scientia Navalis*, written in 1738, published in 1749] And so, no doubt should remain on this subject.

Physics, the science that studies all the phenomena of nature, was bereft of all manifest utility. Nevertheless, all men who loved truth still attached an elevation and grandeur to this science. And for this alone, all the sciences which give more scope and perfection to physics ought to be of extreme importance in our eyes. But physics is the deepest source of useful results in ordinary life. What would we say of higher mathematics if I should prove that, without it, no progress would be possible in physics? To begin with, most phenomena that we know how to explain belong as much to mathematics as to physics: these are the phenomena explained by mechanics, hydrostatics, aerodynamics, optics and astronomy. Next, in all phenomena where we observe some modification of matter, do we not need to take motion into account? To see how and by what matter it is produced? To what variations it is susceptible? etc. All these are studies which depend on deep understanding of mechanics, and another even deeper subject is hydrodynamics, on account of fluidity. Now, all modification of matter observed in nature is due to motion. It is therefore clear that mechanics, that is the science of motion, is necessary to explain even the simplest changes that occur in the universe. If we observe closely the phenomena that seem to be the simplest, and if we wish to account for them by the laws of mechanics, they present to us so much complexity that it is impossible to explain them, even when making use of higher analysis. This case is presented principally in physiology, which studies the movements of living beings. In its current state, this branch of physics gives us phenomena which are impossible to explain and which depend on complete notions concerning the motions of solids and liquids, along with a deep understanding of higher analysis. Who would dare to undertake, without sufficient resources, research on the movement imparted to the blood by the heart, and on the flow of the blood in the arteries and the veins? Before arriving at such an explanation, one must arrive at a solution to numerous and difficult problems, the solutions for which higher analysis is still powerless, even as advanced as it is. All of this will seem clearer than the day, if we are willing to read the authors who have tried to give rational explanations of the phenomena of physics and physiology. I will content myself to cite the book of Borelli [1681—*De motu animalium*] on the movement of living things. One sees on almost every page how he needs the full force of analysis to arrive at his goal, and frequently, when he lacks this aid, he stops, discouraged, and he does not know how to find a lever. Borelli was nevertheless well learned in the mathematics of his times; but he did not receive, until too late, the developments necessary for research of this type.

I believe I have amply attained the goal which I set for myself, to make clear the extreme utility of higher analysis. Other arguments in great number might have confirmed my demonstration. I could have proved that analysis gives more vigor to the spirit and renders it more able in research of the truth. But the enemies of mathematics would find here material to dispute. My first arguments are irrefutable, and I will stop here.

Topically Related Articles

Within topics, articles are arranged in the order in which Euler wrote them. This is the order in which the articles appear in this book and is not necessarily the order in which they were originally published.

Series

E 19
E 20
E 25
E 43
E 41
E 46
E 47
E 55
E 71
E 72
E 74
E 122
E 123
E 125
E 128
E 130
E 63
E 158

Geometry and curves

E 1
E 3
E 5
E 42
E 48
E 51
E 73
E 23
E 129

Calculus of variations

Elliptic integrals

Differential equations

Number theory

Theory of equations

Topology

Philosophy

Index

About the Author

C. Edward Sandifer received his AB from Dartmouth College 1973, and his MA, PhD, from the University of Massachusetts, Amherst 1975, 1980. He was Chair of the Northeastern Section of the MAA, 1998–2000, and Secretary of The Euler Society, 2002–present. He is also a member of the American Mathematical Society, the Canadian Society for the History and Philosophy of Mathematics, and the British Society for the History of Mathematics. He is one of the founding members of The Euler Society, and on the charter committee at the founding of the History of Mathematics Special Interest Group of the MAA (HoMSIGMAA). He has run 34 consecutive Boston Marathons and won the 1984 Northeastern (USA) Regional Marathon Championship.